D1210358

GEORG CANTOR

GEORG CANTOR

His Mathematics and Philosophy of the Infinite

Joseph Warren Dauben

PRINCETON UNIVERSITY PRESS
PRINCETON, NEW JERSEY

Published by Princeton University Press,
41 William Street, Princeton, New Jersey 08540

First Princeton Paperback printing, 1990
Reprinted by arrangement with the author
and Harvard University Press

Library of Congress Cataloging-in-Publication Data
Dauben, Joseph Warren, 1944–
Georg Cantor: his mathematics and philosophy of the infinite/
Joseph Warren Dauben.
p. cm.
Reprint. Originally published: Cambridge, Mass.: Harvard
University Press, 1979.
Includes bibliographical references and index.
ISBN 0-691-08583-8 (cloth)
ISBN 0-691-02447-2 (paper)
1. Set theory—History. 2. Numbers, Transfinite—History.
3. Cantor, Georg, 1845–1918. 4. Infinite. I. Title.
QA248.D27 1990
511.3′22′09—dc20 90-8579

Princeton University Press books are printed on acid-free paper,
and meet the guidelines for permanence and durability of the
Committee on Production Guidelines for Book Longevity of the
Council on Library Resources

10 9 8 7 6 5 4 3 2 1
10 9 8 7 6 5 4 3 2 1 (pbk.)

Printed in the United States of America by
Princeton University Press, Princeton, New Jersey

FOR MY PARENTS,
FOR ELAINE,
AND FOR DAVID

Acknowledgments

Henry Adams once wrote that if Harvard College gave nothing else, it gave calm. Though some might assess Harvard's recent past somewhat differently, I am nevertheless indebted to Harvard University for six years of study, teaching, and research, made possible in part by a Harvard Graduate Prize Fellowship. I am further indebted to Harvard for providing me with a University Traveling Fellowship, and to the National Science Foundation for a dissertation grant which supported a year of archival research in Germany, Sweden, and Italy during the academic year 1970–71. Generous support from the Faculty Research Foundation of the City University of New York and a grant from the Mellon Foundation have also helped make this book possible.

Of the many librarians and archivists who gave willingly of their time and help, special appreciation is due to Dr. Herta Battré of the Akademie der Wissenschaften der DDR (Berlin) for her patience in guiding me through the early and perplexing adjustment to problems of German handwriting and the idiosyncrasies of individual penmanships. Similarly, I am grateful to Dr. Schippang of the Handschriftenabteilung and the Darmstaedter Sammlung of the Staatsbibliothek Preussischer Kulturbesitz (Berlin, Dahlem), and to Dr. Haenel of the Handschriftenabteilung of the Niedersächsische Staats- und Universitätsbibliothek, Göttingen, for their help and interest in my archival research. In Djürsholm, Sweden, Dr. Lennart Carleson generously arranged accommodations for the month I spent at the Institut Mittag-Leffler, including an office in the institute, which allowed maximal use of the archive and library there. I am also grateful to the Biblioteca Apostolica Vaticana and to the American Academy in Rome for the use of their libraries and research facilities. In particular, I am happy to thank Janice and Norman Rosenthal of New York

City, who offered peace and quiet in Porto Ercole, where the final version of this manuscript was written.

For permission to consult the documents surviving in the estate of Georg Cantor, I am pleased to acknowledge the kind cooperation of Oberstudienrat Wilhelm Stahl of Bad Godesberg, Germany. Professor Herbert Meschkowski, who was always willing to discuss matters of Cantor's life and work, offered me materials from his personal library, including two dozen letters written by the Halle mathematician Eduard Heine to H. A. Schwarz, several of which are reproduced in the appendixes. I am also grateful to Mrs. Lily Rüdenberg for allowing me to examine the correspondence between her father, Hermann Minkowski, and David Hilbert, before they were made available in the edition recently published by Professor Hans Zassenhaus. Additional acknowledgment of archival and library sources used in the course of research for this book can be found at the beginning of the appendixes.

In one form or another, parts of the typescript have been read by Professors Kurt-R. Biermann, the late Carl Boyer, Richard Brauer, I. Bernard Cohen, Thomas Hawkins, John Murdoch, and Imré Toth. Their comments and suggestions have helped sharpen at many points the acumen of both my expression and my exposition. For help with translation, I am especially indebted to Professor John D. Poynter, who first taught me German and who has been a continuous source of encouragement and inspiration from the time I was one of his students at Claremont. I am grateful for his help in adding precision and literacy to the English translations supplied throughout this book.

Similarly, Professors John Murdoch and Judy Grabiner were the major forces guiding my preparation in the history of mathematics, and their help and encouragement define in great measure the direction and nature of my interests generally. I also owe special thanks to my two readers, Professor Erwin Hiebert of Harvard University and Professor Emeritus Dirk J. Struik of the Massachusetts Institute of Technology, for overseeing the general development of my doctoral dissertation ("The Early Development of Cantorian Set Theory," Harvard University, 1972), which was devoted to material comprising much of the first six chapters of this book.

A major debt I shall never be able to repay belongs to Dr. Ivor Grattan-Guinness, Hertfordshire, England, who has always been generous with his time, merciless with his criticisms, and indispensable with his encouragement. I appreciate in particular his continuing interest and advice and his willingness to provide me with copies of his own papers long before they were published. Equally valuable has been the advice and criticism of my colleague Professor Esther Phillips of Herbert H. Lehman College of the City University of New York. Her scrutiny of the final typescript has saved me from many an oversight or imprecision.

One last expression of my appreciation must be added, one of special importance to me, and one I have made before. In 1972 I dedicated my dissertation to Professor and Mrs. I. Bernard Cohen, to Professor Dr. and Mrs. Kurt-R. Biermann, and to Dr. and Mrs. Ivor Grattan-Guinness, for their continuing interest, support, and friendship. Over the years since leaving Harvard, my gratitude to them has only deepened, and I am happy to acknowledge their special importance to me once again.

A final word to Vivian H. Breitel and to David F. Grose, who patiently followed the entire course of this book from beginning to end. That it is finally finished is due as much to their efforts as to anyone's. They proved to be indefatigable sources of inspiration, solace, and sympathy, and with boundless affection and appreciation, my ultimate thanks are to them.

Contents

Cantor with his wife, Vally, about 1880.
In the possession of Egbert Schneider.

In re mathematica ars proponendi quaestionem
pluris facienda est quam solvendi.

—Georg Cantor

Introduction

Georg Cantor (1845–1918), the creator of transfinite set theory, is one of the most imaginative and controversial figures in the history of mathematics. Toward the end of the nineteenth century his study of continuity and the infinite eventually forced him to depart radically from standard interpretations and use of infinity in mathematics. Because his views were unorthodox, they stimulated lively debate and at times vigorous denunciation. Leopold Kronecker considered Cantor a scientific charlatan, a renegade, a "corrupter of youth," but Bertrand Russell described him as one of the greatest intellects of the nineteenth century.[1] David Hilbert believed Cantor had created a new paradise for mathematicians, though others, notably Henri Poincaré, thought set theory and Cantor's transfinite numbers represented a grave mathematical malady, a perverse pathological illness that would one day be cured.[2] Both in his own time and in the years since, Cantor's name has signified both controversy and schism. Ultimately, transfinite set theory has served to divide mathematicians into distant camps determined largely by their irreconcilable views of the nature of mathematics in general and of the status of the infinite in particular.

Like many controversial figures in history, Cantor was often misunderstood, not only by his contemporaries, but by later biographers and historians as well. This is particularly clear from the myths which have arisen concerning his personality and his nervous breakdowns. In his own day Cantor was regarded as an eccentric, if exciting, man, who apparently stimulated interest wherever he went, particularly among younger mathematicians. But it was the mathematician and historian E.T. Bell who popularized the portrait of a man whose problems and insecurities stemmed from Freudian antagonisms with his father and whose relationship with his archrival Leopold Kronecker was exacerbated because both men were Jewish.[3] In fact, Cantor was not Jewish. He was born and baptized a Lutheran and was a devout Christian during his entire life.[4]

Equally unreliable are Bell's assertions about Cantor's mental illness. Bell's interpretation deserves to be reevaluated not only because it has been uncritically accepted by so many mathematicians and historians, including the late Bertrand Russell, but because newly discovered evidence makes it possible to assess the nature and significance of Cantor's breakdowns more accurately and in a manner consistent with his biography and intellectual development.[5] In fact, Cantor's cycles of manic depression contributed in a unique and heretofore unsuspected way to his own interpretation of the nature of transfinite set theory itself (see Chapter 12).

In order to justify this revised interpretation of Cantor's personality, to explain the significance of his manic depression, and to suggest how both were intimately connected to very deep theological preoccupations which were always uppermost in Cantor's mind, it is necessary to go beyond materials currently available in print concerning his life and work. Regrettably, the unpublished documents relating to his family and career are less abundant than might be expected, considering the fact that his importance as one of the world's leading mathematicians was recognized in his own lifetime and well before his death in 1918. Nevertheless, shortly after World War I, his library was sold. Some of his papers and letters remained in the hands of his children, who continued to live in the family house on Händelstrasse in Halle, Germany (DDR). Then, during World War II, much of Cantor's literary estate was lost. Before the war, for example, there were twenty letter-books in which he had drafted his correspondence. But in 1945 Cantor's house was occupied, the family was forced to leave, and following the occupation many of Cantor's papers were missing, and only three of the twenty letter-books were to be found.[6]

This is particularly unfortunate, since it is largely through his correspondence that it is possible to study the evolution of Cantor's mathematical ideas before publication. Loss of the letter-books, however, is not so grave as it might be. Often the mathematicians to whom Cantor wrote kept his letters, and many of these are now preserved in various archives and private collections. It is particularly fortunate that in virtually every productive period of Cantor's career, there was at least (and usually only) one mathematician to whom he would write in detail about his work and in whom he would confide. It is doubtless a reflection on Cantor's personality that his friendships were often intense but of relatively short duration. For example, his friendship with H.A. Schwarz came to a premature end, although letters which the two exchanged in the early 1870s demonstrate that Schwarz gave Cantor a good deal of encouragement and even some fundamental techniques which were successfully applied in Cantor's earliest work on trigonometric series. Equally valuable in documenting the earliest development of Cantorian set theory are letters written by Cantor to Dedekind between 1872 and 1879. Their friendship, however, fell by the way shortly thereafter, apparently over Cantor's resentment of Dedekind's refusal to accept a position at the university in Halle.[7] But by then the Swedish mathematician Gösta Mittag-Leffler had become Cantor's confidant. He was one of the first mathematicians to take an active interest in Cantor's set theory, since he had found Cantor's work essential for results of his own. More important for the future of transfinite set theory, Mittag-Leffler did all he could as editor of the journal *Acta Mathematica* to promote Cantor's work and through translations to help make it known outside Germany. But even this very important relationship was not to last. By 1887 Cantor had ended their professional collaboration and even refused to publish in *Acta Mathematica* because of Mittag-Leffler's suggestion that he not print a premature version of his general theory of order types.[8] Thereafter, Cantor began to diversify his interests and included among his correspondents theologians, philosophers, and even a circle of literati interested in proving that Bacon was the true author of Shakespeare's plays. He wrote at length to students like Franz Goldscheider and to mathematicians like Felix Klein, editor of *Mathematische Annalen*. It was Klein who published Cantor's last major work, the *Beiträge Zur Begründung der transfiniten Mengenlehre* (1895–1897). Cantor's letters to Klein have been particularly helpful in documenting many of the philosophical and mathematical developments that occurred in this very important phase of Cantor's work, preceding publication of the *Beiträge*. In addition to these, drafts of Cantor's letters to figures like Giuseppe Peano, Charles Hermite, Philip Jourdain, and Grace Chisholm Young (among others) provide valuable information as to how and why Cantor worked as he did.

Until recently, no study of Cantor's life and work went beyond Adolf Fraenkel's lengthy obituary, written in 1930 for the Deutsche Mathematiker-Vereinigung.[9] Thereafter, most historians of mathematics have been content to follow Fraenkel's account, or worse, to believe the stories written by E.T. Bell, who based much of his work on Fraenkel's but with additional ill-founded embellishments of his own.[10]

There are two notable exceptions, however, and the research presented here owes a good deal to both. In 1967 Herbert Meschkowski published his book *Probleme des Unendlichen: Werk und Leben Georg Cantors,* easily the most informative guide to Cantor's life and work available since the outdated efforts of Jourdain and Fraenkel.[11] Shortly thereafter, Ivor Grattan-Guinness published his discovery of a previously unknown Cantor manuscript and followed that with his article "Towards a Biography of Georg Cantor."[12] The corpus of Cantor's writings, published by Zermelo in a collected edition, has been available to German readers since 1932. But English readers have for the most part been unaware of the very interesting works, in part philosophical and theological, which preceded Cantor's best known publication, his *Beiträge* of 1895 and 1897, translated into English by P. E. B. Jourdain in 1915 and provided with a lengthy informative preface.

As Jourdain realized, Cantor's mind was a fertile source of provocative ideas which profoundly influenced the history of modern mathematics. As a historically minded mathematician, Jourdain exchanged a number of letters with Cantor and probed specifically the motives and origins of Cantor's set-theoretic research.[13] For anyone concerned with intellectual history, in fact, the development of Cantorian set theory may be regarded as a microcosm in which the nature of the creation and development of a significant new idea of science may be studied. It provides a model that is ideal in many respects. Cantor's revolution of the mathematical infinite was created almost single-handedly, in the space of a few years. Original opposition and rejection of his work, not only by mathematicians, but by philosophers and theologians, eventually gave way to acceptance by some and to wholly new theories and domains of study undertaken by others.

The eminent historian of science Alexandre Koyré liked to emphasize the difficulties in conception and philosophy that accompanied the revolutionary shift in thinking required of the Renaissance thinkers.[14] Their transition in advancing from the closed world of Aristotle's universe to the infinite world of the post-Copernican era was in many respects a painful and traumatic one, but profound in its implications for the subsequent history of Western thought. The development of Cantorian set theory offers a close comparison of similar events and responses, though in another place and in another time. While none of the major participants in the modern attempt to move from a closed mathematical universe into a surprisingly and complexly infinite one was ever burned at the stake, George Cantor, in a less dramatic way, faced inquisition and repudiation at the hands of many of his contemporaries. University and Church were to weigh the evidence as carefully as any Galilean trial, and though the verdicts were never formalized, Cantor's work was given careful scrutiny and equally rigorous criticism and interpretation (see Chapter 6).

It is well to remember that this book is neither the biography of a man, nor even the history of a single idea, although it does focus on the mathematics of Georg Cantor and specifically upon the background, emergence, and development of his theory of sets and transfinite numbers. Perhaps the best way to describe what I have attempted is to say that this book represents a study of the pulse, metabolism, even in part the psychodynamics of an intellectual process: the emergence of a new mathematical theory. To be sure, the history of set theory and Cantor's own work provide the major foundations for what follows, but I have attempted to go beyond names, dates, and theorems to show how and why a new scientific theory emerges, the problems it faces, and the evolution, resolution, changing assumptions necessary for its eventual acceptance into the larger body of legitimate theory. At the same time, it is a tribute to the mathematician whose mind and imagination made transfinite set theory possible, and more than possible, something very real, bearing the stamp of his own perspectives and personality. To borrow the phraseology of the critic John Ciardi, the ultimate and underlying motivation of the present analysis has always been an answer to the admittedly broad question, *how* does a theory *mean*? The present investigation of the ways in which the mathematics and philosophy of set theory have come to mean what they do today concentrates upon the early efforts and inspiration of one man in particular: Georg Cantor.

CHAPTER 1

Preludes in Analysis

Georg Cantor's creation of transfinite set theory was an achievement of major consequence in the history of mathematics. Among his earliest papers was a series of articles that began to appear in 1870 and that dealt with a major problem concerning trigonometric series.[1] Functional analysis, in fact, spurred Cantor's interest in point sets and inspired his discovery of transfinite numbers. Set theory, at least in part, was produced in response to Bernhard Riemann's highly fertile investigations of trigonometric series and the related study of discontinuous functions. Interest in such functions led in a very natural way to the examination of point sets over domains of definition where discontinuities of various kinds occurred.

Cantor, however, was not the first to introduce special point sets for the sake of such investigations. Dirichlet, Riemann, and to a greater extent, Lipschitz and Hankel, had all written on various aspects of the subject, especially upon sets of points for which certain functions were either discontinuous or for which questions of convergence became difficult. But it was only Cantor who systematically developed the myriad implications of point sets in general, and who produced an entirely new field of mathematical research in the process.

GUSTAV PETER LEJEUNE DIRICHLET, 1805–1859

In 1829 Dirichlet, who was among the first contributors to Crelle's *Journal*, published an article dealing with the convergence problem of Fourier's series.[2] Joseph Fourier, in his celebrated studies on the conductivity of heat, had established that arbitrarily given functions could be represented by trigonometric series with coefficients of a specified type, and subsequently these became known as Fourier series. This discovery startled mathematicians and opened a new era of research in analysis. Though Fourier brought greater rigor to his mathematics than is generally recognized, his work nevertheless raised more questions than he was interested in answering or capable of solving.[3] Following Fourier, A. L. Cauchy advanced results of his own, but Dirichlet was not satisfied. He was anxious to establish the theory of Fourier series with greater rigor and clarity.

Cauchy, for example, had written the only article known to Dirichlet that attempted to deal with Fourier series in general terms. This had appeared in the *Mémoires de l'Académie des Sciences de Paris* for the year 1823. Though Cauchy had to admit that he had failed with some functions for which convergence was nevertheless incontestable, Dirichlet found that certain arguments were insufficient even in cases for which Cauchy had thought his methods were successful.[4] For example, although Cauchy had shown that the terms of a given series $\varphi(n)$ were decreasing, Dirichlet stressed that this was far from proving that the same terms formed a convergent series. Cauchy had established that the ratio of the nth term of the series and the quantity $A\,\frac{\sin nx}{n}$, where A was a constant determined by the extreme values of the function, differed from $+1$ by a quantity which diminished indefinitely as n became increasingly large. From this result and the conclusion that $\sum A\,\frac{\sin nx}{n}$ was convergent, Cauchy concluded that the general trigonometric series was also convergent. But as Dirichlet was able to show, particularly in the case of alternating series, this was false. It was not difficult to produce two series, one convergent, the other divergent, although the ratio of their nth terms approached $+1$ as n became very large. The two series which Dirichlet had in mind were those with general terms

$\frac{(-1)^n}{\sqrt{n}}$ and $\frac{(-1)^n}{\sqrt{n}}\left(1+\frac{(-1)^n}{\sqrt{n}}\right)$; the first was convergent, the second was not.

Subtracting the two produced the divergent series $-1-\frac{1}{2}-\frac{1}{3}-\frac{1}{4}\ldots$, even though the ratio of nth terms of the two series, $1\pm\frac{1}{\sqrt{n}}$ converged as n increased without limit.[5] Consequently, Cauchy's method for establishing the convergence of trigonometric series was far from acceptable.

Related to such difficulties was the question of alternating series. In fact, Dirichlet was apparently led to study the behavior of trigonometric series upon his realization that nondivergent series were of two types, conditionally convergent and absolutely convergent.[6] He found that certain alternating series could be shown to converge to different sums, or even to diverge, depending only upon the arrangement of their terms. He offered as an illustration the two series:

$$1-\frac{1}{2}+\frac{1}{3}-\frac{1}{4}+\frac{1}{5}-\frac{1}{6}+\ \ldots$$

$$1+\frac{1}{3}-\frac{1}{2}+\frac{1}{5}+\frac{1}{7}-\frac{1}{4}+\ \ldots.$$ [7]

Riemann, in assessing the significance of Dirichlet's work on trigonometric series, stressed the importance of this distinction between conditionally convergent and absolutely convergent series. Since convergent Fourier series were not necessarily *absolutely* convergent, it was impossible to establish the convergence of such series solely from the fact that the terms of the series continually diminished, which was one of the arguments Cauchy had offered. Dirichlet, however, reduced the representation problem to consideration of the integral:

$$\int_0^h \frac{\sin i\beta}{\sin \beta} f(\beta) d\beta,$$

where i was integer-valued, and abstracting sign (Dirichlet's expression for absolute value), the terms were decreasing.[8] In thus appreciating the necessity of absolute convergence to ensure unique summations, Dirichlet was also underscoring the fact that Fourier's convergence problem was only one part of a much larger problem. Convergence could assume many forms, and absolute convergence was but one, as mathematicians in the nineteenth century were to discover.

In his effort to establish the conditions for convergence of Fourier series, Dirichlet began with a simple case, to which others he chose to consider might be reduced. Narrowing his attention to the interval $(0,\pi/2)$, where $0 < h \leq \pi/2$, and assuming $f(\beta)$ to be continuous on $(0,h)$, monotonically decreasing yet always positive, he considered the behavior of:

$$\int_0^h \frac{\sin i\beta}{\sin \beta} f(\beta) d\beta.$$

By insisting that i be positive, and introducing a particular partition of $(0,h)$, he was eventually able to establish, for $0 < g < h \leq \pi/2$, that

$$\int_g^h f(\beta) \frac{\sin i\beta}{\sin \beta} d\beta \tag{1.1}$$

converged to a definite limit as i increased without limit. The limit was always zero, unless $g = 0$, when the integral converged to a value of $(\pi/2)f(0)$.

With this result in hand, Dirichlet could then prove the convergence of the Fourier series representing arbitrary functions within given limits. First he restricted attention to the first $2n + 1$ terms of the resulting representation:

$$\frac{1}{2\pi}\int \varphi(\alpha) d\alpha + \tag{1.2}$$

$$\frac{1}{\pi}\left\{\begin{array}{l} \cos x\int \varphi(\alpha)\cos \alpha\, d\alpha + \cos 2x\int \varphi(\alpha)\cos 2\alpha\, d\alpha + \ldots \\ \sin x\int \varphi(\alpha)\sin \alpha\, d\alpha + \sin 2x\int \varphi(\alpha)\sin 2\alpha\, d\alpha + \ldots \end{array}\right\}.$$

The Dirichlet partial sum could then be rewritten:

$$\pi\int_{-\pi}^{\pi}\varphi(\alpha)d\alpha[\tfrac{1}{2}+\cos(\alpha-x)+\cos(2\alpha-2x)+\ldots+\cos n(\alpha-x)].$$

The cosine series then assumed the well-known form:

$$\frac{1}{\pi}\int_{-\pi}^{\pi}\frac{\varphi(\alpha)\sin(n+\tfrac{1}{2})(\alpha-x)}{2\sin\tfrac{1}{2}(\alpha-x)}d\alpha.$$

The central question was thus reduced to evaluating the limit of this integral as n increased indefinitely. Again, by careful attention to partitioning, Dirichlet was able to refer his argument back to the earlier conclusion for (1.1). In the final analysis, he showed that the series did in fact converge and was equal to $\frac{1}{2}(\varphi(x+\epsilon)-\varphi(x-\epsilon))$ for all x on $(-\pi,\pi)$, and at the endpoints assumed the value $\frac{1}{2}(\varphi(\pi-\epsilon)+\varphi(-\pi+\epsilon))$.

Thus, for Dirichlet, a given function $f(x)$ was completely represented by its Fourier series whenever the function was continuous, with possible exception at the endpoints, where $f(\pi)=f(-\pi)$ would ensure representation. Otherwise, at points of discontinuity, or at an endpoint $\pm\pi$, the function was represented by the series only when certain additional conditions were fulfilled.

The important question of *which* functions actually admitted such representations led to the formulation of what are today known as Dirichlet's conditions, and his paper of 1829 was concerned with *only* those functions satisfying these conditions:

> The preceding considerations prove in a rigorous manner that if the function $\varphi(x)$ (for which all values are supposed finite and determined) only presents a finite number of discontinuities between the limits $-\pi$ and π, and if in addition it does not have more than a determined number of maxima and minima between these same limits, then the series (1.2) (where the coefficients are the definite integrals depending on the function $\varphi(x)$) is convergent and has a value expressed generally by
>
> $$\tfrac{1}{2}(\varphi(x+\epsilon)+\varphi(x-\epsilon)),$$
>
> where ϵ designates an infinitely small number.[9]

While Dirichlet's conditions reflected the success with which he had solved the question of convergence of Fourier series and the sufficient conditions for ensuring representability, they simultaneously characterized, quite succinctly, the limitations beyond which there were still no answers. Dirichlet, we know, was unhappy about the restrictions: "But in order to establish this with all the clarity one could desire," he wrote, "there are some details concerning fundamental principles of infinitesimal analysis which will be treated in another note, in which I shall also consider some other equally remarkable properties of the series."[10] Though he never managed to produce a further, more general and systematic exposition, he did offer certain extensions, both in print, and in his private correspondence. Two advances are especially noteworthy.

In an addition to a paper of 1837, "Sur les séries dont le terme général dépend de deux angles, et qui servent à exprimer des fonctions arbitraires entre des limites données,"[11] Dirichlet explained how his earlier assumptions concerning the function $f(\beta)$ could be relaxed. Earlier he had required that $f(\beta)$ be continuous and monotonically increasing or decreasing between the limits of integration. The more general case allowed the function to be discontinuous and suspended the monotonic requirement. Furthermore, he showed how the validity of his restrictions were unaffected by allowing isolated infinities. Supposing c to be such a singularity in $(0,h)$, where $c \neq 0$, Dirichlet considered the four integrals:

$$\int_0^{c-\epsilon} f(\beta)\frac{\sin k\beta}{\sin \beta}d\beta,\ \int_{c-\epsilon}^{c} f(\beta)\frac{\sin k\beta}{\sin \beta}d\beta,\ \int_{c}^{c+\epsilon} f(\beta)\frac{\sin k\beta}{\sin \beta}d\beta,$$

$$\int_{c+\epsilon}^{h} f(\beta)\frac{\sin k\beta}{\sin \beta}d\beta.$$

Earlier results established directly that the first integral converged in value to $\frac{1}{2}\pi f(0)$, the fourth to zero. For arbitrarily small ϵ,

$$\left|\int_{c-\epsilon}^{c} f(\beta)\frac{\sin k\beta}{\sin \beta}d\beta\right| < \frac{F(c)-F(c-\epsilon)}{\sin(c-\epsilon)}$$

$$\text{and}\quad \left|\int_{c}^{c+\epsilon} f(\beta)\frac{\sin k\beta}{\sin \beta}d\beta\right| < \frac{F(c+\epsilon)-F(c)}{\sin c}.$$

Since Dirichlet assumed $f(x)$ to be continuous, and $c \neq 0$, he could easily conclude that the two integrals were negligible.

Even more impressive was the way in which Dirichlet was able to weaken the restrictions concerning the number of maxima and minima occurring in any given interval. In a letter to Gauss, sent from Berlin in 1853, he showed how it was possible to admit more than a finite number.[12] It was, in fact, an easy proposition, so long as the case of infinitely many maxima/minima at $\beta = 0$ was excluded. For then:

$$\lim_{k\to\infty}\int_0^{c} f(\beta)\frac{\sin k\beta}{\sin \beta}d\beta = \frac{\pi}{2}f(0),\quad \lim_{k\to\infty}\int_b^{c} f(\beta)\frac{\sin k\beta}{\sin \beta}d\beta = 0, \tag{1.3}$$

where $0 < b < c \leq \pi/2$.

The essential element of proof involved Dirichlet's introduction of what today would be termed accumulation points and sets of zero measure:

The extension of these theorems to the case just mentioned, where $f(\beta)$ presents an infinite number of maxima and minima, is very easy, at least for the second theorem, and just as easy for the first, if only infinitely many maxima and minima do not lie in the immediate neighborhood of $\beta = 0$ [the reference to Theorems I and II concerns the two integrals in (1.3), Theorem I dealing with the interval $(0,c)$, Theorem II with (b,c)]. From the integrals one may separate only the intervals [*Theile*] within which such an accumulation [*Anhäufung*] of maxima and minima occurs, and which can be arranged so that, if the limits of these partial integrals approach one another sufficiently closely, these remain arbitrarily small, no matter how large k becomes.[13]

Dirichlet thus required that such sets of singular points be distributed so as to fall into intervals whose dimensions could be made arbitrarily small. This idea might have inspired valuable extensions had it been addressed to a wider audience. As it was, Dirichlet's letter to Gauss was first printed in 1897, in the second volume of Dirichlet's collected works. But the essential features of his idea were not to be lost. The naturalness of the approach, and the generality it afforded, were too striking to remain unnoticed.

In addition to his study of trigonometric series, Dirichlet also responded to Fourier's results by defining the concept of function as any correspondence between given domains, a definition which expressed the arbitrariness that mathematicians had begun to recognize as essential if their work were to embrace complete generality. But in his own work, Dirichlet's general definition was never seriously developed. The functions he dealt with were invariably more circumscribed, in fact bounded, monotonic, and in general continuous. In assessing the terrain Dirichlet had covered, the interesting and suggestive realm of pathological functions had scarcely been explored.

As Dirichlet brought his 1829 paper to a close, he returned to the conditions for representation which he had managed to obtain. Besides requiring continuity in general, with only finitely many singularities allowed, he added an additional observation. $\varphi(x)$, he recalled for his readers, had to be well-behaved to the extent that, for any values a and b such that $-\pi < a < b < \pi$, one could always find values r and s, $a < r < s < b$, where $\varphi(x)$ would be continuous over (r,s). Otherwise it would be impossible to establish the veracity of his results.[14] Dirichlet's conditions, in fact, signaled the point from which Riemann was to initiate his own investigations of the Fourier problem.

GEORG FRIEDRICH BERNHARD RIEMANN, 1826–1886

Turning directly to Dirichlet for inspiration, Riemann decided to devote his *Habilitationsschrift* to the representation problem in the theory of trigonometric series and to complete the task Dirichlet had left undone.[15] Riemann was determined to go beyond the simple class of functions useful in physics, which had provided the focus for Dirichlet's analysis, to include more complex and interesting functions of wider application in pure analysis. Number theory in particular required the introduction of highly discontinuous functions, and while Dirichlet had described a function which was discontinuous at every rational point in its domain of definition, Riemann was able to formulate in even more precise detail a similar function.[16] Riemann consequently wondered about the extent to which Dirichlet's results might be generalized. With Dirichlet's set of *sufficient* conditions in mind, Riemann began the search for conditions that were *necessary* in order to represent a given function by trigonometric series.

Riemann's *Habilitationsschrift* began with a critical narrative of work done in the previous two centuries concerning the mathematical treatment of numerous physical problems by means of trigonometric series.[17] The study of such series, he emphasized, was intimately linked with the principles of analysis, and a careful study could serve to bring greater clarity and definition to the understanding and development of these principles. Not only did he show how the concept of function was made more precise by Dirichlet's work, but he also set the tone for his own investigations. Dirichlet had gone no further than to establish sufficient conditions for representation, and there the matter stood. History pointed to the next logical step: what conditions could be shown as necessary for representation to succeed?

Riemann's historical introduction also made it clear that he was breaking new ground. Just as Fourier's results had inspired Dirichlet to reformulate the concept of function, Riemann similarly insisted that the time had come to reconsider the foundations of the integral as well.[18] What was to be understood by the expression:

$$\int_a^b f(x)dx?$$

Riemann wanted to sharpen not only the definition of the integral, but to broaden the extent to which its application was meaningful. The motivation for such generalization stemmed directly from Riemann's hope of admitting at least a large class of discontinuous functions, if not all discontinuous functions, to the class of functions representable by trigonometric series. Since the integrability of such functions was essential, some extension of the concept had to be made.

Cauchy, assuming the continuity of the function $f(x)$, defined the integral over (a,b) in terms of the sum:

$$\int_{x_0}^{x} f(x)dx = s = (x_1 - x_0)f(x_0) + (x_2 - x_1)f(x_1) + \ldots + (x - x_{n-1})f(x_{n-1}).$$ [19]

Riemann, on the other hand, considered the sum:

$$S = \delta_1 f(a + \epsilon_1\delta_1) + \delta_2 f(x_1 + \epsilon_2\delta_2) + \ldots + \delta_n f(x_{n-1} + \epsilon_n\delta_n),$$
$$\delta_n = x_n - x_{n-1}.$$ [20]

The Riemann sum thus depended upon both δ and ϵ. The integral was said to exist if the sum S approached a definite limit, where $\delta_i = x_i - x_{i-1}$, $0 < \epsilon_i < 1$, so that $x_{i-1} < x_{i-1} + \epsilon_i\delta_i < x_i$, however δ and ϵ might be chosen, as the $\delta_i \to 0$. Riemann's analysis advanced considerably beyond Cauchy's in its treatment of the necessary conditions for integrability. Riemann approached this problem in a remarkable way. In any subinterval of (a,b), allowing D_n to represent the maximum variation of $f(x)$ between x_{n-1} and x_n, where $\delta_n = (x_{n-1}, x_n)$, Riemann required that the sum:

(1.4) $$\delta_1 D_1 + \delta_2 D_2 + \ldots + \delta_n D_n \to 0 \quad \text{as} \quad \delta \to 0.$$

If $\delta < d$, Riemann then assumed that the sum (1.4) could achieve at most the value Δ. Δ was thus a function of d:

$$\delta_1 D_1 + \delta_2 D_2 + \ldots + \delta_n D_n \leq \Delta(d).$$

Consequently, as $d \to 0$, $\Delta \to 0$.

Denoting by s the sum of all intervals in which $f(x)$ was of variation greater than σ, Riemann could argue that

$$\delta_1 D_1 + \delta_2 D_2 + \ldots + \delta_n D_n \geq \sigma s.$$

It then followed that $\sigma s \leq \Delta$ and $s \leq \frac{\Delta}{\sigma}$.

Whenever σ was given, by appropriate choice of d, $s \to 0$ as $d \to 0$. If the function $f(x)$ was always finite, Riemann formulated the necessary and sufficient conditions for integrability as follows. Considering the total magnitude s of all intervals in which the variation of $f(x)$ was greater than a given σ, if $s \to 0$ as $\delta \to 0$, then the sum S converged as $\delta \to 0$. Riemann was able to generalize these conditions in order to admit an even larger class of functions to integrability. It was possible to allow the functions in question to become infinite, but only for isolated (*einzelne*) values of the variable. Such functions were then integrable, if there was a definite limit value for S as the limits of integration approached infinitely near to such isolated infinities of the function.[21]

The advances which conditions for integrability in this form allowed were made clear almost immediately. Riemann had assumed only the boundedness of the functions in question. His focus had been almost exclusively limited to the behavior of the sum function, and the conditions necessary to ensure its convergence to a unique limit. Subsequently Riemann turned to specific examples of functions which, under his newly given definition, were integrable.

What could be said for functions that were infinitely often discontinuous between any two limits, however close they might be? Riemann was able to construct such a function, one that was discontinuous for a certain dense set of rational values of x. He produced the function

$$f(x) = \frac{(x)}{1} + \frac{(2x)}{4} + \frac{(3x)}{9} + \ldots = \sum_{1}^{\infty} \frac{(nx)}{n^2},$$

where (x) denoted the difference between x and the nearest integer, or was taken to be zero if x were to fall midway between two consecutive integers. Letting n denote any integer, p an odd number, then $f(x)$ converged for all values of x, except for those values $x = \frac{p}{2n}$ (where p and n were assumed to be relatively prime). At such points,

$$f(x+0) = f(x) - \frac{1}{2n^2}\left(1 + \frac{1}{9} + \frac{1}{25} + \ldots\right) = f(x) - \frac{\pi^2}{16n^2},$$
$$f(x-0) = f(x) + \frac{1}{2n^2}\left(1 + \frac{1}{9} + \frac{1}{25} + \ldots\right) = f(x) + \frac{\pi^2}{16n^2}.$$

Thus for every rational value of x with even denominator, the function was discontinuous, although the number of discontinuities for which $|f(x+0) - f(x-0)| \geq \sigma$ was always finite. Consequently, $f(x)$ was Riemann integrable, despite the infinite number of discontinuities on an everywhere-dense set.[22]

Riemann was similarly able to deal with cases where, at a given point, the function became infinite, but the phenomena which eluded him were those involving functions with infinitely many maxima and minima. In such cases, Riemann was at a loss to show that for any arbitrarily given value of σ, d could be produced to establish the conditions ensuring integrability of the discontinuous function in question. Consequently, Riemann's work was incomplete. Total generality was lacking, and this may well have convinced him not to publish the paper in his own lifetime.[23] Nevertheless, it is undeniable that Riemann had laid the essential foundations for a new era in the theory of functions of a real variable. Innovations however were by no means exhausted with Riemann's reconsideration of the problem of integrable functions. The importance of his *Habilitationsschrift* lay as much in its contribution to the theory of trigonometric series as in its reformulation of the definite integral. Here the question of convergence was of prime importance.

Convergence of trigonometric series may be considered in two ways, either for all values x in a continuous domain of definition, or for convergence only at certain points x. There are two additional distinctions to be made. One can either consider, first, a given function, and ask what properties it must have in order to admit a representation either at a particular value or over an entire interval, *or* one can begin with the trigonometric series itself, and ask what can be concluded about convergence in terms of properties of the coefficients.

Dirichlet had followed the first approach and was able to show that $S_n(x) \to f(x)$ for all x interior to $(-\pi,\pi)$, where $f(x)$ was continuous. At points of discontinuity the series was shown to converge to $\frac{1}{2}(f(x+0)+f(x-0))$, (which may not equal $f(x)$). At the endpoints of the interval, the series converged to $\frac{1}{2}(f(\pi-0)+f(-\pi+0))$.

Riemann turned the entire problem around. This was precisely what he meant when he said that the direct path Dirichlet had followed was impossible to pursue further. The nature of the problem itself was more complicated, and Riemann was consequently forced to try a more indirect route, one which, as it turned out, was to be of limited success. Even so, Riemann's approach was a very significant departure from Dirichlet's, for he did not make the restrictive assumption that the coefficients a_n, b_n of his series be given by Fourier's integral formulae. Instead of dealing only with Fourier series, Riemann dealt with trigonometric series in general, and he was in fact the first to emphasize this distinction.

Assuming that a given function was representable, Riemann began with the trigonometric series:

$$(1.5)\quad f(x) = \tfrac{1}{2}b_0 + a_1 \sin x + b_1 \cos x + a_2 \sin 2x + b_2 \cos 2x + \ldots,$$

and emphasized that $f(x)$ could be said to exist for only those values where the series converged. If the coefficients $a_n, b_n \to 0$, then the terms of the series (1.5) converged throughout an entire interval; otherwise, convergence could occur only for special values of x, and Riemann was careful to divide the two cases.

The pivotal element of Riemann's approach to the study of trigonometric series was his introduction of an auxiliary function $F(x)$, which became the focus of investigation both of the uniqueness of representation and of convergence. It is of special importance to note that, arbitrarily given a trigonometric series, it is possible to decide both questions in terms of this auxiliary function. Given a trigonometric series:

$$\tfrac{1}{2}b_0 + \sum a_n \sin nx + b_n \cos nx,$$

the Riemann function $F(x)$ arises from twice formally integrating the series (1.5), and assumes the form:

$$F(x) = C + C'x + A_0 \frac{x \cdot x}{2} - A_1 - \frac{A_2}{4} - \frac{A_3}{9} - \ldots,$$

where $A_0 = \frac{1}{2}b_0$; $A_n = a_n \sin nx + b_n \cos nx$.[24]

The usefulness of this complete shift in perspective is made clear by advantages gained in studying such series in terms of the Riemann function $F(x)$, which is not only continuous, but convergent as well, uniformly and absolutely (though Riemann did not make these distinctions himself). To clarify further the behavior of the function, Riemann offered three lemmas which in turn prepared the way for theorems establishing the necessary conditions for representation.[25]

Lemma 1: If the series (1.5) converges, and if α and β become infinitely small so that their ratio remains finite, then

$$\frac{F(x+\alpha+\beta)-F(x+\alpha-\beta)-F(x-\alpha+\beta)+F(x-\alpha-\beta)}{4\alpha\beta}$$

converges to the same value as the series.

Lemma 2:
$$\frac{F(x+2\alpha)+F(x-2\alpha)-2F(x)}{2\alpha}$$

becomes infinitely small as α becomes infinitely small.

Lemma 3: If one denotes by b and c two arbitrary constants, the greater by c, and by $\lambda(x)$ a function whose first differential-quotient is always continuous between b and c and is equal to zero at the boundaries, and whose second differential-quotient does not have infinitely many maxima and minima, then the integral

$$\mu\mu \int_b^c F(x) \cos\ \mu(x-a)\,\lambda\,(x)dx$$

becomes arbitrarily small as μ becomes infinitely large.

Armed with these specific characteristics of the function $F(x)$, Riemann devoted all of Section 9 of his *Habilitationsschrift* to the problem of representability.[26] In each case it was assumed that the terms of the general series (1.5) converged to zero:

Theorem I: If a periodic function $f(x)$ with period 2π is to be represented by a trigonometric series whose terms eventually become infinitely small for all values of x, then there must be a continuous function $F(x)$ so that

$$\frac{F(x+\alpha+\beta)-F(x+\alpha-\beta)-F(x-\alpha+\beta)+F(x-\alpha-\beta)}{4\alpha\beta}$$

converges to $f(x)$ if α and β become infinitely small and their ratio remains finite. Furthermore,

$$\mu\mu \int_b^c F(x) \cos\ \mu(x-a)\,\lambda\,(x)dx$$

must become infinitely small as μ becomes infinitely large, if $\lambda(x)$ and $\lambda'(x)$ are zero at the limits of integration, and $\lambda''(x)$ does not have infinitely many maxima and minima.

Theorem II: Conversely, if these two conditions are fulfilled, then there exists a trigonometric series in which the coefficients eventually become infinitely small, and which represents the function wherever it converges.

Theorem III: Let $b < x < c$. Let $\rho(t)$ be a function such that $\rho(t) = \rho'(t) = 0$, for $t = b$ and $t = c$, and between these values let $\rho(t)$ and $\rho'(t)$ be continuous. Suppose $\rho''(t)$ does not have infinitely many maxima and minima, and moreover, for $t = x$, let $\rho(t) = 1$, $\rho'(t) = 0$, $\rho''(t) = 0$, $\rho'''(t)$ and $\rho^{iv}(t)$ be finite and continuous; then the difference between the series $A_0 + A_1 + \ldots + A_n$ and the integral

$$\frac{1}{2\pi}\int_b^c F(t)\,\frac{dd\,\dfrac{\sin\dfrac{2n+1}{2}(x-t)}{\sin\dfrac{x-t}{2}}}{dt^2}\,\rho(t)\,dt$$

eventually becomes infinitely small as n increases without limit. Thus the series $A_0 + A_1 + A_2 + \ldots$ will converge or not, depending upon whether or not

$$\frac{1}{2\pi}\int_b^c F(t)\,\frac{dd\,\dfrac{\sin\dfrac{2n+1}{2}(x-t)}{\sin\dfrac{x-t}{2}}}{dt^2}\,\rho(t)\,dt$$

converges to a fixed limit with increasing n.

The first of Riemann's three theorems outlined the necessary and sufficient conditions for representation to take place. Theorem II was concerned with the conditions under which a trigonometric series might exist such that the sum would be that of a predescribed function. In the course of his proof no use was made of convergence of the trigonometric series at any particular point. No stipulation was made that the series involved be a Fourier series. The theorem held for all trigonometric series in general, whatever form the coefficients might assume. Theorem III gave the necessary and sufficient conditions for a trigonometric series to converge at a particular point x.[27]

The third and final aspect of Riemann's paper to be considered here concerns his treatment of discontinuous functions; of special importance are those which possess an infinite number of maxima/minima. Dirichlet found such functions particularly troublesome, and Riemann had to content himself with only partial success. Nevertheless, it would be no exaggeration to claim that one of the major stimuli to refinements in analysis during the nineteenth century were discontinuous functions, and Riemann was the first to give them any sort of systematic treatment.

Riemann had redefined the integral in order to allow evaluation of a larger class of discontinuous functions than previously had been possible under Cauchy's definition. It was no minor contribution to have given such functions prominent and separate consideration in their own right: functions displaying maxima and minima were given special attention at the end of the *Habilitationsschrift*.[28] First, Riemann turned to the problem of functions with only finitely many maxima/minima for which Dirichlet's investigations were incomplete. Assuming the functions under consideration were bounded, Riemann had already shown, earlier in his paper, that they were integrable. In fact, as Riemann had noted, dismissing Dirichlet's unnecessary assumption that the function be continuous, Dirichlet's proof was perfectly applicable, granting his supposition that

$$\int_x^{x+b} f(t) \frac{\sin \frac{2n+1}{2}(x-t)}{\sin \frac{x-t}{2}} dt$$

converged to $\pi f(x+0)$. Riemann summarized the necessary and sufficient conditions for the representability of functions displaying discontinuities:

> Except for functions which have infinitely many maxima and minima, it is therefore necessary and sufficient for the representability of the function $f(x)$ by a trigonometric series (with coefficients converging to zero) that if the function becomes infinite for $x = a$, then $f(a+t)t$ and $f(a-t)t$ become infinitely small with t, and $f(a+t) + f(a-t)$ be integrable up to $t = 0$.[29]

Nevertheless, functions with infinitely many maxima/minima continued to elude Riemann's grasp. As he brought his investigations to a conclusion, he turned to consider the most problematic, yet the most suggestive (as Hankel was to show) of the discontinuous functions—those with an infinite number of oscillations in any given finite interval.

Riemann began by stressing that such functions were sometimes integrable, even though they were not representable by a Fourier series. He generated a specific example in considering the case where

$$\int f(x)dx = \varphi(x) \cos \psi(x),$$

where $\varphi(x)$ was assumed to be arbitrarily small, $\psi(x)$ arbitrarily large, for infinitely small values of x, and where these functions were also assumed to be continuously differentiable without infinitely many maxima/minima. Despite integrability, the Fourier series did not converge.[30] On the other hand, Riemann had to add that the converse was no better: it was entirely possible that, though a given function might be completely impossible to integrate over any arbitrarily small interval, it was still possible that there were infinitely many values x in that same interval for which the trigonometric series (1.5) would converge. Riemann offered his example of the function

$$\sum \frac{(nx)}{n},$$

where (x) denoted the difference between x and the nearest integer value. If x lay midway between two integers, (x) was taken to be zero. The function

$$\sum \frac{(nx)}{n}$$

was defined for all rational values of x and was represented by the trigonometric series

$$\sum_{1,\infty}^{n} \frac{\Sigma^{\theta} - (-1)^{\theta}}{n\pi} \sin 2\pi nx,$$

where $\Sigma^{\theta} - (-1)^{\theta}$ was understood as the sum of positive and negative units so that for even values of n the term was negative, and for odd values of n the term was positive. But the function was not integrable in Riemann's sense, since it was bounded in no arbitrarily small interval by finite limits.[31] With two examples, Riemann had exhibited emphatically the diverse and perplexing questions which had yet to be resolved. Among these, one question was of particular interest: given a function represented by a trigonometric series, was there only *one* such series? Eduard Heine emphasized the importance of this problem in a paper on the subject in 1870, and through his encouragement, Cantor was led to produce his very important proof of the uniqueness of such representations in 1872. Before turning to Cantor's major contribution, however, it is necessary to say something of two additional developments.

RUDOLPH OTTO SIGISMUND LIPSCHITZ, 1832–1904

The paper which Lipschitz published in Crelle's *Journal* for 1864 made clear his esteem for Dirichlet and demonstrated the strong impression Dirichlet's approach to the study of Fourier series had made upon Lipschitz.[32] Though Riemann's *Habilitationsschrift* was the immediate successor to Dirichlet's work on trigonometric series, Lipschitz's memoir of 1864 was really much closer in spirit and in form to the subject as Dirichlet had developed it. Thus, when compared with Riemann's results, Lipschitz's approach actually serves to underscore the ingenuity and importance of Riemann's analysis. But Riemann's paper was not generally available until 1867, the year in which Dedekind published it, and Lipschitz seems to have lacked any knowledge of its existence. In fact, Lipschitz regarded his own work as providing the sequel to Dirichlet's paper of 1829, a sequel which Dirichlet had promised but had never produced.[33] According to Lipschitz, it was possible to work from hints scattered throughout Dirichlet's publications in order to produce a reconstruction, almost, of the paper Dirichlet had failed to write. Though Dirichlet had insisted that the cases admitting infinite numbers of discontinuities or maxima/minima could be reduced to the cases he had previously treated with success, he also warned that there were difficulties lying at the very foundations of analysis which required prior consideration. Armed with these guides and caveats, Lipschitz set out to resolve problems about which Dirichlet had only conjectured.

In attempting to deal more completely with the question of *which* functions could be represented by trigonometric series, Lipschitz worked entirely within the compass of procedures developed by Dirichlet. Given a function $\varphi(x)$, it could fail to satisfy Dirichlet's conditions in several ways.[34] It might assume infinite values at isolated points on the interval $(-\pi,\pi)$, or display an infinite number of discontinuities. The most interesting exception concerned yet another alternative and involved functions with an infinite number of maxima/minima. Lipschitz emphasized the special character of these functions by giving them separate consideration. The first two cases, however, were treated together, following in large measure Dirichlet's procedures outlined in his paper of 1837. Considering an interval (a,b) in $(-\pi,\pi)$, it sufficed to probe the behavior of the integral

$$\int_a^b \varphi(x)dx.$$

Under what circumstances could the more discontinuous functions $\varphi(x)$ in question preserve the integral as a finite and continuous function between a and b? Given $c_1, c_2, \ldots, c_\mu$ as points where, in the first case, $\varphi(x)$ might assume an infinite value, Lipschitz enclosed each c_μ in a δ-interval, described as intervals of the *first* form.[35] What remained of the original interval, discounting first form intervals $(c_n - \varphi_n,\ c_n + \varphi_n)$, where the values φ_n could be taken as arbitrarily small, were intervals of the *second* form, over which $\varphi(x)$ satisfied the Dirichlet conditions.

For the second case, involving functions with an infinite number of discontinuities, a similar procedure sufficed, provided that in every interval (a,b) it was possible to find numbers r and s such that the function $\varphi(x)$ continued to satisfy Dirichlet's conditions over (r,s). Essentially Lipschitz required that the points of discontinuity be nowhere densely distributed (a condition which had its origin in Dirichlet's earlier paper of 1829).[36] It was then easy for Lipschitz to establish that for arbitrarily small φ, any value x for which the function neither became infinite, nor displayed a discontinuity, could be made to fall entirely within an interval of the second form, where $\varphi(x)$ was continuous and finite. The argument could then be reduced to showing that first form integrals, those defined over first form intervals, assumed only the value zero, while those over second form intervals remained defined, continuous, and finite.

But what of the third case, where $\varphi(x)$ was allowed to assume an infinite number of maxima/minima in any finite interval? Lipschitz emphasized that here he was forced to proceed with care, because there were no real aids to be had from Dirichlet.[37] The difficulties lay with the behavior of such infinitely often oscillating functions in arbitrarily small neighborhoods where the continuity of the function could no longer be assured. For the study of the two earlier cases of functional discontinuity, Lipschitz had relied ultimately upon a theorem that was no longer applicable in the case of infinitely many oscillations in a given interval. Previously, Lipschitz could be satisfied with a value in the limit for

$$\int_g^h f(\beta)\frac{\sin k\beta}{\sin\beta}\,d\beta,$$

assuming among other things that $f(\beta)$ was finite, continuous, and constant or at least monotonic over (g,h). But the assumptions were too strong to be applied successfully to the case of infinitely often oscillating functions.

Before introducing modifications suitable for the problem at hand, Lipschitz pondered the various ways in which maxima/minima could be distributed, infinitely often, over a finite interval.[38] Oscillations could occur at a given isolated point r in (a,b), whereby there would be infinitely many maxima/minima in any arbitrarily small δ-neighborhood $(r-\delta, r+\delta)$, but only finitely many maxima/minima elsewhere in the interval. Oscillations could also be densely distributed over (a,b), so that no intervals (r,s) could be found containing only a finite number of maxima/minima. Combinations of these two cases offered a third possibility for the distribution of oscillations over (a,b).

Following Dirichlet's means of approach, Lipschitz cast the problem in terms of the series (1.2) and wondered what conditions were necessary to ensure convergence. The analysis was subsequently reduced to consideration of the integral representing the first $2n+1$ terms of (1.2):

$$\frac{1}{\pi}\int_{-\pi}^{\pi}\varphi(\alpha)\frac{\sin(n+\frac{1}{2})(\alpha-x)}{\sin\frac{1}{2}(\alpha-x)}\,d\alpha.$$

The heart of Lipschitz's argument was his Theorem II:

> *Theorem II:* Let us designate by g and h quantities such that $0\leq g< h\leq\pi/2$, and let $f(\beta)$ be a function which remains between positive and negative values of some constant in the interval (g,h) such that the difference $f(g+\delta)-f(g)$ converges to zero with δ, and such that

> $f(\beta+\delta)-f(\beta)$, for $g<\beta<h$, has a value absolutely less than the product of a constant by any positive power of δ; then the integral
>
> $$\int_g^h f(\beta)\frac{\sin k\beta}{\sin \beta}\,d\beta$$
>
> has a limit as k increases indefinitely. This limit is zero if g is positive and has the value $\frac{\pi}{2}f(0)$ if g is zero.[39]

What is most often remembered about Lipschitz's memoir of 1864 is not the distinction it made between intervals of the first and second form, but the condition it introduced for integrability,[40] namely, that the following inequality for a constant B and any positive power of δ must hold:

$$\left| f(\beta+\delta) - f(\beta) \right| < \delta^{\alpha}B.$$

What impresses anyone interested in set theory, however, is Lipschitz's classification of intervals into two forms, and his use of concepts equivalent to both everywhere- and nowhere-dense sets. The fact that he even bothered to offer special terms for certain types of point distributions was an advance. But Lipschitz's designations revealed a conceptual prejudice as prevalent in his mind as in the minds of his predecessors: values in the domain of definition were of interest only insofar as the behavior of an oscillating function at such points became critical. Lipschitz's only concern for the form or structure of the domain of definition was revealed in his requirement that certain infinite numbers of singularities be distributed in a nowhere-dense way. This example, drawn from Lipschitz's analysis of functions with infinitely many discontinuities, is particularly revealing, since the distinctions go no further than to ensure functional continuity *over certain specified intervals*. Confusion between nowhere-dense and negligible sets persisted. Until mathematicians began to shift their frame of reference from functional behavior in the range, to structural properties in the domain, the various point sets under consideration could not begin to constitute the elements of a general theory.

Lipschitz never really came to terms with his exceptional point sets of the second form. His analysis, similar to the one Dirichlet had outlined in his 1853 letter to Gauss, was at least somewhat more general. His short-term interest had been the completion of Dirichlet's program, and this he had accomplished. But on comparison with Riemann's *Habilitationsschrift*, written ten years before Lipschitz's paper, Lipschitz's results appear to be little more than direct extensions of work Dirichlet had already done.

Hankel, on the other hand, went well beyond either Riemann or Lipschitz in the matter of discontinuous functions. Like Lipschitz, he appreciated functions displaying infinitely many maxima/minima, but he also advanced to even more complicated examples of considerable heuristic value. Though Lipschitz's paper of 1864 failed to develop even the most immediate implications of the newly described sets of the first and second forms, Hankel's Tübingen *Universitätsprogramm* went significantly further.

HERMANN H. HANKEL, 1839–1873

On March 6, 1870, Hankel delivered his *Universitätsprogramm* at the university in Tübingen.[41] Ostensibly concerned with research devoted to infinitely often discontinuous and oscillating functions, the paper revealed its major aim in an appended subtitle: "A Contribution to the Determination of the Concept of Function in General." Recalling Dirichlet's reformulation of the concept of function in the face of Fourier's provocative work, Hankel emphasized several obvious shortcomings. Dirichlet, he believed, had offered a definition of function in such general terms that it was scarcely of value in mathematics and did nothing to suggest a more useful way of thinking about functions themselves.[42] Hankel sought to remedy this situation and found the basis for a solution in Riemann's work, specifically within the domain of complex analysis. This aspect, unquestionably the guiding principle behind Hankel's memoir, is not of direct consequence for the subsequent development of set theory. Nevertheless, Hankel's underlying reasons for investigating the infinitely often discontinuous functions must be kept in mind for later reference. For now, attention is directed more specifically to one section of Hankel's memoir where his indebtedness to Riemann and Lipschitz is clearest, and where his advances in producing new results are the most significant.

Systematically proceeding from a preliminary clarification of the function concept, the meaning of continuity and the nature of continuous functions, Hankel outlined in Section 4 his method of the condensation of singularities.[43] This method, in summary, employed a function with a given singularity. In turn, this singularity was used to construct another function yielding an infinite number of such singularities. Beginning with a function $\varphi(y)$ on $(-1,+1)$ which admitted expansion into a Taylor series of the form

$$\varphi(x+\delta) = \varphi(y) + \delta\varphi'(y) + \tfrac{1}{2}\delta^2\varphi''(y) + \ldots$$

where δ was arbitrarily small, Hankel supposed that $\delta(y)$ had a singularity at $y=0$, which precluded expandability at $y=0$. Hankel further assigned the definite value $\varphi(0)=0$. The secret of Hankel's method lay in its move to impregnate this function's singularity at $y=0$ into another function whose behavior ensured an infinite number of singularities in any finite interval. Consequently, Hankel defined

$$f(x) = \sum_{n=1}^{\infty} \frac{\varphi(\sin n x\pi)}{n^s}.$$

Closer investigation revealed that $f(x)$ was not only continuous for every irrational value of x, but that a finite derivative existed at every point. For the singular value $y = 0$, however, and for every rational point x where this singularity again appeared, $f(x)$ possessed no derivative. Hankel underscored the procedure and its significance in a single sentence: "The infinitely many singularities which $\varphi(\sin y\,\pi)$ possesses according to the assumed properties of the function φ at discrete points $y = 0, 1, 2, 3 \ldots$, are imposed upon the function $f(x)$, so to speak, on a finite interval, and I therefore have named the principle of going from such functions φ to functions f as that of the *condensation of singularities*."[44]

The generation of such a function was only of heuristic interest for Hankel. Uppermost in his mind were the consequences his new method could have for the further clarification of the function concept. More than half of the paper was devoted to investigating these consequences.

After a short discussion of continuous functions possessing infinitely many singularities, in particular the function $\varphi(y) = y \sin \dfrac{1}{y}$, Hankel turned to the definition and examples of linear, discontinuous functions. A linear, discontinuous function was simply a function with an infinite number of points of discontinuity in a finite line.[45] In developing a theory and classification of such functions, Hankel made some necessary distinctions:

> If an aggregate [*Schaar*] of points having a certain property lies on a line, then I say that these points fill up the line if no interval, however small, can be given in which there is not at least one point of that aggregate; on the other hand, I say that this aggregate of points does *not* fill up the line, but that the points lie *loosely* [*zerstreut*] on the line, if between any two points, however close, an interval can always be given in which there is no point of that aggregate.[46]

From these distinctions two classes of linear, discontinuous functions emerged:

> *Class I:* Point-wise discontinuous [*punktirt unstetigen*] functions are those which are only linearly discontinuous in loosely distributed points, more precisely: *Point-wise discontinuous* functions are those linearly discontinuous functions for which the points with jumps occur only in loose distribution and fill up *no* interval [*Strecke*], however small the value σ greater than zero may be.[47]

> *Class II:* Totally discontinuous functions are those which are completely discontinuous in entire intervals, or more precisely: *Totally linearly discontinuous* functions are those for which points with jumps which exceed a certain finite magnitude fill up entire intervals.[48]

Turning to Riemann's definition of the definite integral, Hankel concluded (though erroneously) that the functions of Class I were always integrable, while those of Class II were not.

Hankel's primary interest in discontinuous functions was directed toward their pathology as a means of elucidating the function concept. What was the extent, the limit to which discontinuous functions could be meaningful, analytically operable, before only complete arbitrariness was left? Hankel's introduction of the linear, discontinuous functions of Class II, those with infinitely many discontinuities in every interval, were intended to answer this question:

> As for totally discontinuous functions, since they cannot be subjected to the analytical operations of differentiation and integration, the conjecture arises that they might evade representation by analytical forms, and that this may be the limit which separates functions of interest to the analyst from the transcendental functions, i.e., as the only conceivable ones for practical purposes. This view however is not confirmed, for I believe I have found an unassailable example of a totally discontinuous function through a modification of the principle of condensation.[49]

The example Hankel offered depended on his earlier example, where

$$\sum_{n=1}^{\infty} \frac{1}{n^s} \left[\frac{1}{\varphi(\sin nx\pi)} \right]^2$$

took the value $\sum \frac{1}{n^s}$ on all irrational values of the independent variable, but became infinitely large for all rational values of x. Constructing the function

$$f(x) = \frac{\sum \frac{1}{n^s}}{\sum \frac{1}{n^s} \left[\frac{1}{\varphi(\sin nx\pi)} \right]^2},$$

he could conclude that it was fully determined for all values of x, and was everywhere discontinuous, assuming the value 1 for all irrational x, the value 0 for all rational x. He could then push to his conclusion, a comprehensive critique both of Euler's definition of the function concept in its narrowest terms, and of Dirichlet's in its most general.[50]

Hankel described Euler's definition as limiting functions to expressions connecting constants and variables through the four basic arithmetic operations. This might seem to ensure that all functions under Euler's definition would share certain identifiable properties: continuity, differentiability, integrability, and so on, but as Hankel showed, "Eulerian" functions could be defined which were discontinuous for an infinite differential quotient. This seemed as unsatisfactory as Dirichlet's definition of function, which permitted a completely arbitrary correspondence between all points of a function's range and domain of definition. In short:

> Both concepts (of function) are thus insufficient to provide a foundation for a general theory of functions. For the task of all mathematics is to derive from given relations which concern a domain [*Gebiet*] of objects, new relations in which one brings into application the general properties of concepts which underlie all those objects. But if there are no such general properties, then every transformation of the given relations remains an empty tautology.[51]

If the function concept was to be defined suitably for the purposes of functional analysis, Hankel believed that his "principle of the permanence of formal rules,"[52] advanced in his theory of complex variables, was the appropriate solution. This principle was suggested by the way in which the real numbers were traditionally extended from the four basic arithmetic operations for the integers. To solve problems of division, rational numbers were introduced. To solve quadratic equations, irrational numbers and complex numbers were needed. The complex numbers, in fact, seemed to provide "an organic conclusion" to the domain of numbers.

Just as the system of such numbers and arithmetic laws were "typical and would be shown to be permanent," the same was true for functions where algebraic expressions could be regarded as derived directly from the four arithmetic operations and produced similarly "permanent laws." Transcendental functions, however, were not exhausted by only finitely many such operations but required an infinite number for their complete determination. Hankel took the transcendental functions as analogous to the irrational and complex numbers, and just as alternative complex numbers were possible which did not follow the basic laws of arithmetic, so too functions were possible which did not follow the "permanent formal rules" of algebraic functions in general. These alternative functions Hankel termed "illegitimate." The existence of illegitimate complex numbers (Hamilton's noncommutative quaternions, for example) demonstrated that the domain of number was not complete with the introduction of complex numbers, and similarly Hankel argued that: "Actual proof of the existence of functions of the most illegitimate kind, which I have endeavored to provide in this paper, has been necessary to show beyond a doubt that the legitimacy of functions does not come to us arbitrarily dictated from some mysterious, inflexible necessity which lies in the 'Nature of Things,' as one often hears, but is rather a wise, convenient and conventional limitation which we impose on ourselves."[53]

> *Definition:* Legitimate functions are those which, for all real values of the argument with the exception of only a finite number in any finite interval,

> are determined, finite, and continuous, and whose differential quotients all have the same properties.[54]

The purpose of Hankel's *Universitätsprogramm* was an exact analysis of the extent to which the concept of function could be pushed, and still remain legitimately useful in analysis. But having carefully described the essential differences between the "legitimate" and "illegitimate" functions, there was yet another step to be taken. In order to resolve the problem of extending a legitimate function over and beyond a singular point, the tools of real analysis were insufficient. Rather than "jump over them," as he put it, such points could be avoided by allowing the variable to include complex as well as real values. In his final summary he argued that the most general, simple law unifying and describing the entire class of legitimate functions was given by Riemann's monogenic functions:

> *Definition:* a variable complex number w is a monogenic function of $(x + yi)$ if it varies with x,y such that in general (for all but individual points of the line) the differential quotients $\frac{\partial w}{\partial x}, \frac{\partial w}{\partial y}$ are determined and finite and fulfill the conditions $i\frac{\partial w}{\partial x} = \frac{\partial w}{\partial y}$.[55]

"Now this is the general, simple rule," wrote Hankel, "under which all legitimate, i.e., monogenic functions are subsumed."[56] The concept of function was thus pressed to its most general form, more comprehensive than Euler's but short of disintegrating into the most arbitrary and consequently impractical definition given by Dirichlet. Hankel had at last reached his goal and had shown that the most general, still practical, and useful definition of the concept of function was to be found only by considering complex variables. But for the history of set theory, Hankel's method of condensing singularities and his examples of functions that were Eulerian in form but not in character, functions continuous but nondifferentiable, or completely discontinuous, nonintegrable, were particularly stimulating.

What, then, was the meaning for Hankel of his condensations and the notion of "nowhere-dense" distributions as he used them, for example, to characterize linearly discontinuous, point-wise discontinuous functions?

> *Theorem:* The necessary and sufficient condition that a linearly discontinuous function be point-wise discontinuous is that the totality of all intervals in which oscillations $>\sigma$ occur, for however small a $\sigma > 0$, can be made arbitrarily small.[57]

Though he attempted to describe structural differences between the exceptional sets defined by point-wise discontinuous functions as opposed to totally discontinuous functions in terms of measure-theoretic properties, the result was incorrect. Point sets more complicated than any Hankel had considered could be nowhere dense, and yet of positive measure. In the final analysis, he seemed to have insufficient feeling for what the structural properties of the sets of singular points were like. After all, he was interested in the behavior of the *function* at singular points. That these could be condensed, or made to appear in any arbitrarily small interval, was interesting because it led to extraordinary conclusions concerning the nature and compass of "legitimate functions." But the domain of singularities was scarcely explored. This deficiency was underscored by an English mathematician, H. J. S. Smith, who produced some very important examples of nowhere-dense sets of positive measure in a memoir which appeared in 1875, shortly after Hankel's death:

> The result obtained in the last example deserves attention, because it is opposed to a theory of discontinuous functions, which has recently received the sanction of an eminent geometer Dr. Hermann Hankel. In an interesting memoir Dr. Hankel had laid down the distinction, here adopted from him, between a system of points which completely fill a segment, and a system of points which do not completely fill any segment, but lie in loose order. Dr. Hankel then asserts that, when a system of points is in loose order on a line, the line may be so divided as to make the sum of the segments containing the points less than any assignable line. It must be conceded that this demonstration is rigorous, if the number of points in the system is finite; but the construction indicated ceases to convey any clear image to the mind, as soon as the number of points becomes infinite.[58]

Why did Hankel and his predecessors fail to develop a more general and autonomous theory of sets? Since they, like Cantor, were for the most part at work within the framework of trigonometric series, similarities between their approaches and their results are to be expected. But why did no one, until Cantor, develop an independent theory? Though Hankel's analysis, like that of Lipschitz, or Riemann, or Dirichlet, included a naive sort of real-line topology, it was always secondary, and served only as a means to producing certain theorems in functional analysis. The set theory implicit in their work was neither emphasized in more general terms, nor developed.

Cantor, for reasons to be discussed in the next chapter, was led to focus his attention upon the ways in which point sets with various specified properties might be defined. His approach, moreover, required the development of a rigorous theory of real numbers. This was a necessary step before point sets of complicated structure could be satisfactorily identified, described, and analyzed. He was also the first to realize that there were differences in magnitude that had to be identified among infinite sets, and this discovery proved to be a turning point in his study of infinite sets as a subject wholly independent from the theory of functions (see Chapter 3). The next two chapters illustrate the significant innovations in the conceptualization and the approach to functional analysis that made Cantor's study of singular points and trigonometric series unique among his contemporaries. Though his advances were indebted to the results of his colleagues and predecessors, Cantor was able to modify old ideas and to forge new ones in order to establish the uniqueness of functional representations by means of trigonometric series. In the process, Cantor began to formulate an entirely new mathematical discipline: transfinite set theory.

CHAPTER 2

The Origins of Cantorian Set Theory: Trigonometric Series, Real Numbers, and Derived Sets

On December 14, 1866, Georg Cantor received his *Promovierung,* officially marking the end of his studies at the University of Berlin. With the exception of his first semester as a student in Zürich, and another in Göttingen during the summer of 1866, he had remained in Berlin to study under the greatest mathematicians of the day: Kummer, Kronecker, and Weierstrass. His major interest then was number theory, and his dissertation, "De aequationibus secundi gradis indeterminatis," was lauded by its readers as *docta et ingeniosa,* "learned and clever."[1] He passed his oral examinations before the faculty *magna cum laude,* and when he was finished with his studies at age 22, he taught briefly at a local school for girls in Berlin. In 1868, having passed the formalities of the Prussian *Staatsprüfung,* he joined the prestigious Schellbach Seminar for mathematics teachers.

The following year he left Berlin to accept an appointment to teach as Privatdozent at the university in Halle, not far from Leipzig. Upon his arrival Cantor presented the faculty with his *Habilitationsschrift,* which again discussed a problem in number theory dealing with the transformations of ternary quadratic forms.[2] Like his dissertation, Cantor's *Habilitationsschrift* reflected his early interest in the theory of numbers, though his great creation of transfinite set theory was not indebted to this early work. Instead, responding to suggestions of the mathematician Edward Heine, Cantor shortly found new inspiration in analysis, and in particular, in the study of trigonometric series.

Heine, one of Cantor's senior colleagues at Halle, was then at work on problems in analysis and was completing a study of trigonometric series when he first encouraged Cantor to take up a particularly difficult question: given an arbitrary function represented by a trigonometric series, is the representation necessarily unique? Uniqueness theorems for special classes of functions under certain assumptions had already been given by others. Heine in particular had published early in 1870 the following theorem:

Theorem: A function $f(x)$, continuous in general but not necessarily finite, can be represented in only one way by a trigonometric series of the form:

$$f(x) = \tfrac{1}{2} a_0 + \Sigma\, (a_n \sin nx + b_n \cos nx), \tag{2.1}$$

if the series is subject to the condition of being uniformly convergent in general. The series represents the function in general from $-\pi$ to π.[3]

Requiring uniform convergence and continuity except for at most a finite number of points, Heine's theorem invited direct generalizations. Nevertheless, in the opening paragraphs of his paper, Heine emphasized that even the great minds, including Dirichlet, Lipschitz, and Riemann, had been unable to solve the uniqueness problem. They had only managed to confirm for a number of specific cases, and under certain restrictions, that arbitrary functions could be represented by trigonometric series. But as Heine stressed, they had failed to determine the number of different ways in which such representations could be given.[4]

With this challenge so neatly articulated by Heine in mind, Cantor set out to show that there was only one way in which such a development could occur, assuming that one existed. He was also determined to do away with as many restrictive assumptions as he could, but this made the problem considerably more difficult. For the most general solution he could give, entirely new concepts were needed.

FIRST RESULTS CONCERNING TRIGONOMETRIC SERIES

One important new concept introduced in the nineteenth century was that of uniform convergence. Karl Weierstrass, one of Cantor's professors at Berlin, had defined and stressed the significance of such convergence in his lectures on functional analysis. Cantor himself noted that the uniqueness problem could not be solved, as had been generally assumed, by multiplying each term of the representation series (2.1) by $\cos n(x-t)dx$, followed by term-by-term integration from π to $-\pi$. This process assumed not only the integrability of $f(x)$, but without the uniform convergence of the series, term-by-term integration was also illegitimate. Cantor gave the requirements for integration, term-by-term, as follows:

If one sets

$$f(x) = A_0 + A_1 + \ldots + A_n + R_n, \tag{2.2}$$

> for any given quantity ϵ there must exist an integer m such that, for $n \geq m$, the absolute value of R_n is less than ϵ for all values of x which come into consideration.[5]

Denoting the value m by which uniform convergence was established as a discontinuous function of x and ϵ, that is, $m(x,\epsilon)$, Cantor summarized the difficulty of the traditional approach:

> One doesn't know, for a given ϵ, whether the function $m(x,\epsilon)$ lies within finite limits for all values of x. It is even easy to see that if $f(x)$ is discontinuous for $x = x_1$, the function $m(x,\epsilon)$, for constant ϵ, must take on values which exceed any given limit as x approaches x_1 arbitrarily closely.[6]

Thus it was clear that one could not expect to probe in general the uniqueness of functional representations by trigonometric series in this way. Instead, turning first to Riemann's *Habilitationsschrift*, Cantor showed how a proof sent to him by Schwarz could be used to demonstrate the uniform convergence of the remainder terms of a special series. This result, in turn, enabled him to show that if a function admitted representation by a trigonometric series, then the representation was unique.

Assuming there were *two* trigonometric series representing the *same* function $f(x)$ and convergent to the same value for each value of x, Cantor subtracted the two, thus producing a representation of zero likewise convergent for all values of x:

$$0 = C_0 + C_1 + C_2 + \ldots + C_n + \ldots, \tag{2.3}$$

where

$$C_0 = \tfrac{1}{2}d_0 \quad \text{and} \quad C_n = c_n \sin nx + d_n \cos nx.$$

A month earlier Cantor had already published a preliminary result needed for his proof of the uniqueness theorem:

> *Theorem* (March 1870; known as the Cantor-Lebesgue Theorem): If two infinite sequences $a_1, a_2, \ldots, a_n, \ldots$ and $b_1, b_2, \ldots b_n, \ldots$ are so constituted that the limit of $a_n \sin nx + b_n \cos nx$ is equal to zero with increasing values of n for every value of x in a given interval $(a > x > b)$, then a_n as well as b_n converge to zero as n increases.[7]

Consequently, Cantor could conclude that the representation of zero, actually the starting point for his proof of the uniqueness theorem, involved a trigonometric series whose coefficients c_n and d_n, with increasing index, became arbitrarily small. The uniqueness theorem itself could therefore be established if Cantor could show that the coefficients c_n and d_n were identically zero on all indices.

Following Riemann, Cantor constructed the function[8]

$$F(x) = C_0 \frac{x \cdot x}{2} - C_1 - \ldots - \frac{C_n}{n \cdot n} - \ldots . \tag{2.4}$$

The Riemann function $F(x)$ was not only continuous in the neighborhood of every value of x, but its second derivative:

$$\lim_{\alpha \to 0} \frac{F(x+\alpha) + F(x-\alpha) - 2F(x)}{\alpha \cdot \alpha}, \tag{2.5}$$

approached zero as α diminished without limit.[9]

Cantor realized that the uniqueness theorem would follow easily if only he could show that $F(x)$ was linear. In fact, on February 17, 1870, he wrote to his friend H.A. Schwarz and asked if there might not be some way to show that the Riemann function had to assume the form $F(x) = cx + c'$. Schwarz replied that it was indeed possible to demonstrate the linearity of $F(x)$, and he gave a proof establishing the desired property.[10] Taking $F(x) = cx + c'$, Cantor then rearranged the terms of (2.4):

$$c_0 \frac{x \cdot x}{2} - cx - c' = C_1 + \frac{C_2}{2^2} + \ldots + \frac{C_n}{n^2} + \ldots . \tag{2.6}$$

Rewriting this result as

$$C_0 \frac{x^2}{2} - cx - c' = \sum (a_n \sin nx + b_n \cos nx)/n^2, \tag{2.7}$$

it was clear that since $\Sigma (a_n \sin nx + b_n \cos nx)/n^2$ had a period of 2π, the left-hand side of the equation must also be periodic. This could only happen in equation (2.7) if $C_0 = 0$ and $c = 0$. Cantor immediately reduced (2.7) to the following:

$$-c' = C_1 + \frac{C_2}{2^2} + \ldots + \frac{C_n}{n^2} + R_n, \tag{2.8}$$

an equation for which, given any $\epsilon > 0$, he could then produce a whole number m such that for all values of $n \geq m$, the absolute value of R_n was less than ϵ. This held, as Cantor italicized, "*for all values of x*."[11] Since the convergence of R_n was therefore uniform, he could employ the conclusions from Weierstrass' work on uniform convergence and therefore multiply each term in (2.3) by $\cos n(x-t)dx$.

Performing the now legitimate integration from $-\pi$ to π, Cantor concluded that

$$c_n \sin nx + d_n \cos nx = 0, \tag{2.9}$$

and consequently, that $c_n = 0$, $d_n = 0$. Thus it was established that a representation of zero by a trigonometric series convergent for all values of x was possible only if all coefficients c_n and d_n of the series (2.3) were identically zero. Cantor's uniqueness theorem then followed immediately:

> *Theorem* (April 1870): If a function of a real variable $f(x)$ is given by a trigonometric series convergent for every value of x, then there is no other series of the same form which likewise converges for every value of x and represents the function $f(x)$.[12]

CANTOR'S *NOTIZ* AND THE UNIQUENESS THEOREM FOR FINITE EXCEPTIONAL SETS

Despite so impressive an achievement, this was not the end of Cantor's work on trigonometric series. In a very short time he was able to obtain several valuable refinements, and in January 1871 he published these as a short "Notiz" to his theorems of the previous year.[13] First he noted that Kronecker had offered a suggestion making the uniqueness theorem independent of the March proof of the Cantor-Lebesgue theorem. By taking

$$(2.10) \qquad 0 = C_1 + C_2 + \ldots + C_n + \ldots$$

where $C_n = a_n \sin nx + b_n \cos nx$, one could substitute for x the values $x = u+x$ and $x = u-x$. Adding these produced the series:

$$(2.11) \quad 0 = e_0 = e_1 \cos x + e_2 \cos 2x + \ldots + e_n \cos nx + \ldots,$$

where $e_n = c_n \sin nu + d_n \cos nu$. If u was then replaced by x, the coefficients e_n then coincided with the elements C_n of (2.10). Now Cantor's previously established result for (2.10) could similarly be applied to (2.11), producing $e_n = c_n \sin nu + d_n \cos nu = 0$, and consequently, $c_n = d_n = 0$. Cantor was so impressed with the technique Kronecker had suggested that he used it again, in a fourth paper on trigonometric series, to improve the argument of his first paper of March 1870 so that "it would leave nothing to be desired in the way of clearness and simplicity."[14]

It is worth noting that Kronecker's interest in helping Cantor simplify his proof of the uniqueness theorem shows that the two were still on good terms during Cantor's first years at Halle. But as Cantor began to produce remarkable extensions of his uniqueness theorem, allowing infinitely many points in the domain of definition where exceptions to the representation of the function or to the convergence of the series might be permitted, Kronecker became increasingly uneasy. As Cantor began to press further, developing ideas which Kronecker thoroughly opposed, Kronecker became Cantor's earliest and most determined critic. But in 1871 he was still sufficiently interested in Cantor's work to offer friendly advice. The most significant part of the "Notiz," however, was not the simplification suggested by Kronecker.

Originally, Cantor's uniqueness theorem had required that the representation of zero by a convergent trigonometric series hold for *all* values of x in the function's domain of definition. By devising a simple modification of this initial assumption, however, he managed to transform certain elements of that proof into an embryonic mathematical idea, one which thenceforth grew in Cantor's mind, and which eventually led to his theory of real numbers, to the transfinite numbers, and to set theory itself.

The first generalization of the uniqueness theorem, offered in the "Notiz," was in no way unconventional. As he expressed it himself, the new theorem "allows the assumptions to be modified in the following sense, that for certain values of x either the representation of zero by (2.10) or the convergence of the series is given up."[15] Heine had done as much in his paper on trigonometric series by permitting conditions of continuity and convergence to apply "in general."

Cantor extended the validity of his uniqueness theorem by assuming the existence of an infinite sequence of values $\ldots, x_{-1}, x_0, x_1, \ldots$, increasing with index such that for these x_n, either convergence or representation of zero by the series (2.10) was given up. The important restriction, that these x_n occur in finite intervals in *only finite numbers,* was stressed.

Returning to the Riemann function $F(x)$ introduced in his original proof, Cantor noted that $F(x)$ was again a linear function of the form $F(x) = k_\nu x + l_\nu$ for each interval $(x_\nu \ldots x_{\nu+1})$. He then showed that each of these must be identical over all the intervals $(x_\nu \ldots x_{\nu+1})$. To do this, he took successive pairs of intervals, and hence the neighboring linear functions $k_\nu x + l_\nu$, $k_{\nu+1} x + l_{\nu+1}$, and then applied a method Heine had already exploited in his paper "Über trigonometrische Reihen."[16] The continuity of $F(x)$, plus Lemma II from Riemann's *Habilitationsschrift* for the value $x_{\nu+1}$ of x, gave

$$F(x_{\nu+1}) = k_\nu x_{\nu+1} + l_\nu, \tag{2.12}$$

and

$$\lim \frac{x_{\nu+1}(k_{\nu+1} - k_\nu) + l_{\nu+1} - l_\nu + (k_{\nu+1} - k_\nu)}{\alpha} = 0 \text{ for } \lim \alpha = 0. \tag{2.13}$$[17]

As Cantor showed, this was only possible if $k_\nu = k_{\nu+1}$, $l_\nu = l_{\nu+1}$, and thus the identity of $F(x)$ over all the intervals $(x_\nu \ldots x_{\nu+1})$ was established for every ν. By appealing to the same reasoning used in the earlier proof of the uniqueness theorem, the linearity of $F(x)$, shown to be identical over all the intervals $(x_\nu \ldots x_{\nu+1})$, led to the conclusion that $c_n = d_n = 0$.

It would be tempting to suggest that this paper marked a transition in Cantor's thinking. In a certain sense it did. It was in the 1871 "Notiz" that he first considered exceptional sets. On the other hand, he was still very much at work within the established traditions of mid-nineteenth century mathematics, tinkering with theoretical tools which he had borrowed from Riemann, Heine, and

Schwarz. Like his theorems published in 1870, the conclusions Cantor reached in 1871 were notable contributions to the mathematical literature of his time, but they were primarily achieved through previously available results. By April 1872, however, he had produced some important innovations which grew directly out of the more general uniqueness theorem of 1871. The "Notiz," however, was only the beginning: "This extension of the theorem is by no means the last," wrote Cantor. "I have succeeded in finding an equally rigorous but much broader extension which I shall communicate when I have an opportunity to do so."[18] The opportunity was not long in coming.

CANTOR'S REVIEW OF HANKEL'S *UNIVERSITÄTSPROGRAMM*

Early in 1871 Cantor published a review of Hankel's Tübingen *Universitätsprogramm* of 1870. Cantor told readers of the *Deutsches Zentralblatt* that he was particularly impressed with the author's "principle of the *condensation of singularities.*"[19] Coupled with the fact that Hankel's work was stimulated by Riemann's *Habilitationsschrift* on trigonometric series, Hankel's method of creating functions which were discontinuous on all rational points of a given domain of definition must have stimulated Cantor's imagination. It may have been Hankel's memoir, in fact, which prompted Cantor to search for certain novel condensations of singular points distributed in such a way that he might extend the results of his uniqueness theorem beyond merely finite numbers of exceptional points.

With Hankel's method of the condensation of singularities in mind, Cantor's "Notiz" provided both a point of departure and a valuable guide to the possible argument that a deeper proof of the uniqueness theorem might follow. Working with Riemann's function $F(x)$, Cantor had already shown that it had to be identical over two intervals separated by a singularity x_0. Finitely many such singularities in a given interval (α,β) posed no additional difficulty, but what might happen if an infinite number of such exceptional points x_ν were allowed in (α,β)? The Bolzano-Weierstrass theorem ensured there was at least one condensation point in any neighborhood of which an infinite number of the singular points x_ν were to be found. Suppose there were only *one* such accumulation point x' in (α,β). Considering only the interval (α,x'), any proper subinterval (s,t) could contain only a *finite* number of the exceptional points x_ν, otherwise there would have been another condensation point in (s,t), contrary to the assumption that x' was the only point of condensation in all of (α,β). For all intervals (s,t) however, and their strictly finite number of singularities x_ν, Cantor's result from the "Notiz" ensured that $F(x)$ was linear and identical over each one. Since $F(x)$ was also continuous, and since the end-points s and t might be brought arbitrarily close to (α, x'), it was possible to conclude that $F(x)$ was also linear over (α, x'). The same argument was clearly true if a finite number of such condensation points $x_0', x_1', \ldots, x_n'$ were admitted. Thus it was possible to establish Cantor's uniqueness theorem even for infinitely many points of exception, so long as they were distributed in this specified way. Moreover, there was no reason to stop with only a finite number of such points of condensation x_ν'. If there were infinitely many of these distributed in (α,β), then again the Bolzano-Weierstrass theorem ensured there would be at least one condensation point of the infinitely many x_ν'. Suppose there were only one such accumulation point, denoted x''. It would then follow, as before, that only a finite number of the points x_ν' could occur in any sub-interval (s,t) of (α,x''). But $F(x)$ was linear over such intervals, as previously demonstrated. By letting the endpoints s and t approach α,x'' as nearly as one liked, it was a direct conclusion that $F(x)$ must be linear over (α,x''). The argument was easily extended any finite number of times, to higher and higher levels of condensations of condensation points of singularities.

But technical difficulties intervened. The idea was one thing; making it work was quite another. How could Cantor describe such a procedure in a clear and rigorous way? How might he disentangle the various levels of singularities condensed one atop another? Given such complex point sets, how was it possible to identify their elements and to distinguish one from another in a mathematically precise manner? The only satisfactory solution was to proceed arithmetically, but Cantor discovered that to do so required a rigorous theory of real numbers. This in turn raised the difficulty of relating the arithmetic continuum of real numbers with the geometric continuum of points on a line. Thus before any significant extension of his uniqueness theorem was possible, there were a number of fundamental problems that had to be resolved.

CANTOR'S THEORY OF IRRATIONAL NUMBERS

The first section of Cantor's paper of 1872 was devoted to the real numbers, in particular to the irrationals. In a paper published somewhat later, he stressed explicitly the objections he had to previous attempts to define irrational numbers in terms of infinite series.

> Here there would be a logical error, since the definition of the sum Σa_ν would first be won by equating it with the finished number b, necessarily defined beforehand. I believe that this logical error, first avoided by Weierstrass, was earlier committed quite generally, and was not noticed because it belongs to those rare cases in which real errors can cause no significant harm in calculations.[20]

Thus Cantor set himself the task of developing a satisfactory theory of the irrationals which in no way presupposed their existence. He followed Weierstrass in beginning with the rational numbers.[21] The set of all rationals (A), including zero, thus provided a foundation for all further concepts of number. Cantor then was explicit that whenever he spoke of a numerical quantity in any further sense he did so above all in terms of infinite sequences of rational numbers $\{a_n\}$:

> *Definition:* The infinite sequence
>
> (2.14) $$a_1, a_2, \ldots, a_n, \ldots$$
>
> is said to be a *fundamental sequence* if there exists an integer N such that for any positive, rational value of ϵ, $|a_{n+m} - a_n| < \epsilon$, for whatever m and for all $n > N$.[22]

If a sequence $\{a_n\}$ satisfied this condition, then Cantor said that "the sequence (2.14) has a definite limit b."[23] This had a very special meaning for Cantor and must be taken as a convention to express, not that the sequence $\{a_n\}$ actually had the limit b, or that b was presupposed as the limit, but merely that with each such sequence $\{a_n\}$ a definite symbol a was associated with it.[24] At this point in his exposition Cantor was explicit in using the word "symbol" (*Zeichen*) to describe the status of b. He then defined the order relations between the fundamental sequences as follows: If $\{a_n\}$ was associated with b, $\{a_n'\}$ with b', then if for all n greater than some arbitrarily large N, $a_n - a_n' < \epsilon$, one could say that $b = b'$; if $a_n - a_n' > \epsilon$, $b > b'$; if $a_n - a_n' < -\epsilon$, $b < b'$. Thus the sequence (2.14), associated with the limit b, could have only one of three relations with a rational number a (where a could be defined by the constant sequence $\{a\}$, which was clearly a fundamental sequence and for which the value a was actually the value reached by the sequence in the limit). Cantor noted, by comparing the sequences, that either $b = a$, $b < a$ or $b > a$, and concluded:

> From these and the definitions immediately following it follows that if b is the limit of the sequence (2.14), then $b - a_n$ becomes infinitely small as n increases, whereby, *incidentally*, the designation "limit of the sequence (2.14)" for b finds a certain justification.[25]

The definitions to which Cantor referred were those for which arithmetic operations might be extended from the domain of rational numbers to the new numbers b (Cantor now called them numbers [*Zahlengrössen*] instead of symbols). Given three numbers b, b', b'' of the domain B, he considered the infinite sequences $\{a_n\}$, $\{a_n'\}$, and $\{a_n''\}$, respectively, associated with each. He then took the formulae $b + b' = b''$, $b \cdot b' = b''$, $b/b' = b''$ as expressing the following relations:

$$\lim\ (a_n + a_n' - a_n'') = 0,$$

$$\lim\ (a_n \cdot a_n' - a_n'') = 0,$$

$$\lim\ (a_n/a_n' - a_n'') = 0.$$

With elementary arithmetic operations thus extended to the elements of B, and having stressed the naturalness of describing them as "limits" of fundamental sequences of rational numbers, Cantor felt justified in abandoning his earlier use of "symbol" and now designated the elements of B as "numbers."[26] Certainly there was something reassuring in being able to refer to the familiar real numbers of B as *numbers,* rather than to some (the rationals) as numbers, and to others (the irrationals) as symbols. But despite the return to familiar terminology, there was a philosophical problem concerning the nature and existence of the domain B generated from A. Cantor regarded the numbers of B as meaningless in themselves. On this point he was explicit and emphatic, arguing that he had built a theory "in which the numbers (above all lacking general objectivity in themselves) appear only as components of theorems which have objectivity, for example, the theorem that the corresponding sequence has the number as limit."[27] Hence the numbers of domain B received whatever objectivity they might have from the sequences to which they were related. Clearly their objectivity was of a very different kind from that enjoyed by the rational numbers of domain A. Consequently an element $b \in B$ should be considered as a number only for the sake of convenience. Ultimately it only represented a fundamental sequence.[28]

Just as Cantor had succeeded in generating the numbers of B from fundamental sequences of numbers from A, he likewise built the domain C in terms of fundamental sequences of elements from B. With each fundamental sequence $\{b_n\}$ there was associated a definite limit c. The entirety of all such numbers c constituted the domain C. While for every $a \in A$ there was a corresponding $b \in B$, the converse was false. However, not only was it possible to equate each $b \in B$ with an element $c \in C$, but every $c' \in C$ also had its equal $b' \in B$. Although the elements of the domains B and C could thus be equated, Cantor underscored their very important conceptual difference. This difference was most easily explained by the fact that even in the domain B the "identity" of two elements $b=b'$ only expressed a certain relation between the sequences to which the limits b, b' were related.[29]

Cantor proceeded to define higher-order domains from C. Proceeding through λ such constructions, he reached the domain L, which was closed under the arithmetic operations and upon which order relationships were defined as above for the domain B. Just as B and C were in a special sense reciprocally identifiable, so too was the domain L reciprocally coincident with all the generated domains $B, C, \ldots, K$. A of course was still excepted, and though A could be described as imbedded in L, L was much richer in elements than the merely rational set A. Given an element l in the domain L, reached by λ successive generalizations from the domain A, Cantor explained: "It is given as a number, value, or limit of the λth kind, by which it is evident that I use the words *number, value* and *limit* in general as synonymous in meaning."[30] The equivalence of Cantor's vocabulary was justifiable, again by the special way in which he had taken the elements of the λth domain L as numbers. Though he stressed that these λth domains were valuable conceptually,[31] he reserved the opportunity for discussing the absolute extensions he had in mind for a later occasion.[32] Clearly he had already considered the possibility of extending his definitions and operations for the benefit and advancement of analysis itself, as he put it, though he was well aware that this would require a more careful approach than that furnished in the 1872 paper. Nevertheless, as he had best described it himself, the germ of a new and significant development for mathematics was here laid out, embryonically, for the first time.

Cantor next turned to the task of identifying numbers with points of the geometric continuum. For the rational numbers this was not a difficult problem. He knew that given a point on the line, if it had a rational relation from the origin o to the unit abscissa, then it could be expressed by an element from the domain A. Otherwise, it could be approached by a sequence of rational points (2.15)

$$a_1, a_2, a_3, \ldots, a_n, \ldots,$$

each of which corresponded to an element in A. Furthermore, Cantor could take $\{a_n\}$ as a fundamental sequence which approached the given point arbitrarily closely. He expressed this condition as follows: "The distance of the point to be determined from the point o (the origin) is equal to b, where b is the number corresponding to the sequence (2.15)."[33] Since every element from A was uniquely embedded in the domain B, the uniqueness of this representation of the points of the line and the domain B followed easily. Cantor was unable to show, of course, the converse, that for every element of the domain B there corresponded a unique point of the real line. Thus he had to invoke the well-known axiom: "that also conversely, to every number there corresponds a definite point of the line, whose coordinate is equal to that number."[34] This axiom served two functions for Cantor. It gave a sense of symmetry and completeness to the relation between points of the line and real numbers, a relation which would have been one-sided otherwise, with no guarantee that for all elements of the domain B there corresponded unique elements of the line itself. Equally important, it identified geometric points with the arithmetically derived domains $B, C, \ldots$, thereby giving the elements of these domains a further objectivity. Cantor stressed, however, that the numbers in these various domains remained entirely independent of this geometric identification, and the isomorphism served, really, as an aid in thinking about the numbers themselves.[35] But above all, Cantor had devised a scheme whereby it was possible to identify with precision and rigor complicated configurations of points distributed in certain specified ways on the geometric continuum. He was at last in a position to introduce one of his most innovative concepts: derived sets of the first species.

DERIVED SETS OF THE FIRST SPECIES

Though the geometric character of a point set might make it somewhat easier to visualize, Cantor cautioned that whenever he used the word "point," the associated numerical value of the given point should always be kept in mind. He noted that a given finite or infinite collection of numbers was termed a *Wertmenge;* correspondingly, a finite or infinite collection of points on the line a *Punktmenge*.

> *Definition:* By a "limit point of a point set P" I mean a point of the line for which in any neighborhood of same, infinitely many points of P are found, whereby it can happen that the (limit) point itself also belongs to the set. By a "neighborhood of a point" is understood an interval which contains the point in its interior. Accordingly, it is easy to prove that a point set consisting of an infinite number of points always has at least *one* limit point.[36]

This last conclusion was the Bolzano-Weierstrass theorem that any infinite bounded point set must have at least one limit point. With this result, Cantor could describe the immediate relationship between a given point set P and any point of the real line, namely, that every point of the line was either a limit point of the set P, or it was not. Thus with any point set P there was always given, conceptually, its set of limit points, which Cantor denoted as P' and called "the *first derived point set* of P."[37] If P were the set of all rational points of the line corresponding with the set of all rational numbers A, then P' was the set of all points of the linear continuum corresponding with the set B of all real numbers. Just as he had been able to generate from B an entire system of λ-domains, he did the same with P'. If P' was an infinite point set, then it gave rise to a second derived point set P'', and so on, until "one finds through ν such transitions the concept of the νth derived point set $P^{(\nu)}$ of P."[38]

Cantor had succeeded in translating his fundamental sequences $\{a_n\}$, which had served to define the domain B, into the language of point sets, where the limit points corresponded to the elements of B and were defined by an infinite set of points (corresponding to the fundamental sequence) in any arbitrarily small neighborhood of that point. The case which then became interesting for Cantor's uniqueness theorem was the one in which, after ν repetitions, the derived point set $P^{(\nu)}$ consisted only of a finite number of points, thus making the extension to further derived sets impossible, that is, $P^{(\nu+1)}$ did not exist.[39] When this occurred, Cantor designated the original set P as being of the νth kind. All point sets of the νth kind, for finite ν, constituted the derived sets of the first species.[40]

Cantor showed that such point sets existed by appealing to a point whose abscissa was determined by a number of the νth kind. Retracing his steps from

$P^{(\nu)}$ to $P^{(\nu-1)}$, $P^{(\nu-2)}$, . . . , he finally reached a set Q in the domain A which produced an infinite set of rational numbers. Translated back into the language of point sets, the point set Q was then a set of the νth kind, and $Q^{(\nu+1)}$ did not exist.[41]

Thus Cantor had achieved his goal, the determination of infinite point sets with structural properties neatly identified by his numbers of the νth kind and tailored specifically to the requirements of his extension of the uniqueness theorem. The proof of the theorem itself was an easy matter once the derived sets of the first species had been introduced.

THE UNIQUENESS THEOREM FOR INFINITE EXCEPTIONAL SETS

Theorem: If an equation is of the form

$$0 = C_0 + C_1 + C_2 + \ldots + C_n + \ldots, \tag{2.16}$$

where $C_0 = \frac{1}{2}d_0$, $C_n = c_n \sin nx + d_n \cos nx$ for all values of x with the exception of those which correspond to the points of a given point set P of the νth kind in the interval $(0 \ldots 2\pi)$, where ν denotes any integer, then $d_0 = 0$, and $c_n = d_n = 0$.[42]

The proof itself proceeded in a fashion similar to the methods used in the 1870 uniqueness theorem and the first extension of that theorem to finite exceptional sets in 1871. Considering the Riemann function,

$$F(x) = C_0 \frac{xx}{2} - \frac{C_2}{4} - \ldots - \frac{C_n}{nn} - \ldots, \tag{2.17}$$

Cantor showed that given the nature of a set P of the νth kind, there must be an interval $(\alpha \ldots \beta)$ in which no points of P occurred. On such an interval $(\alpha \ldots \beta)$, by virtue of the assumed convergence of the series, $\lim (c_n \sin nx + d_n \cos nx) = 0$ for all values of x in $(\alpha \ldots \beta)$. Thus, by appealing to his 1871 paper on trigonometric series, it followed that $\lim c_n = 0$ and $\lim d_n = 0$. Now, by introducing certain results from Riemann's *Habilitationsschrift*,[43] Cantor claimed the following properties for $F(x)$:

1. $F(x)$ was continuous in the neighborhood of any value of x.
2. $\lim \dfrac{F(x+\alpha) + F(x-\alpha) - 2F(x)}{\alpha \cdot \alpha} = 0$ if $\lim \alpha = 0$ for all values of x, with the exception of those values which corresponded to points of the set P.
3. $\lim \dfrac{F(x+\alpha) + F(x-\alpha) - 2F(x)}{\alpha} = 0$ if $\lim \alpha = 0$ for all values of x without exception.[44]

He then claimed that $F(x)$ was a linear function, namely that $F(x) = cx + c'$ over the entire interval $(0 \ldots 2\pi)$. For sub-intervals $(p \ldots q)$ which contained only a finite number of points $x_0, x_1, \ldots, x_r$ from P, he considered the collection of sub-intervals into which (p, q) was divided by these exceptional points. The *linearity* of $F(x)$ over each sub-interval followed from properties 1 and 2, just as he had argued in his first uniqueness proof of 1870. The *identity* of the linear functions $F(x)$ on each sub-interval was established from properties 1 and 3, just as he had argued in his "Notiz" of 1871. Consequently,

> (A) If $(p \ldots q)$ is any interval in which only a finite number of points of the set P lie, then $F(x)$ is linear in this interval.[45]

Cantor then considered intervals $(p' \ldots q')$ which contained only a finite number of points $x_0', x_1', \ldots, x_\nu'$ of the *first derived set* P' of P. Here again he needed to show that not only was $F(x)$ linear over each of the sub-intervals $(x_\nu' \ldots x_{\nu+1}')$, but that it was in fact identical over the entire interval $(p' \ldots q')$.

He began with the sub-interval $(x_0' \ldots x_1')$. Since, in general, $(x_0' \ldots x_1')$ contained an infinite number of points from the originally given point set P, the "Notiz" result (A) could not be applied. However, each sub-interval $(s \ldots t)$ falling entirely within $(x_0' \ldots x_1')$ could have only a finite number of points from P, otherwise there would be a limit point in $(s \ldots t)$. But it had been assumed that there was no limit point of P between x_0 and x_1. Now the result (A) was applicable and the function $F(x)$ had to be linear over $(s \ldots t)$. Since the endpoints s and t could be taken arbitrarily near the endpoints x_0' and x_1', it followed that the continuous function $F(x)$ was also linear over $(x_0' \ldots x_1')$. Cantor applied this conclusion over each sub-interval $(x_\nu' \ldots x_{\nu+1}')$ of $(p' \ldots q')$ and deduced as before the following:

> (A') If $(p' \ldots q')$ is any interval in which only a finite number of points of the set P' lie, then $F(x)$ is linear in this interval.[46]

The proof then proceeded analogously, considering intervals $(p^{(k)} \ldots q^{(k)})$ which contained only finitely many points of the kth derived point set $P^{(k)}$ of P. $F(x)$ was linear over all such intervals. Since P was assumed to be of the νth kind, however, Cantor could press his conclusion. After a finite number of steps he reached the intervals which contained points of $P^{(\nu)}$. There could only be a finite number of such points, and there were no derived sets beyond $P^{(\nu)}$. Thus he concluded that $F(x)$ was linear in any arbitrarily given interval $(a \ldots b)$. $F(x)$ could therefore be given the form $F(x) = cx + c'$, and the proof followed exactly as it had in both of the previous versions of the uniqueness theorem which Cantor had given.

DERIVED SETS OF THE SECOND SPECIES

In order to prove his uniqueness theorem over infinite exceptional sets, Cantor was restricted to the consideration of point sets of the first species, those for which $P^{(\nu)} = 0$ for some finite $\nu < \infty$. But it is clear, even from his own remarks, that by 1872 he had proceeded well beyond the simple sequence of derived sets of the first species, $P^{(1)}$, $P^{(2)}$, . . . , $P^{(\nu)}$, to consider the entire sequence of derived sets of the second species as well, those sets for which $P^{(\nu)} = 0$ for *no* finite value of ν, however large. Taking ∞ as the least infinite number beyond all the finite numbers, hence $\nu < \infty$ for all finite values of ν, sets of the second species P were consequently those for which $P^{(\infty)} \neq 0$. In fact, it was possible to proceed beyond $P^{(\infty)}$ to even higher orders of derived sets of the second species, including:

$$P^{(\infty)}, P^{(n\infty^{\infty})}, P^{(\infty^{\infty}+1)}, P^{(\infty^{\infty}+n)}, P^{(\infty n^{\infty})}, P^{(\infty^{\infty^{n}})}, P^{(\infty^{\infty^{\infty}})}, \ldots . \tag{2.18}$$

Cantor was explicit about his earliest formulation of this sequence. He stressed as a note to a later paper of 1880 (a note which Zermelo in his edition of Cantor's *Gesammelte Abhandlungen* failed to include) that "I succeeded [in producing the sequence] ten years ago, and I took the opportunity in a proper presentation of the concept of number (Cantor, 1872) to hint at the discovery."[47]

In fact, a reference Cantor made originally in the 1872 paper was not so oblique as to make his meaning unclear: "The number concept, in so far as it is developed here, carries within it the germ of a necessary and absolutely infinite extension."[48] What Cantor, looking back in 1880 to his paper of 1872, described as a *dialektische Begriffserzeugung*,[49] was the conceptual generation of derived sets of the *second species* which had grown out of his general idea of derivation in a *necessary* and *absolute* way. Thus the derived set of second species $P^{(\infty)}$, where $P^{(\infty)} \neq 0$, led to a derived set of $P^{(\infty)}$ itself (whenever $P^{(\infty)}$ contained an infinite number of points), namely $P^{(\infty')}$, which Cantor denoted $P^{(\infty+1)}$. The process could easily be continued without difficulty to increasingly higher orders of derived sets of the second species.

In discussing his construction of the irrational numbers, Cantor produced the set B from the rationals A. From B proceeded the unending sequence of further derived sets $B,C,D,\ldots,L,\ldots$, a development which Dedekind, from his own perspective, found quite unnecessary.[50] Cantor of course had noted that further derivations beyond B produced no new elements in the sense that each domain derived from B was equivalent to B. Thus, as Dedekind remarked, these "higher order real numbers" contributed nothing that the first derived set B from A did not already contain. But Cantor in 1872 was interested in more than just an analysis of the real number system and a rigorous construction of the irrationals. In order to prove the uniqueness theorem over infinite exceptional sets, the derived sets of the first species were essential. While Dedekind would have left the sets of higher order derived from B to collapse by virtue of their isomorphisms to the set B itself, Cantor stressed what for him were important distinctions.[51] These differences were significant because they enabled him to specify certain sets of points which might be excepted in the proof of his uniqueness theorem. But they were also important in identifying the sets derived from B, considered dialectically in the case of *point sets of the second species,* for these led directly to Cantor's germinal idea: the sequence (2.18).

This idea itself, however, did not mature immediately, nor was it transformed into the creation of Cantor's transfinite numbers for more than a decade. In the 1870s his interest focused primarily upon point sets. Following the remarkable success of his most general proof of the uniqueness theorem, he began to study the distinctions to be made between discrete and continuous domains. Such studies, in fact, were to be major steps toward establishing an independent theory of sets.

CONCLUSION

Having shown that his uniqueness theorem could be proven even on certain infinite sets of exceptional points, Cantor knew that the nature of these sets and the derived sets themselves posed problems closely related to the structural properties of the set of all real numbers. His development of the concepts of limit point and of derived sets clearly revealed the direct association between the two. Since the nature of Cantor's proofs of the uniqueness theorem depended heavily on the nature of point relationships on the line, and only to a lesser extent on the trigonometric functions themselves, it would have been natural for Cantor to turn his attention to the former. In fact, his next published paper dealt specifically with the differences between denumerably infinite sets like the rationals and continuous sets like the reals.[52] The discoveries he was about to make demonstrated his gift for posing incisive questions, and for sometimes finding unexpected, even unorthodox answers.

The evolution of the uniqueness theorem displayed Cantor's facility for knowing how to tailor a theorem to the mathematical tools at his disposal. As he developed each new extension of the theorem, he marshalled with increasing success the previous work of Riemann, Heine, and Schwarz. He modified lemmas or the details of proof to obtain, in a clear and natural way, the uniqueness theorem in considerable generality. But it was only in the final stage of this work that he displayed the most important feature of his mathematical ability: the capacity for creating new forms and concepts when existing approaches failed. At this point Cantor established himself as an innovative member of his own generation of mathematicians, capable of forging new concepts and tools which could be used to obtain radically new kinds of solutions to problems in the theory of functions.

Thus growing out of the specific problems posed by the nature of trigonometric representations, set theory soon gained an autonomy of its own. But just as Fourier had shocked Lagrange and the mathematicians of his day with his results on the trigonometric representation problem, Cantor's work was to provide equally distressing dilemmas. Yet in his early papers on the nature and uniqueness of such representations, there was no hint of the paradoxes to come. Instead, Cantor had managed to find impressive solutions to problems which were well known to mathematicians of his own time. As he turned away from the strictly trigonometric context of his early work, a new era in mathematics was about to emerge.

CHAPTER 3

Denumerability and Dimension

Je le vois, mais je ne le crois pas.

—Cantor to Dedekind, June 29, 1877

Transfinite set theory did not follow immediately from Cantor's paper of 1872 on trigonometric series. Though his introduction of derived sets was almost immediately recognized as profitable in applications to analysis (Ulysse Dini and Gösta Mittag-Leffler were among the first to make use of Cantor's new methods), the state of his thinking in 1872 did not allow any significant progress toward a theory of infinite sets.[1] What soon brought transfinite set theory to life was Cantor's discovery that the set of all real numbers was nondenumerable.

Between 1872 and the first publication of his theory of linear point sets in 1879, Cantor made some of his most important mathematical discoveries. Not only did he establish the existence of different magnitudes of infinity, but he produced the startling result that spaces of arbitrary dimension could be uniquely mapped onto the one-dimensional line of real numbers. While this raised for a time some concern about the nature and definition of dimension, it also helped determine the character of Cantor's later research. Above all, his studies of continuity and dimension represented the first steps toward the development of a comprehensive theory of infinite point sets which Cantor began to publish in 1879.

RICHARD DEDEKIND (1831–1916): CANTOR AND CONTINUITY

Cantor was not alone in studying the properties of continuous domains in rigorous detail. Most notably, in the same year (1872) that Cantor's theory of real numbers was outlined in his paper on trigonometric series, Richard Dedekind published a short memoir: *Stetigkeit und die Irrationalzahlen.*[2] Cantor had mailed a copy of his own paper of 1872 to Dedekind, which Dedekind acknowledged in his preface. In return, Dedekind forwarded his publication to Cantor. In carefully reading Dedekind's short monograph, Cantor agreed heartily with its results, though he was never completely convinced of the naturalness of Dedekind's approach, nor of its easy application in analysis.[3] Nevertheless, their exchange of papers marked the beginning of a friendly exchange of ideas, both by letter and through discussion during vacations spent together in the Harz Mountains. Dedekind was to have an important influence upon Cantor's thinking during one of its most creative periods. Indeed, for a time Dedekind assumed the roles of mentor, counsel, and friend.

Though Dedekind did not publish his study of continuity and the real numbers until 1872, he nevertheless had made his discoveries much earlier. In 1858 he had accepted an appointment to teach at the Polytechnic School in Zürich. In preparing his lectures on the elements of the differential calculus, he was particularly struck by the lack of secure foundations for analysis, and in attempting to remedy this situation, he was led to one of his best-known contributions to mathematics: the "Dedekind cut." He first made his discovery, as he carefully recorded in one of his notebooks on November 24, 1858, and declared that he had produced in the process a true definition of continuity.[4]

Because his new theory of real numbers was not simple, and because it did not seem very promising at the time, he did little to circulate the results of his Zürich lectures. But Heine's presentation of Cantor's theory of irrational numbers in "Die Elemente der Functionenlehre," reinforced by Cantor's paper of 1872 (both came to Dedekind's attention within days of each other, in March 1872), persuaded him to present his own theory of the irrationals and the continuum.[5] For Dedekind, the nature of limits, continuity, and an arithmetic theory of the real number continuum shared important connections, just as they had in Cantor's proof of the uniqueness theorem in 1872. Moreover, he felt that his introduction of the "cut" had the advantage of being simpler and clearer, though Cantor later disputed the appropriateness of Dedekind's "cuts" for use in functional analysis. Nevertheless, there were important similarities between the approaches each had taken. Section II of Dedekind's paper compared the rational numbers with the points of a straight line. When Dedekind then turned in Section III to discuss continuity and to emphasize that in any line segment there were infinitely many points which corresponded to no rational numbers, he wrote suggestively of Cantor's next great advance: "The line L is infinitely richer in point-individuals," wrote Dedekind, "than is the domain R of rational numbers in number-individuals."[6] In this statement lay part of the secret to continuity. Though Dedekind had absolutely no idea how much richer the continuum really was in such point individuals, he had specified a feature of the real numbers that had begun to interest Cantor greatly.

Unfortunately, the earliest set-theoretic ideas appearing in Cantor's paper of 1872 had yet to be given any substantial autonomy of their own. Though the necessary elements for the transfinite numbers are recognizable in the sequence of derived sets

$$P', P'', \ldots, P^{(\nu)}, \ldots, P^{(\infty)}, P^{(\infty+1)}, \ldots,$$

Cantor had no basis for any articulate conceptual differentiation between $P^{(\nu)}$ and $P^{(\infty)}$. In 1872 he had not yet developed any precise means for defining the first transfinite number ∞ following all finite natural numbers n. Nor could he begin to make meaningful progress until he had realized that there were necessary distinctions to be made in orders of magnitude between discrete and continuous sets. Before the end of 1873 he did not even *suspect* the possibility of such differences.

After his success in applying first species sets to the uniqueness theorem for trigonometric series representations, Cantor must have been intrigued by the reasons for the validity of the result. The proof had permitted countably infinite sets for which representation was given up, with the points of exception being distributed in a carefully specified way. Nevertheless, such points could occur infinitely many times, and this fact must have raised an allied question in his mind concerning the relation between infinite sets like the natural numbers, and continuous sets like the reals.

On November 29, 1873, a year after their first meeting in Switzerland, Cantor wrote to Dedekind.[7] In his letter, Cantor raised a straightforward question which had been suggested by his analysis of irrational numbers. Though brief, the letter is significant because it reveals how Cantor had begun to reexamine the nature of continuity:

> Take the collection of all positive whole numbers n and denote it by (n); then think of the collection of all real numbers x and denote it by (x); the question is simply whether (n) and (x) may be corresponded so that each individual of one collection corresponds to one and only one of the other? At first glance one might say no, it is not possible, for (n) consists of discrete parts while (x) builds a continuum; but nothing is won by this objection, and as much as I am inclined to the opinion that (n) and (x) permit no such unique correspondence, I cannot find the reason, and while I attach great importance to it, the reason may be a very simple one.[8]

Cantor had good reasons for suspicion. One might have supposed, on purely intuitive grounds, that both the rationals and the algebraic numbers would be nondenumerable. Nevertheless, each formed a countable set. But what was the nature of the continuum? The search for an answer to this question was to haunt Cantor for the rest of his life.

How could one characterize the important difference between the rationals and the reals? The rationals were dense, between any two there was always a third, but they were not continuous. It was perhaps natural to suspect that there were *more* irrationals than rational numbers, but what did that mean? Though intuition might disagree, Cantor noted that both the rational and the algebraic numbers were countable. However, Liouville had established the existence of nonalgebraic numbers, the transcendentals, and this fact was also one for Cantor to ponder.[9] But despite his best efforts, he could find no reason to establish or to deny the countability of the reals. Cantor wondered if he might not be making some simple blunder which obscured an otherwise trivial or obvious solution, and he therefore turned to Dedekind for help. But Dedekind was in no better a position to suggest an answer, as Cantor himself indicated:

> It was a rare pleasure today to receive your reply to my last letter. I raised the question because I have considered it for a number of years and have always found myself doubting whether the difficulty it gave me was subjective or whether it was due to the subject itself. Since you write that you are also in no position to answer it, I may assume the latter.[10]

Thus it seemed that no obvious solution was forthcoming. Nevertheless, Cantor's interest in raising the question was a necessary step, and it suggests that he had begun to consider the possibility that some infinite sets were larger, in absolute or cardinal terms, than others. But what mechanism could show this in a direct way?

NONDENUMERABILITY OF THE REAL NUMBERS

The story of Cantor's discovery of the nondenumerability of the real numbers is by no means an easy one to unravel. As his interests turned away from trigonometric series to an analysis of continuity, an important element was missing: the idea of denumerability. In so far as the concept of denumerability led to the still more general, and correspondingly more useful ideas of equivalence and power, its origin in Cantor's thinking deserves careful attention.

Both Cantor and Dedekind, in formulating their axioms of continuity of the real numbers, were emphasizing the same point: the real numbers were far richer in number and in properties than the rationals. Somehow the irrational numbers managed to fill the gaps, and consequently transcended the set of rational numbers both in the sense of completeness and of continuity. But what was to account for this inherent difference? Shortly before Christmas 1873 Cantor found the secret to answering the question he had put to Dedekind more than a year earlier. It was impossible to establish a one-to-one correspondence between the natural numbers N and the real numbers R.

Cantor's original proof that the real numbers were nondenumerable was developed during December 1873 and published early in 1874.[11] The title, "On a Property of the Collection of All Real Algebraic Numbers," implied that the major accomplishment of his new methods was the proof that the set of all algebraic numbers was countable.[12] The significant result, however, which established the nondenumerability of R, was introduced in a curious way. The proof was given, Cantor claimed, in order to demonstrate an *application* of the conclusion that the set of all algebraic numbers A was countable. The application, as it turned out, was a corroboration of Liouville's proof that in any given interval (α,β) there were infinitely many transcendental numbers. Since A was countable, and R was not, Cantor could claim the same result.

In proving that R was nondenumerable, Cantor chose to argue by means of contradiction. Assuming that the real numbers *were* countable, it followed that they could be sequenced on an index of the natural numbers N:

$$\omega_1, \omega_2, \omega_3, \ldots, \omega_\nu, \ldots. \tag{3.1}$$

Cantor went on to claim that it was possible, given any interval $(\alpha, \beta) \subset R$, to find at least one real number $\eta \in R$ such that η failed to be listed as an element of (3.1). This Cantor set out to establish. Assuming that $\alpha < \beta$, he picked the first two numbers from (3.1) which fell within the interval (α, β). Denoted α', β', respectively, these were used to constitute another interval $(\alpha' \ldots \beta')$. Proceeding analogously, Cantor produced a sequence of nested intervals, reaching $(\alpha^{(\nu)} \ldots \beta^{(\nu)})$, where $\alpha^{(\nu)}, \beta^{(\nu)}$ were the first two numbers from (3.1) lying within $(\alpha^{(\nu-1)} \ldots \beta^{(\nu-1)})$. There were then two possibilities.

If the number of intervals thus constructed were finite, then at most only one more element from (3.1) could lie in $(\alpha^{(\nu)}, \beta^{(\nu)})$. It was easy in this case for Cantor to conclude that a number η could be taken in this interval which was not listed in (3.1). Clearly any real number $\eta \in (\alpha^{(\nu)}, \beta^{(\nu)})$ would suffice, as long as η was not the one element possibly listed in (3.1).

On the other hand, if the number of intervals $(\alpha^{(\nu)}, \beta^{(\nu)})$ were not finite, Cantor's argument shifted to consider two alternatives in the limit. Since the sequence $\alpha, \alpha', \ldots, \alpha^{(\nu)}, \ldots$ did not increase indefinitely, but was bounded within (α, β), it had to assume an upper limit which Cantor denoted α^∞. Similarly the sequence $\beta, \beta', \ldots, \beta^{(\nu)}, \ldots$ was assigned the lower limit β^∞. Were $\alpha^\infty < \beta^\infty$, then, as in the finite case, any real number $\eta \in (\alpha^\infty, \beta^\infty)$ was sufficient to produce the necessary real number not listed in (3.1). However, were $\alpha^\infty = \beta^\infty$, Cantor reasoned that $\eta = \alpha^\infty = \beta^\infty$ could *not* be included as an element of (3.1). He designated $\eta = \omega_\rho$. But ω_ρ, for sufficiently large index ν, would be excluded from all intervals nested within $(\alpha^{(\nu)}, \beta^{(\nu)})$. Nevertheless, by virtue of the construction Cantor had given, η had to lie in every interval $(\alpha^{(\nu)}, \beta^{(\nu)})$, regardless of index. The contradiction established the proof: R was nondenumerable.

This proof, however, was a streamlined version of a more complicated draft version sent early in December 1873 to Dedekind.[13] Subsequently modified for the proof which appeared in Crelle's *Journal*, the original draft suggests more clearly the roots of Cantor's thinking. More directly, it also bears the marks of what may be the last remnants of his attempt to prove that the real numbers, like the rational and algebraic numbers, were actually countable.[14]

Like his published proof, the argument Cantor developed in a letter to Dedekind of December 7, 1873, first assumed that the real numbers could be counted, thus producing a sequence in which every real number would appear:

(3.2) $$\omega_1,\ \omega_2,\ \ldots,\ \omega_n,\ \ldots.$$[15]

Considering (3.2) and beginning with ω_1, he then removed an infinite sequence by taking ω_α as the next number in (3.2) larger than ω_1; ω_β was then picked as the number next larger than ω_α. Continuing, analogously, Cantor changed the subscripts and added superscripts:

(3.3) $$\omega_1 = \overset{'}{\omega_1};\ \omega_\alpha = \overset{2}{\omega_1};\ \omega_\beta = \overset{3}{\omega_1};\ \ldots.$$

What emerged was an infinite sequence:

(3.4) $$\overset{1}{\omega_1},\ \overset{2}{\omega_1},\ \overset{3}{\omega_1},\ \ldots,\ \overset{n}{\omega_1},\ \ldots.$$

Removing this first sequence (3.4) from the sequence (3.2), an infinite number of elements remained. Denoting $\overset{'}{\omega_2}$ as the first term then left in (3.2), Cantor in like fashion produced the sequence:

(3.5) $$\overset{1}{\omega_2},\ \overset{2}{\omega_2},\ \overset{3}{\omega_2},\ \ldots,\ \overset{n}{\omega_2},\ \ldots.$$

In fact, he was able to generate an infinite array:[16]

(3.6)
$$\begin{array}{ll}
(1) & \overset{1}{\omega_1},\ \overset{2}{\omega_1},\ \ldots,\ \overset{n}{\omega_1},\ \ldots \\
(2) & \overset{1}{\omega_2},\ \overset{2}{\omega_2},\ \ldots,\ \overset{n}{\omega_2},\ \ldots \\
(3) & \overset{1}{\omega_3},\ \overset{2}{\omega_3},\ \ldots,\ \overset{n}{\omega_3},\ \ldots \\
\vdots & \\
\vdots & \\
(k) & \overset{1}{\omega_k},\ \overset{2}{\omega_k},\ \ldots,\ \overset{n}{\omega_k},\ \ldots \\
 & \text{where } \overset{\lambda}{\omega_k} < \overset{\lambda+1}{\omega_k}.
\end{array}$$

Attention subsequently shifted to an interval (p,q) on $(0,1)$ containing no element from (1). For example, (p,q) in $(\overset{1}{\omega_1},\overset{2}{\omega_1})$. In general there would always be some kth sequence, terms of which lay in (p,q). Otherwise the theorem would follow immediately by picking $\eta \in (p,q)$. Assuming such a kth sequence, there had to be a sub-interval $(p'\ldots q')$ in (p,q) containing no points on the kth sequence. Repeating this argument, there had to be a sequence k' with terms in $(p\ldots q)$. But Cantor could pick another sub-interval $(p''\ldots q'')$ which contained no points of k'. Thus he produced an infinite sequence of nested intervals:

(3.2) $$(p\ldots q),\ (p'\ldots q'),\ (p''\ldots q''),\ \ldots.$$

The proof then proceeded much as it did in the published version of 1874, showing that there must exist some $\eta \in R$, where $\eta \in (\alpha^{(\nu)}, \beta^{(\nu)})$ for any ν. This established the result directly, since η, were it indexed in (3.2), could be excluded from $(\alpha^{(\nu)},\beta^{(\nu)})$ for sufficiently large ν.

Almost immediately Cantor realized that his proof was more cumbersome than it needed to be, and two days later he introduced several modifications.[17] Eliminating the construction of the array (3.6), he stated simply that for any given interval $(\alpha\ldots\beta)$ in (0,1), he could show the existence of a number $\eta \in (\alpha,\beta)$ which was not included in a sequence:

$$\omega_1, \omega_2, \ldots, \omega_n, \ldots$$

assumed to exhaust (0,1). That this conclusion was the key to many problems, Cantor was certain:

> It thus follows directly that the collection (x) cannot be uniquely corresponded with the collection (n), and I conclude therefrom that among collections and sets of values distinctions exist which until recently I could not establish.[18]

Cantor had dramatically changed his opinion, and in less than a week. Earlier he had said that the matter was of no practical interest and therefore hardly worth much attention. Suddenly he saw more clearly the emergence of wholly new systems which previously he had scarcely noticed.

Cantor hastened to add that his two proofs, taken together, showing that the set of algebraic numbers was countable while the set of real numbers was not, provided an independent and totally different sort of corroboration of Liouville's proof that there were an infinite number of transcendental numbers in any given interval (α,β) of reals. But this was hardly the most significant part of Cantor's conclusion. As he described it, without particular emphasis:

> Moreover, the theorem in Section 2 represents the reason why aggregates of real numbers, which constitute a so-called continuum (say the entirety of real numbers, which are ≥ 0 and ≤ 1), cannot be uniquely corresponded with the aggregate (ν); thus I found the clear difference between a so-called continuum and an aggregate like the entirety of all real algebraic numbers.[19]

Dedekind had insisted upon the nature of continuity in his concept of the *Schnitt;* Cantor had already given his axiom of continuity in the 1872 paper. Now he had contributed another insight, equally valuable. Everyone had long suspected that continuity was somehow involved with the fact that there were "more" real numbers than natural numbers. But then it had long been assumed that there were more integers than even numbers, still more fractions, and certainly more transcendental and real numbers. It was Cantor's merit to have shown how the problem could be approached quantitatively in absolute terms.

The first step required that he develop some way of counting infinite collections, for which he developed the idea of one-to-one correspondences. Then it was a matter of discovering which aggregates were countable and how such a process might actually be described. He did this for the algebraic numbers and then showed why the same could not be done for the reals. In answering these questions with such directness and simplicity, he simultaneously indicated the direction in which his new research would develop. Armed with the tools and techniques of point sets, derived sets, and one-to-one correspondences, Cantor set out to map the unknown territory that was to become set theory.

LINES AND PLANES

With the nondenumerability of the real numbers finally established, Cantor sent a new puzzle to Dedekind on January 5, 1874:

> Is it possible to map uniquely a surface (suppose a square including its boundaries) onto a line (suppose a straight line including its endpoints), so that to each point of the surface one point of the line and reciprocally to each point of the line one point of the surface corresponds?[20]

Cantor cautioned that the solution was one of great difficulty, though one might be tempted to say that the answer was clearly "no," and that a proof was superfluous. In fact, when Cantor mentioned the same problem to friends while visiting Berlin in the spring of 1874, they were astonished at the absurdity of the question, since they were convinced that two independent variables could not be reduced to one.[21]

Apparently Dedekind was in no hurry to answer. More than three years passed before Cantor again raised the problem of mapping lines and surfaces. In the meantime, he had met with distractions of his own. Through his sister Sophie, he was introduced to Vally Guttmann.[22] Early in 1874 the two were engaged, and following their wedding in mid-summer, the newly married Cantors spent their first vacation together, meeting Dedekind in the Harz mountains. It is quite possible that the early adjustments of marriage and settling into a new way of life in Halle might have taken time away from Cantor's work. Whatever the reasons for the apparent hiatus in correspondence, Cantor wrote to Dedekind in 1877 to say that, contrary to prevailing mathematical opinion, the "absurd" correspondence between lines and planes was not impossible. Continuous spaces of ρ-dimensions, Cantor claimed, had the same power (*Mächtigkeit*) as curves. He was aware that the result raised fundamental questions for geometry, as he explained:

> This view seems opposed to that generally prevailing in particular among the advocates of the new geometry, since they speak of simply infinite, two, three, . . . , ρ-dimensional [*fach*] infinite domains. Sometimes one

will even find the idea that the infinity of points of a surface may be produced so to speak by squaring, that of a solid by cubing the infinity of points of a line.[23]

Cantor's proof as he sketched it for Dedekind was simple, but it also had a flaw, luckily one he was later able to avoid. Beginning with ρ independent variables defined on [0,1]:

$$x_1, x_2, \ldots, x_\rho,$$

he then considered a $(\rho + 1)^{th}$ variable y on [0,1], and then asked whether it was possible to map ρ numbers $x_1, x_2, \ldots, x_\rho$ to one y, so that to every ρ-tuple $(x_1, x_2, \ldots, x_\rho)$ a definite value y, and reciprocally, to every definite value y one and only one ρ-tuple $(x_1, x_2, \ldots, x_\rho)$ corresponded?

Despite years of having believed the contrary, he admitted that the only possible answer must be given in the affirmative. Any number $x \in [0,1]$ could be given a unique decimal expansion:

$$x = \alpha_1 \frac{1}{10} + \alpha_2 \frac{1}{10^2} + \ldots + \alpha_\nu \frac{1}{10^\nu} + \ldots$$

If one then considered a point x in a space of dimension ρ, its position was determined by the ρ variables:

$$x_\rho = \alpha_{\rho,1} \frac{1}{10} + \alpha_{\rho,2} \frac{1}{10^2} + \ldots + \alpha_{\rho,\nu} \frac{1}{10^\nu} + \ldots$$

Cantor then considered the single point:

$$y = \beta_1 \frac{1}{10} + \beta_2 \frac{1}{10^2} + \ldots + \beta_\nu \frac{1}{10^\nu} + \ldots \tag{3.8}$$

and devised a correspondence which uniquely mapped the x_ρ variables onto y in terms of four equations:

$$\begin{aligned} \alpha_{1,n} &= \beta_{(n-1)\rho+1} \\ \alpha_{2,n} &= \beta_{(n-1)\rho+2} \\ \alpha_{\sigma,n} &= \beta_{(n-1)\rho+\sigma} \\ \alpha_{\rho,n} &= \beta_{(n-1)\rho+\rho} \end{aligned}$$

The process was also reversible. Given a value y expressed as a nonterminating decimal, it was an easy matter of pulling the terms of (3.8) apart in order to form the ρ "independent" variables x_ρ.

Cantor was so unprepared for this result that it prompted him to exclaim, with unabashed surprise, "I see it, but I don't believe it!"[24] He was particularly impressed by the richness which his discovery revealed in the domain of real numbers. It showed, he said, "what wonderful power there is in the ordinary real (rational and irrational) numbers, since one is in a position to determine uniquely, with a single coordinate, the elements of a ρ-dimensional continuous space."[25]

But there was a difficulty in Cantor's first proof which Dedekind saw immediately.[26] In order to avoid the representation of one and the same value x twice, the assumption had to be added that no representation be allowed which, after a certain index, was always zero. Otherwise a number x of the following form would have two representations:

$$x = 3.0000\ldots$$

$$x = 2.9999\ldots$$

The only exception to the above restriction would of course be the representation of zero itself. Under these conditions, Cantor's mapping was necessarily incomplete. Considering merely the two-dimensional case, for

$$x_1 = \alpha_1, \alpha_2, \ldots, \alpha_\nu, \ldots$$

$$x_2 = \beta_1, \beta_2, \ldots, \beta_\nu, \ldots$$

the corresponding y under Cantor's map would be:

$$y = \alpha_1, \beta_1, \alpha_2, \beta_2, \ldots, \alpha_\nu, \beta_\nu, \ldots$$

Nevertheless, under y, a representation of the form

$$y = \alpha_1, \beta_1, \alpha_2, \beta_2, \ldots, \alpha_\nu, \beta_\nu, 0, \beta_{\nu+1}, 0, \beta_{\nu+2}, 0, \ldots$$

was necessarily inadmissible, since it would have decomposed as:

$$x_1 = \alpha_1, \alpha_2, \ldots, \alpha_\nu, 0, 0, 0, \ldots$$

$$x_2 = \beta_1, \beta_2, \ldots, \beta_\nu, \ldots$$

where such x_1's were by assumption beyond consideration.

Cantor realized his error at once, and sent off a postcard in complete agreement with Dedekind's warning: "Unfortunately your objection is entirely correct; fortunately it affects only the proof, not the conclusion."[27] Two days later he had managed to repair the proof but was forced to admit that the simplicity of the earlier approach had to be sacrificed in the process. And despite his original claim that the difficulty was only an error in presentation, on June 25 Cantor was saying that the result itself could only be achieved by means of unfortunate complexity. His new proof was long and lacked the directness of the earlier formulation. Cantor passed over the burdens of the new approach by hoping that with the theorem once established, he might eventually be able to simplify the proof. With Dedekind's help, he had earlier brought considerable simplification to his proof that the real numbers were nondenumerable, and it was certainly possible that together he and Dedekind might prevail once more. After all, he had managed the proof itself, and that was what mattered:

On a postcard which I sent you yesterday I acknowledged the gap you found in my proof, and also remarked that I am in a position to fill it, even though I cannot help expressing a certain regret that the matter cannot be resolved without more complicated considerations; this however is the nature of the subject and I must console myself; perhaps later it will turn out that the error in that proof can be avoided more easily than it is possible for me to do at present.[28]

Having answered one question, Cantor also recognized that he had raised another. If planes and lines could be corresponded uniquely, then the long-accepted basis for defining the dimension of a given space seemed entirely inadequate. He recognized that he, like most of his fellow mathematicians, had assumed all along that the elements of any n-dimensional continuous space were determined by n independent real coordinates. He was quick to add, however, that he had always regarded that assumption as a theorem greatly in need of a proof. In light of his new discovery, he charged "that all philosophical or mathematical deductions which make use of that erroneous assumption, are inadmissible."[29]

Dedekind's response was mixed. He congratulated Cantor on the results of his theorem, but took exception to the conclusions Cantor had drawn from the proof itself. Dedekind criticized Cantor's haste in discrediting all work, both philosophical and mathematical, which had assumed the invariance of dimension. In spite of the proof, or perhaps because of certain features of dimensionality it brought prominently to light, Dedekind was not so negative as Cantor. Furthermore, he was convinced, "that the dimension of a continuous space is the first and most important invariant (of that space), and I must defend all earlier authors on this point."[30] Dedekind naturally conceded that the proof required more than the tacit assumption of its correctness. The additional assumption, even more significant, was the universal belief that continuity played the decisive role in determining the invariance of dimension. Dedekind emphasized this second assumption by formulating it explicitly:

If a reciprocally unique and complete correspondence between the points of a continuous domain A of a dimensions on the one hand and the points of a continuous domain B of b dimensions on the other is possible, then this correspondence is necessarily completely discontinuous if a and b are *unequal*.[31]

To Dedekind, it was clear that his colleague in Halle was far too hasty in condemning those who had all along assumed invariance in terms of continuity. As he stressed, Cantor had not given any real grounds upon which to challenge that traditional assumption. Dedekind doubted that he could. The closing paragraph from his letter of July 2 was a stern reminder that Cantor should not rush his result, full of provocative but inaccurate criticism, into print: "I hope I have expressed myself clearly enough; the point of my writing is only to request that you not polemicize openly against the universal beliefs of dimension theory, heretofore held true, without first giving my objections a thorough consideration."[32]

Cantor was naturally pleased that Dedekind had found no errors in his proof, but he reserved comment on Dedekind's criticisms until he had found the time to consider more carefully his position. Two days later his answer was in the mail: it was all a misunderstanding.

Cantor was anxious to clarify his earlier criticism of those who had merely assumed the invariance of dimension. Actually, he declared that the assumption they were guilty of making was *not* that of the invariance of dimension, but that "of the determinateness of the independent coordinates, whose number is assumed by certain authors to be equal in all cases to the number of dimensions." On the contrary, declared Cantor, "if one considers the concept of coordinates *in general,* without any assumption about the nature of the intervening functions, the number of independent, unique, complete coordinates, as I've shown, can be brought to any arbitrary number."[33]

Cantor conceded his own personal belief, similar to Dedekind's, that the secret to the problem involved the *continuity* of the mapping of various dimensions onto one another. But this, too, he rightly refused to accept without proof. The new situation, however, was an exceedingly complex one. He was unwilling to guess at how one would have to interpret "in general continuous correspondence." What constituted a generally continuous domain? What limitations were necessary and sufficient in terms of the continuity of the mapping functions themselves? Cantor may have had his eye on the proofs both he and Heine had produced for functions continuous "in general," where certain results were valid even on infinite sets of excepted points. This complicated matters considerably, making the proof of the invariance of dimension one of great difficulty. Whatever the final solution, Cantor wanted to reassure Dedekind that he had no desire to stir up any sort of polemic. Instead, he hoped his new result would help to further explanations concerning the invariance of dimension.

CANTOR'S *BEITRAG* OF 1878

Though the Cantor-Dedekind correspondence between May and October 1877 is helpful in recounting the immediate steps Cantor took in developing the technical apparatus for his proof of that year, we are left to conjecture the process by which he must have been led originally to formulate the problem of dimension. Three years earlier he had established that R was nondenumerable. Following this remarkable discovery, certain new avenues of inquiry must have appeared simultaneously. If, in terms of cardinality, there were more real numbers than natural numbers, were there further quantitative distinctions to be made between and beyond them? How did the new discovery contribute to the explanation and understanding of continuity? Was it possible, perhaps, that higher dimensions would produce point sets of increasingly larger cardinality? If continuity of the linear set R was somehow related to its nondenumerable character, could still higher orders of infinity account for continuity of physical space, for example? As Cantor raised the question of mapping lines and planes, such questions seem to have been uppermost in his mind. The entire matter, however, took a completely unexpected turn when he discovered the equivalence map between sets of arbitrarily different yet positive dimension. Simultaneously, the result convinced him that the study of continuous domains was possible in terms of the real numbers alone. Cantor's subsequent concentration on the characteristics of linear point sets reflected his early contention that among point sets, none was to be found of an infinite order greater than that of R itself. In the "Beitrag zur Mannigfaltigkeitslehre," published in 1878, he followed in large measure his reasoning as he had outlined it originally to Dedekind.[34]

The idea of the one-to-one correspondence, implicit in the paper of 1874 achieved its full generality in the *Beitrag* of 1878. Its use also marked the appearance of an important and essential set-theoretic concept, that of the *power* of a set. Sets were said to be of equal power if they could be identified by a unique, reciprocal one-to-one correspondence. The problems of determining the powers of sets and of determining the sequence and hierarchy of such powers were to be a major early preoccupation of Cantor's work with the new theory.

The first theorems Cantor outlined in the *Beitrag* dealt with the class of all denumerable sets. As he remarked, this class was a particularly rich one, including not only the integers and rationals, but as he had also demonstrated, the algebraic numbers too. The class also contained those sets described by Cantor in 1872 as sets of the nth kind, and in general any simply infinite sequence or n-fold sequence, by virtue of their index over the set of natural numbers N. He also declared that it was not difficult to see that "the sequence of positive whole numbers ν constitutes, as is easy to show, the least of all powers which occur among infinite aggregates."[35] But these dense, though nowhere-continuous sets were not the focus of Cantor's interest: "In the following only the so-called continuous, n-dimensional spaces with respect to their power will be studied."[36]

Cantor emphasized the significance of his research by raising the question of the invariance of dimension and the unsubstantiated assumptions previously made by mathematicians concerning dimension. Apparently Cantor had taken Dedekind's admonitions to heart, for he chose not to criticize the work of Helmholtz and Riemann in as much detail as he had in the letters to Dedekind.[37] But he was nevertheless careful to make the seriousness of the continuity assumption quite clear:

> Apart from making the *assumption,* most are silent about how it follows from the course of this research that the correspondence between the elements of the space and the system of values $x_1, x_2, \ldots, x_n$ is a *continuous* one, so that any infinitely small change of the system $x_1, x_2, \ldots, x_n$ corresponds to an infinitely small change of the corresponding element, and conversely, to every infinitely small change of the element a similar change in its coordinates corresponds. It may be left undecided whether these assumptions are to be considered as sufficient, or whether they are to be extended by more specialized conditions in order to consider the intended conceptual construction of n-dimensional continuous spaces as one ensured against any contradictions, sound in itself.[38]

By making no assumptions concerning the nature of the correspondence between a set and its coordinates, Cantor emphasized even more strongly that this was the real crux of the problem, whatever others might regard as the essential feature of dimension. He described the surprising result he had achieved as follows:

> As our research will show, it is even possible to determine uniquely and completely the elements of an n-dimensional continuous space by a single real coordinate t. If no assumptions are made about the kind of correspondence, it then follows that the number of independent, continuous, real coordinates which are used for the unique and complete determination of the elements of an n-dimensional continuous space can be brought to any arbitrary number, and thus is *not* to be regarded as a unique feature of a given space.[39]

Knowing that two continuous sets of equal dimension could be uniquely mapped onto one another, Cantor reduced his investigation to the following theorem:

> *Theorem A:* If $x_1, x_2, \ldots, x_n$ are n independent variable real numbers for which it can be assumed that all are ≥ 0 and ≤ 1, and if t is another variable with the same domain $(0 \leq t \leq 1)$, then it is possible to correspond the one number t with the system of n numbers $x_1, x_2, \ldots, x_n$ so that to every definite value of t a definite system of values $x_1, x_2, \ldots, x_n$, and conversely, to every definite system $x_1, x_2, \ldots, x_n$ a certain value of t corresponds.[40]

In order to prove Theorem A, Cantor was prompted to investigate the set of irrational numbers and its relation to the real line. The basis of his argument rested upon the representation of irrational numbers by infinite continued fractions, establishing a unique correspondence between the irrationals and the set of infinite sequences defined on N, where

$$e = \cfrac{1}{\alpha_1 + \cfrac{1}{\alpha_2 + \ddots}}$$

was denoted as $e = (\alpha_1, \alpha_2, \alpha_3, \ldots)$. Furthermore, he considered only such $e \in (0,1)$, and assumed that all the $\alpha_i \in N$. Given a finite two-dimensional array of such irrationals:

$$e_1 = (\alpha_{11}, \alpha_{12}, \ldots)$$
$$e_2 = (\alpha_{21}, \alpha_{22}, \ldots)$$
$$\vdots$$
$$e_n = (\alpha_{n1}, \alpha_{n2}, \ldots)$$

he was able to show that from these n irrationals a new irrational number could be uniquely determined. He denoted this new irrational as $d = (\beta_1, \beta_2, \beta_3, \ldots, \beta_\nu, \ldots)$ and added that the required relationship between the two-dimensional array and d was given by:

$$\beta_{(\nu-1)n+\mu} = \alpha_{\mu,\nu} \begin{cases} \mu = 1,2, \ldots, n \\ \nu = 1,2, \ldots \end{cases}$$

Beginning with a given irrational number $d = (\beta_1, \beta_2, \ldots, \beta_\nu, \ldots)$, one could also work backwards to determine uniquely an array of n infinite sequences, each determining uniquely an irrational number. Cantor set this result apart as Theorem C:

Theorem C: If $e_1, e_2, \ldots, e_n$ are n independent variables each one of which can assume all irrational values in the interval $(0,1)$, and if d is another variable with the same domain, then it is possible to correspond uniquely and completely the one number d and the system of n numbers $e_1, e_2, \ldots, e_n$.[41]

Section 3 of Cantor's paper was devoted to showing how the proof of Theorem A could be reduced to a proof involving only mappings between the set of irrationals E and the set of reals R. In fact, once he had established a correspondence $E \leftrightarrow R$, he could immediately conclude the correspondence of elements from his Theorem A, $x_1, x_2, \ldots, x_n$ and t with the elements from his Theorem C, $e_1, e_2, \ldots, e_n$ and d. In this sense the results (C) and (A) may be regarded as equivalent. The entire paper therefore reduced to the single Theorem D:

Theorem D: A variable quantity e which can assume all irrational values of the interval $(0,1)$ can be corresponded uniquely to a variable x which includes all real, i.e., rational and irrational values which are ≥ 0 and ≤ 1, so that to every irrational value $0 < e < 1$ one and only one real value of $0 \leq x \leq 1$ corresponds, and conversely, to every real value of x a certain irrational value of e corresponds.[42]

With the problem thus reformulated, Cantor briefly stopped to introduce some nomenclature intended to simplify the somewhat lengthy and intertwining argument which followed. He first noted that the rest of the paper would be limited to considerations of *linear* sets only. This was the first hint he offered suggesting that all the relevant matters of set theory and the nature of continuity could be considered on the model of R.

Cantor also introduced the following partition of linear sets, which today would be described as a disjoint union:

$$a \equiv \{a', a'', \ldots, a^{(\nu)}, \ldots\}.$$

The symbol $\equiv$ was used to indicate that the intersection of any two of the elements $a^{(\nu)}$, $a^{(\nu')}$, was always empty. Thus a could be considered the disjoint union of linear sets $a^{(\nu)}$, while from any set a a series of mutually disjoint sets $a^{(\nu)}$ could always be formed.[43] Finally, Cantor also introduced a symbol to express his new concept of equivalence and wrote, whenever two domains a and b were of equal power, that $a \sim b$. Of particular note is the theorem which Cantor then gave concerning the relation of two infinite sequences of disjoint elements:

> *Theorem E:* If $a', a'', \ldots, a^{(\nu)}, \ldots$ is a finite or infinite sequence of variables or constants which have no pair-wise connection (the elements of the sequence are all different or disjoint), and if $b', b'', \ldots, b^{(\nu)}, \ldots$ is another sequence of the same character, then every variable $a^{(\nu)}$ of the first sequence corresponds to a definite variable $b^{(\nu)}$ of the second, and if these corresponding variables are always equivalent to one another, i.e., if $a^{(\nu)} \sim b^{(\nu)}$, then it is also always true that $a \sim b$ if
>
> $$a \equiv \{a', a'', \ldots, a^{(\nu)}, \ldots\}$$
> $$b \equiv \{b', b'', \ldots, b^{(\nu)}, \ldots\}.$$ [44]

Once Theorem E had been established, Cantor set out to prove Theorem D, that the set of irrationals and the set of reals were of equal power. He began by considering the set of all rational numbers on the closed interval $[0,1]$. By first demonstrating that they could be arranged as an infinite sequence designated by $\varphi_1, \varphi_2, \ldots, \varphi_\nu, \ldots$, he further noted that the denumerability of Q, aside from the construction he gave, could be concluded immediately from his 1874 result, since the rationals were included automatically in the set of algebraic numbers.

Defining e as the variable in Theorem D, such that it could assume all the real values on $(0,1)$ except the rationals φ_ν, Cantor considered an infinite sequence of irrationals e_ν, assuming only that $e_\nu < e_{\nu+1}$ and that $\lim e_\nu = 1$.

Dieser letzte Satz (D) läßt sich nun durch die Betrachtung der nebenstehenden, merkwürdigen, Curve als wahr erkennen. Die Coordinaten eines laufenden Punktes in derselben sind meine Größen x, y von welchen die eine eine eindeutige Function der andern ist; während aber x alle Werthe des Intervalls $(0 \ldots 1)$ annimmt, ist der Spielraum von y der nämliche mit alleiniger Ausnahme des Werthes 0. — Die Curve besteht aus den unendlich vielen, immer kleiner werdenden parallelen geraden Strecken $\overline{ab}$, $\overline{a'b'}$, $\overline{a''b''}$, ... und aus dem Punkt c. Die Endpunkte $b, b', b'', \ldots$ werden als nicht zur Curve gehörig angesehen.

Die Längen sind:

$$\overline{op} = \overline{pc} = 1\,; \quad \overline{ob} = \frac{1}{2}\,; \quad \overline{bb_1} = \frac{1}{4}\,, \quad \overline{b_1b_2} = \frac{1}{8}\,, \quad \overline{b_2b_3} = \frac{1}{16}\,, \ldots$$

$$\overline{oa} = \frac{1}{2}\,; \quad \overline{a'd'} = \frac{1}{4}\,; \quad \overline{a''d''} = \frac{1}{8}\,; \quad \overline{a'''d'''} = \frac{1}{16}\,, \ldots$$

A page from Cantor's letter to Dedekind dated June 25, 1877.
From the Cantor-Dedekind correspondence
in the archives of the University of Evansville, Indiana.

The reasons for this insistence will become clear in a moment. He then introduced another variable f over (0,1) which assumed all real values except those e_ν; g a variable over (0,1) where both e_ν and φ_ν were excluded from its domain on (0,1). He could easily claim, from $e=\{g, e_\nu\}, f=\{g, \varphi_\nu\}$, that $e \sim f$, because $g \sim g$ and $e_\nu \sim \varphi_\nu$, since both were denumerable.

But this was only the first half of the result. Having shown that the set of irrationals was equivalent to the set of reals, excepting a countably infinite number of irrationals, he further showed that such a set of reals, excepting a countably infinite number of irrationals, was equivalent to the *entire* set of real numbers. Once this Theorem F was established, the conclusion of Theorem D easily followed, since $e \sim f \sim x$.

> *Theorem F:* A variable f which can assume all values of the interval (0,1) with the exception of the values of a sequence e_ν for which $e_\nu < e_{\nu+1}$ and $\lim e_\nu = 1$ for $\nu = \infty$, may be corresponded uniquely and completely to a variable x which assumes all values ≥ 0 and ≤ 1; in other words: $f \sim x$.[45]

Cantor was not prepared to approach his Theorem F immediately, but first offered three auxiliary theorems G, H, and J. Since the results were straightforward, their proofs do not warrant discussion in detail. It is, however, essential to appreciate the nature of the argument by which he established the first of these last three theorems. Defining y as a variable over $(0,1]$, meaning that y could assume all values between 0 and 1 except for the value 0, while including the endpoint 1; and x on $[0, 1]$, defined for both endpoints as well, he proved that $x \sim y$ by introducing a geometric mapping.[46]

The figure consisted of an infinite number of diminishing parallel lines $\overline{ab}$, $\overline{a'b'}, \overline{a''b''}, \ldots$, and the point c. The endpoints $b, b', b'', \ldots$ were not included. Cantor constructed the parallels in such a way that he could list the following lengths, which can be read directly from the bottom of the page as reproduced:

$$\overline{op} = \overline{pc} = 1; \quad \overline{ob} = \frac{1}{2}; \quad \overline{bb_1} = \frac{1}{4}; \quad \overline{b_1b_2} = \frac{1}{8}; \quad \ldots$$

$$\overline{oa} = \frac{1}{2}; \quad \overline{a'\delta'} = \frac{1}{4}; \quad \overline{a''\delta''} = \frac{1}{8}; \quad \ldots$$

(in the original letter, one can still see the points of the compass with which arcs were thrown off to construct the bisectors b, b', b'', ... and the parallels $a^{(\nu)}b^{(\nu)}$).

Cantor further generalized this result, that $(0,1] \sim [0,1]$, to the form $(a, b] \sim [a, b]$ by introducing a suitable transformation. Then it was only a step to the conclusion that $(a,b) \sim [a,b]$. Essentially he had proven that one could map open intervals onto closed ones. With all this as background, he was at last prepared to complete the proof of Theorem F. The task at hand was to show that $f \sim x$, or that the set of irrational numbers, excepting a countable collection of irrationals, was of a power equal to that of the reals. By the method he chose, this was almost immediate. Considering the sequence of irrationals expected from the domain of f, $e_1, e_2, \ldots, e_\nu, \ldots$, he partitioned the set of variables f disjointly over the intervals $(e_1, e_2), (e_2, e_3), \ldots$, such that f was variable over $(e_{\nu-1}, e_\nu)$ with the exception of endpoints; $x^{(2\nu)}$ was a variable assuming all values over the closed interval $[e_{2\nu-1}, e_{2\nu}]$. It was then clear that:

$$f \equiv \{f^{(1)}, f^{(2)}, \ldots, f^{(\nu)}, \ldots, 1\};$$
$$x \equiv \{f^{(1)}, x^{(2)}, f^{(3)}, x^{(4)}, \ldots, f^{(2\nu-1)}, x^{(2\nu)}, \ldots, 1\}.$$

Having shown the equivalence of continuous open and closed intervals, he could conclude directly that $f^{(2\nu)} \sim x^{(2\nu)}$; it was obvious that $f^{(2\nu-1)} \sim f^{(2\nu-1)}$; $1 \sim 1$; by Theorem E it was immediate that $f \sim x$.

Though the published paper of 1878 followed in almost verbatim faithfulness the draft that Cantor had sent to Dedekind, there were nevertheless some interesting revisions.[47] He realized that the proof sent to Dedekind was not so direct as it might have been, and consequently he decided to reformulate at least the major theorem of the paper, about which the entire argument revolved. The simplification made even clearer the fact that he was dealing essentially with the problem of mapping closed and open intervals onto one another, though his terminology in 1878 was still too primitive to have formulated the problem in this way. Letting φ_ν denote any sequence of rationals in $[0,1]$, η_ν any sequence of irrationals in $(0,1)$, he then offered the following decompositions for x, defined for all real values on $[0,1]$, and e defined for only the irrationals on $(0,1)$:

$$x \equiv \{h, \varphi_\nu, \eta_\nu\},$$

where h was defined over the interval $(0,1)$ excepting the sequence of rational numbers φ_ν and the irrational sequence η_ν,

$$e \equiv \{h, \eta_\nu\}.$$

Interposing the step

$$e \equiv \{h, \eta_{2\nu}, \eta_{2\nu-1}\},$$

he saw that $h \sim h$, $\varphi_\nu \sim \eta_{2\nu}$, $\eta_\nu \sim \eta_{2\nu-1}$, which led easily to the conclusion of Theorem D that $e \sim x$.

As Cantor reached the end of his paper, he concluded on a strong and forward-looking note. The question immediately arising from the *Beitrag* of 1878 concerned the infinite variety of sets contained in any real continuum, and the problem of how they could be divided according to their powers. Since any continuous domain, whatever its dimension, could be considered as power-equivalent to any continuous linear point set, the study of continuity could be reduced to the real number continuum R. Linear sets, upon this justification, became the subject of Cantor's next intensive subject of research. Having already shown that the natural numbers formed the basis for the denumerable class, and the real numbers the basis for continuous classes, Cantor further alleged that among continuous linear domains a process of induction led to the conclusion that these were, in fact, the only ones possible:

> Thus the linear aggregates would consist of two classes of which the first includes all spaces which can be given the form of a function of ν (where ν runs over all positive whole numbers), while the second class takes on all those aggregates which are reducible to the form of a function of x (where x can assume all real values ≥ 0 and ≤ 1). Corresponding to these two classes, therefore, would be only two powers of infinite linear aggregates; the exact study of this question we put off for another occasion.[48]

Thus Cantor promised further investigations. The research that followed was directed in large measure to the real line. The major heuristic importance of the paper of 1878 was its justification of the narrower concentration. The real line was a model through which the essential features of continuity might be studied. In limiting his attention to single dimensional point sets, Cantor was to make important progress in developing his new theory of sets and transfinite numbers in a very short period of time.

LEOPOLD KRONECKER (1823–1891): EARLY OPPOSITION TO CANTOR'S WORK

Despite the great advances Cantor had made with his study of the real numbers and the properties of dimension, his work was not well received by all mathematicians. No one could have been more opposed to Cantor's ideas, nor have done more to damage his early career, than Leopold Kronecker. As one of Cantor's professors in Berlin, Kronecker had praised his work in number theory, and had even suggested a helpful modification of the first version of Cantor's uniqueness theorem in 1871. But later, Kronecker did not respond so generously to Cantor's work. Not only did he attempt to prevent the publication of Cantor's paper of 1878 on dimension, but he railed vigorously against the transfinite numbers, and was well known for his emphatic polemics against Cantorian set theory.[49]

Kronecker's initial opposition to Cantor's work came shortly after Cantor had begun work on the representation problem. In April 1870 Schwarz wrote to encourage his work despite Kronecker's antagonism. Schwarz was delighted that Weierstrass had found Cantor's result "entirely correct" and added that he felt Cantor's discovery was "a significant advance in the theory of trigonometric series, one that makes me even happier than if I had made it myself."[50] Heine's letters to Schwarz in the early 1870s make it equally clear that Kronecker was already questioning, and opposing, the use in rigorous mathematics of the Bolzano-Weierstrass theorem, upper and lower limits, and the irrational numbers in general.[51] At one point Kronecker had even tried to dissuade Heine from publishing his "Über trigonometrische Reihen," but Heine was not deterred, and recorded his satisfaction over the paper in unreserved terms:

> My little work "On trigonometric series," the publisher's proofs of which I have at this moment in my hands, and which, moreover, had been greatly debated with Kronecker who wanted to persuade me to withdraw it, has appeared in the current (71st) volume of the journal, and has made me very happy. I sent it to Borchardt in February. Kronecker saw it, and kept it without my knowing until I came to Berlin.[52]

All this is particularly interesting in terms of Cantor's original proof of the nondenumerability of the real numbers and the changes he deliberately made in presenting that result in 1874. One striking peculiarity of the published version of that proof was its title: "On a Property of the Collection of All Real Algebraic Numbers." The *algebraic* numbers? Surely the nondenumerability of the *real* numbers was the problem uppermost in Cantor's mind at the time, as his correspondence with Dedekind shows. Why then did the title of his paper fail to indicate the most significant discovery of the article? The answer to this question depends upon the nature of Cantor's proof and of Kronecker's objections to certain basic features of analysis in general.

In order to press his conclusion that the set of real numbers was nondenumerable, Cantor argued that "if $\alpha^\infty=\beta^\infty$ (a case which always occurs for the collection (ω) of all real algebraic numbers), then it is easy to see, if one only refers to the definition of intervals, that the number $\eta=\alpha^\infty=\beta^\infty$ *cannot* be contained in our sequence."[53] Clearly, to assure the exactness of the assertion, the axioms of continuity and his theory of irrationals were indispensable. Dedekind recognized this in drafting his version of the proof, which Cantor reproduced nearly verbatim in the published version of his paper.[54] But with one important exception. Dedekind, in his notebook for December 7, 1873, reminded himself that:

> On December 8 I received this letter from Cantor and answered it that same day with congratulations for his pleasant success. I also recast the heart of the proof (which was still very complicated) in greater simplicity; this version is carried over almost word-for-word in Cantor's article (Crelle's *Journal, 77*); of course my use of "the principle of continuity" is avoided at the relevant place (p. 261, lines 10–14).[55]

Lines 10–14 are precisely those in which Cantor asserted that $\alpha^\infty=\beta^\infty=\eta$, the key to the entire proof.[56] Though Dedekind had chosen to make explicit the part the axiom of continuity was playing in the proof, Cantor deliberately struck any reference from his public presentation. His reasons for doing so were doubtless related to the objections he feared might be raised against the line of reasoning basic to his entire argument.

In 1886 Kronecker described his misgivings with respect to the attempts both Cantor and Dedekind had made to establish a general theory of irrational numbers.[57] The most critical remarks were included in a detailed footnote:

> The above considerations seem to me opposed to the introduction of those concepts of Dedekind like "module," "ideal," etc.; similarly [they are opposed to] the introduction of various concepts by the help of which it has frequently been attempted in recent times (but first by *Heine*) to conceive and establish the "irrationals" in general. Even the *general* concept of an infinite series, for example one which increases according to definite powers of variables, is in my opinion only permissible with the reservation that in every special case, on the basis of the arithmetic laws of constructing terms (or coefficients), just as above, certain assumptions must be shown to hold which are applicable to the series like finite expressions, and which thus make the extension beyond the concept of a *finite* series really unnecessary.[58]

The details of Kronecker's program of arithmetization, basing all of mathematics on a finite number of operations involving only the integers, was laid out more fully in his article "Ueber den Zahlbegriff."[59] Nevertheless, his opinions were known to most mathematicians, particularly those who had studied in Berlin, by the 1870s. Cantor, who had written both his *Dissertation* and *Habilitationsschrift* in Berlin, could not have been unaware of Kronecker's extreme position. Heine's letters written to Schwarz in the early 1870s make it clear that debate centered primarily upon irrational numbers and the Bolzano-Weierstrass theorem.[60] Kronecker had strong objections to both. As an editor of Crelle's *Journal* he was in a position to refuse any article, and Cantor may have felt it was wisest to minimize those features of his proof that might raise any questions about its acceptability for publication.[61] By making the 1874 paper seem as innocuous as possible, Cantor was perhaps being diplomatic and trying to smooth his own way as best he could. Incorporation of the axiom of continuity and the prominent appeal to his theory of irrational numbers might have seemed unnecessary sources of provocation, ones he could easily forgo.

It may be objected that Cantor's paper of 1874 was so obviously unorthodox from a finitist point of view that Kronecker would have been opposed to it with or without any references to irrational numbers and axioms of continuity. The major result of the article, establishing the nondenumerability of R, was provocative enough by itself. But Cantor was very clever in his presentation, and his letters to Dedekind show that he was purposefully careful. The paper was really about the algebraic numbers, so Cantor said in print. The title indicated nothing more. The paper focused, seemingly, on this one result, and it introduced the proof of the nondenumerability of R only to enable an *application* of the paper's title. With the denumerability of the algebraic numbers established, comparison with the nondenumerable set R provided an independent corroboration of Liouville's proof of the existence of nonalgebraic real numbers. To strengthen support for his proof establishing the nondenumerability of the reals, Cantor added at the very end of his paper that Minnegerode had already offered a specific case of the proposition in question.[62] It was as though Cantor wanted it to appear that his proof was neither new nor particularly special, except for its generality and its verification, in a new way, of Liouville's work. He consequently chose to place no emphasis upon the truly important result of the paper and left readers of Crelle's *Journal* to believe, if they read no further than his title, that he was only writing about *algebraic* numbers. The title was thus carefully chosen to avoid any hint of one of Cantor's most significant discoveries.

Three years later Cantor finished his paper on dimension and sent it off to Crelle's *Journal* on July 12, 1877. Despite the editor's promise to accept it and Weierstrass's efforts to promote its appearance, no steps were taken to prepare the article for press.[63] Suspecting Kronecker's intervention, Cantor became so agitated over the delay that he wrote a bitter letter to Dedekind in which he complained about the treatment of his work and raised the possibility of withdrawing his paper from the journal. Instead, he would ask the publishing house of Vieweg to print it separately. Dedekind, acting upon his own experience in such matters, was able to convince Cantor that he should wait. As it turned out, the *Beitrag* finally did appear the following year.

The customary explanation for Kronecker's reluctance to approve the publication of Cantor's *Beitrag* involves the unusual and unexpected nature of the paper's results.[64] A. Fraenkel once speculated that Kronecker's opposition was based simply upon the apparent paradox of Cantor's conclusion. More recently, Herbert Meschkowski has taken the same view, expressed more eloquently: "Cantor's theorem is thus a beautiful example of a mathematical paradox, of a true statement which seems to be false to the uninformed. And in terms of the problems of set theory, all mathematicians at that time were 'beginners' at best."[65]

But if resistance to unintuitive surprises explains the reluctance of Crelle's *Journal* to print Cantor's *Beitrag,* it is at best only a partial explanation. Despite the counter-intuitiveness of Cantor's result, Kronecker's objections were more specific, and more penetrating. For Kronecker, the proof was meaningless, without any hope of salvation. For example, when Lindemann published his proof of the existence of transcendental numbers, Kronecker could compliment it as a beautiful argument, but one which proved *nothing* because there *were* no transcendental numbers. He could say much the same in Cantor's case.[66] His entire proof rested upon mappings between irrational numbers $e \in (0,1)$ and real numbers $x \in [0,1]$. Kronecker must have been convinced that Cantor was playing with empty concepts, and that consequently he should be prevented from publishing mathematical nonsense.

Though Kronecker did not succeed in suppressing Cantor's paper, the delay in publication represented the first major conflict Cantor was to experience over the acceptance of his work. The *Beitrag* was the first occasion for open hostility. Though the paper was eventually published, he never again permitted his work to appear in the pages of Crelle's highly respected *Journal*. The *Beitrag*, in signaling the lengths to which its adversaries would carry their attempts to defeat it, also reflected the revolutionary ideas with which Cantor had begun to work, ideas that were to unfold systematically over the next twenty years.

INVARIANCE OF DIMENSION

No sooner had Cantor's paper of 1878 established the correspondence between continuous domains of one and two dimensions, than a rush of activity was under way to show that the result failed under the additional condition that the mapping between spaces must be continuous. Both J. Lüroth and J. Thomae published papers in July 1878, followed by E. Jürgens in August and E. Netto in October. Cantor's own response, "Über einen Satz aus der Theorie der stetigen Mannigfaltigkeiten," appeared early in the following year.

Lüroth correctly stressed that the difficulty of resolving the doubts Cantor had raised lay in precisely characterizing the generality of elements involved—of continuous mappings and domains.[67] Specific cases were not so problematic. He was able to show that the two-dimensional plane of points M_2 could not be made to correspond continuously in a one-to-one fashion with the one-dimensional line of points M_1. In fact, by considering more generally the m coordinates of points in a space of dimension m, and by letting $m-2$ coordinates be fixed, Lüroth established that no space of dimension greater than or equal to 2 could be mapped uniquely and continuously onto a straight line. Considering the circumference of a circle defined in M_2, he assumed a unique map for which x and y in M_1 corresponded respectively to a and b and M_2. The elements a,b were assumed to lie on the circumference, and because he had assumed the continuity of the map, there had to be an intermediate point of the line which mapped to a point midway between the two points a and b on the circumference of the circle. Since this point, however, appeared twice on the circumference, Lüroth could support his general conclusion that "for a continuous function of more than one variable a unique inverse mapping is not possible."[68]

However, proceeding more generally became difficult. Lüroth was able to establish that no spaces of dimension $n \geq 3$ could be corresponded uniquely and continuously with the two-dimensional plane, but the argument became so laborious and complex that he went no further.[69]

Jürgens approached the problem much as Lüroth had. Assuming the correspondence between points of the plane and the line to be continuous, he showed that it could not then be one-to-one.[70] But, like Lüroth, he was able to establish this fact only for mappings onto spaces of dimensions one and two; those of higher dimension and the most general nth case still eluded their capabilities. Nonetheless, successful proofs for the most intuitive and visual cases must have strengthened the prevalent belief among mathematicians that the conjecture linking invariance of dimension with continuity was correct in spirit.

Thomae's proof, unlike those of Lüroth and Jürgens, was the first though tentative gesture toward establishing the continuity hypothesis for spaces of arbitrary dimension. But the basis of his presentation erred in its reliance upon what he described as a well-known result of *analysis situs:* "A connected continuous space [*zuzammenhängende continuirliche Mannigfaltigkeit*] M_n of n dimensions cannot be divided into separate parts by one or more spaces of $n - 2$ or lesser dimension."[71] But this presupposed the theorem he was trying to prove. In fact, Lüroth stressed this in commenting on Jürgens' paper read to the section for mathematics and astronomy of the *Gesellschaft Deutscher Naturforscher und Aerzte,* meeting in Cassel in September 1878.[72] Moreover, the flavor of Thomae's method was more one of pure analysis than of topology. As Netto was soon to demonstrate, the problems of dimension theory required, instead, new formulations and approaches.

Netto's paper of 1878, despite its shortcomings, was the most sophisticated and far-reaching.[73] Instead of assuming continuity and then trying to show that the mapping could not be one-to-one, Netto reversed the procedure by assuming the mapping to be one-to-one and then proving the impossibility of continuity. Even Dedekind was impressed by Netto's work, as he admitted to Cantor: "The definitions given by Netto (whose paper pleases me very much, and whose proof, I think, can be made fully correct with a few modifications) constitute a good beginning, but they seem to me capable of simplification and of completion as well."[74]

Netto illustrated his basic idea with the simple correspondence between line and plane that Cantor had used as a focus for his 1878 paper. Assuming a line segment in M_2, the plane, could be mapped in a unique, continuous fashion onto a segment of M_1, the line, he denoted endpoints by a_2, b_2 in M_2; a_1, b_1 in M_1. He then took a point p_1 in M_1 on a_1, b_1, and any point p_1' on the line but not within the segment bounded by a_1, b_1. In order to join p_1, p_1', one of the endpoints, either a_1 or b_1, had to be crossed, but such was not the case for the line in M_2 corresponding to p_1, p_1' in M_1.

The method was simple and admitted easy generalizations to higher domains. Pointing out the general proposition upon which his approach was based, Netto also emphasized the reason for a closer scrutiny of some fundamental concepts:

> The principle of this proof is the following: "In a space of dimension ν every configuration [*Gebilde*] of dimension ν is bounded by another of lesser dimension; in a space of dimension $\nu + 1$ each configuration of

dimension ν coincides with its boundary. In the cases $\nu = 1,2$ this follows in any event from observation or intuition [*Anschauung*]; in general cases, for a more rigorous foundation, the concept of a configuration of dimension ν, of an interior or boundary point of a configuration, etc., must also be determined.[75]

Netto here touched upon a recurring theme in nineteenth-century mathematics: intuition has both its benefits and its limitations. It is easy to imagine Lüroth following the guide of his geometric intuition to put together his first proof. The reasoning was relatively simple for the mappings onto lines, but in the approach he and Jürgens had taken, the geometric visualizations did not seem to suggest, or to allow, any direct extension to higher dimensions. The major virtue of Netto's proof was that it permitted an almost immediate generalization.

Netto also realized that leaving the familiar territory of perceptual three-dimensional space required a sureness of argumentation that mere analogy could not provide. In seeing this, Netto was remarkably unlike his predecessors, who had rushed to show that Cantor's criticism of Riemann's view of dimension was met once the role of continuity was restored to its rightful place. But as Dedekind stressed, everyone had assumed as much all along. Lüroth's success surely came as a surprise to no one, as Cantor's proof, on the contrary, had. Though Cantor's result was unexpected, no one seemed very sensitive to the real questions which had been raised: what, exactly, were the essential characteristics which established the nature of dimension? What factors necessarily and sufficiently determined the dimension of arbitrarily given spaces? Initially, mathematicians seemed content to believe that one-to-oneness and continuity disposed of the dilemma Cantor had raised. But these two conditions only ensured that no dimensions of different order could be mistaken for each other through mappings of a continuous and unique kind. No one had actually said anything about dimension *per se*. No one except Netto.

In the course of his paper, Netto not only defined dimension but went on to give explicit characterizations for what he called regular spaces, interior points, boundary points, connectivity—all basic topological concepts. He also managed to show why the boundary of a given domain must be of a dimension lower than that of the domain itself. He then established his general theorem by induction.

CANTOR'S FAULTY PROOF OF THE INVARIANCE OF DIMENSION

Though Cantor once suggested to Dedekind that there were difficulties with Netto's proof, he never described any of his misgivings specifically. Nor is there any indication that he was aware of Jürgens' critique, which appeared somewhat obscurely in 1879.[76] In fact, when Cantor published his own proof of the invariance of dimension that same year, he did not choose to level any comments against Netto's results. Instead, he argued the superiority of his own analysis in terms of its generality and straightforwardness. Despite the various solutions offered by his colleagues, he could still complain to Dedekind that none of them had been entirely satisfactory.[77]

Cantor told readers of his new proof in the *Göttinger Nachrichten,* that he had attempted to bring new methods to establish the invariance of dimension in a most general form. His new approach incorporated the intermediate value theorem:

> *Theorem I:* A continuous function of a continuous variable t, which assumes a negative value for $t = t_0$, a positive value for $t = t_1$, must be zero at least once (for some value of t between t_0 and t_1).[78]

First he defined spherical surfaces $M_{\mu-1}$ of order $\mu-1$ in a continuous space M_μ. In order to stress explicitly the relation between his new approach and the intermediate value theorem, he noted that the surface $K_{\mu-1}$ in the space M_μ divided the space M_μ into three parts. Having defined such a surface $K_{\mu-1}$ of M_μ by the equation:

$$F = (x_1 - a_1)^2 + (x_2 - a_2)^2 + \ \dots + (x_\mu - a_\mu)^2 - r^2 = 0,$$

he described points *exterior* to $K_{\mu-1}$ as those for which $F > 0$, those *interior* to $K_{\mu-1}$ were those for which $F < 0$, and lastly, those points for which $F = 0$ comprised the surface $K_{\mu-1}$ itself.

Cantor was then able to give a companion reformulation of the intermediate value theorem in terms of interior and boundary points:

> *Theorem II:* If $K_{\mu-1}$ is a spherical surface of order $\mu-1$ lying in M_μ, if a is a point of M_μ interior to $K_{\mu-1}$, b a point of M_μ exterior to $K_{\mu-1}$, and if N is any continuous connected configuration (of any order) within M_μ which contains both points a and b, then there is at least one point c which belongs simultaneously to both $K_{\mu-1}$ and N.[79]

The central theorem, which Cantor proceeded to prove by induction, was then given as follows:

> *Theorem III:* If one has such a correspondence between two continuous configurations M_μ and M_ν such that to every point Z of M_ν *at least* one point z of M_μ corresponds, and furthermore, if this correspondence is a continuous one so that infinitely small variations of Z correspond to infinitely small variations of z, and conversely, then $\mu = \nu$.[80]

Cantor began his proof of the theorem by asserting that if III were *not* correct for $\nu = n$, then one could devise a correspondence between two domains M_μ and M_n, where $\mu < n$, in a continuous way under the assumed conditions. He demonstrated how this would lead to a contradiction, in terms of II and the intermediate value theorem. As he had earlier outlined to Dedekind,[81] the proof proceeded more easily by beginning from points A and B in M_n, considering their images in M_μ, sets Cantor denoted (a') and (b'), with one of the points a in particular from (a') under consideration. K_{n-1} was then described about A, and similarly $K_{\mu-1}$ about a so that its image G in M_n was entirely within K_{n-1}.

Denoting those points z' in M_μ which had images in M_n, and those without image in M_n by z'', he stressed that such z' formed one or more separate continuous parts of $K_{\mu-1}$, and that therefore the image G of these points of M_n had to consist of one or more separate continuous parts. Each z' on $K_{\mu-1}$ had a unique image ζ in G, which each ζ could have many images in $K_{\mu-1}$. But here there was a difficulty that Dedekind drew to Cantor's attention. How could one be certain that, under the conditions of a general map, points ζ and A would not coincide, that G would not be identical with the point A itself? Cantor had overlooked this problem when he originally sketched his proof of the theorem in his correspondence with Dedekind.[82] When the published version appeared, however, Cantor had found an answer:

> In general the configuration G will not, of course, contain the point A, but it may also happen that A is a point of G, namely if points of the collection (a') fall on $K_{\mu-1}$. Now, if ζ' is a point of A different from the point ζ of G, then we take the straight line [*Strahl*] $A\zeta'$ in M_n; when extended, it meets the surface K_{n-1} in a definitely determined point Z'.[83]

The basic idea of the entire theorem, in its earlier form as Cantor had described it to Dedekind and then in its printed version, was to establish a continuous mapping between the surfaces $K_{\mu-1}$ and K_{n-1}. As Dedekind saw it, were A to coincide with G, then there was no way to determine such a map. Cantor, by projecting G onto K_{n-1} by straight lines generated from A, even were A on G, at least achieved the necessary map for all those elements z' of $K_{\mu-1}$ which were not images of A but corresponded to points ζ' different from A.

Cantor now argued the validity of his theorem just as he had before, in his letters to Dedekind. The points z' could not include all points of $K_{\mu-1}$. If so, there would then be a continuous map between $K_{\mu-1}$ and K_{n-1}, were $\mu-1<n-1$, and fulfilling the assumed conditions of Theorem III. But having assumed III to be true for $\nu=n-1$, Cantor's line of argument had reached the desired contradiction. Thus, the all-important conclusion:

> Therefore, there must be points on K_{n-1} with which Z' never coincides; if one connects the point A with one such point P by the straight line AP, then certainly the line AP meets G in no points different from A.[84]

As for the logical soundness of this reasoning, it is significant that Cantor's proof was apparently accepted without challenge until 1899. That originally both Cantor and Dedekind should have overlooked an important logical error in the paper of 1879 reflects both the difficulty and complexity of the problem.

But twenty years later, in the pages of the *Jahresbericht der Deutschen Mathematiker-Vereinigung,* Jürgens mentioned serious *lacunae* in both Cantor's and Netto's attempts to establish the general mapping theorem.[85] Jürgens argued that fundamental questions had been overlooked. What was to be understood by continuity, by variation? He put the entire matter as follows:

> If a ν-fold extension is to be considered at all in *Riemann's* sense of a collection of objects, be they points, colors, functional relations or the like, then the *concept of a continuous correspondence* and simultaneously the *concept of a small variation* of its elements as well must already be certain or at least determinable in a natural and evident way.[86]

And there was yet another problem. What was the meaning of dimension in certain pathological cases? Focusing on specially defined subsets, Jürgens argued that it was not at all clear what dimension should be assigned to certain sub-collections of sets of given dimension. Jürgens discussed certain such subsets which he called "extensionless domains," and criticized Netto's proof for not having taken such difficult possibilities into consideration.[87]

However, these specific criticisms which Jürgens raised are not so interesting as the fact that they appeared after a period of more than twenty years. Jürgens suggested only one reason why no one before him should have bothered to criticize either of Cantor's results:

> Since in mathematics too the printed word easily finds faithful acceptance, particularly if it proceeds from distinguished authors, it is therefore understandable ever since [publication of Cantor's proof], that the entire question has been considered as already solved, and that recently prominent mathematicians still simply cite that work.[88]

But the power of the printed word had not prevented Lüroth from criticizing Thomae's proof almost before it was published, and mathematicians have rarely felt that a proof once in print was necessarily beyond reproach. In offering his own simplifications of Lüroth's original argument, Jürgens (in 1878) like the others must have been convinced of the general validity of the theorem on which he had been working. And while he noted (in 1879) shortcomings in Netto's paper, Jürgens did not publically criticize Cantor's proof of the invariance of dimension until 1898.[89] Whatever may have kept Jürgens from scrutinizing Cantor's arguments more successfully in 1879, it seems that no one was prepared then to raise the sort of questions Jürgens was asking twenty years later. As is so often the case in mathematics, until someone appears with a new viewpoint, insight, or counterexample, difficulties in any complex argument may easily go unnoticed.

Despite its shortcomings, Cantor's proof seems to have succeeded in establishing the invariance of dimension to everyone's satisfaction at the time. Subsequently, Cantor returned to his interest in studying the properties of continuity exhibited by infinite linear point sets. While his early papers on trigonometric series contained the seeds of Cantorian set theory, and the transfinite numbers as well, the papers on dimension were the intermediate steps toward an independent theory. In the same year (1879) that Cantor published his proof purporting to assure the invariance of dimension, he also issued the first in a long series of articles devoted to his new ideas. The next six years would be devoted to publishing the details of his *Punktmannigfaltigkeitslehre*.

CHAPTER 4

Cantor's Early Theory of Point Sets

Between 1879 and 1884 Cantor published six papers designed to introduce the basic elements of his new set theory: "Über unendliche lineare Punktmannigfaltigkeiten."[1] The first four of these were concerned with establishing strictly mathematical results and with providing certain applications of his ideas which he hoped might provoke interest in his work. At the same time, he meant to demonstrate the usefulness of his interesting, if somewhat unorthodox, ideas. But by 1883 Cantor had come to realize that his new theory of the infinite required a more extensive and elaborate defense than his first four papers had provided. Thus, the fifth paper in the series was deliberately constructed to face the philosophical as well as the mathematical issues raised by transfinite set theory, including his answers to objections most commonly raised against any absolute interpretation of infinity. This fifth paper was so important to Cantor that he also arranged to have his publisher, B.G. Teubner, issue the work as a separate monograph, entitled *Grundlagen einer allgemeinen Mannigfaltigkeitslehre*.[2] This, together with the sixth paper in the series (which returned to exclusively mathematical concerns), represented a concise formulation of Cantor's transfinite set theory.

The *Grundlagen,* however, was the product of ideas gestating in Cantor's mind for at least a decade. Initially, evidence of his new ideas had appeared, rather obscurely, in the last paper of 1872 on trigonometric series. In the following years he began to provide a framework and a structure within which new concepts could be formulated. In 1874 he demonstrated not only the existence of infinite sets of distinct and different magnitude, he also showed how infinite sets might be counted, how powers could be determined and equivalence defined. By 1878 he had shown that the analysis of continuity could be restricted to the domain of a single dimension: the real line. In 1879 he presented the first in a series of new papers devoted to the analysis of infinite linear point sets. The present chapter is devoted to showing how this research both conditioned and provoked his writing of the *Grundlagen* as a separate though related enterprise.

THE PAPER OF 1879: CLASSIFICATIONS

Cantor's first in the series of six papers on infinite linear point sets appeared in Crelle's *Journal* in 1879.[3] It was straightforward, predictable, and served to introduce readers unfamiliar with his earlier work to a number of essential, basic concepts. Having decided to concentrate his efforts upon *linear* point sets, Cantor described the fundamental structural properties of derived sets previously noted in 1872 and then began to expand his results, refining details as he went. The major burden of the paper of 1879 was classification, with eventual applications to the study of continuity clearly in mind. Above all, Cantor stressed the importance of derived sets in this undertaking and promised to show later that "the simplest and at the same time the most complete explanation for the determination of a continuum is based upon [the idea of derived sets]."[4]

Cantor began by classifying sets according to the character of their derived sets. Sets P for which derived sets $P^{(\nu)}$ were empty for some arbitrarily large ν had already been introduced in Cantor's paper of 1872 as sets of the first species. Those for which $P^{(\nu)}$ was not empty for any finite value of ν were now introduced as sets of the *second* species.[5] Another concept, by no means new but finally given independent status in the paper of 1879, was the property of a set being everywhere-dense. Cantor had already dealt with such sets in 1872 and more extensively in 1874 when discussing the denumerability of rational and algebraic numbers and the nondenumerability of the reals. Each of these sets, however, was everywhere-dense, and thus the property was of considerable interest.

> *Definition:* If P lies partially or entirely in the interval $(\alpha \ldots \beta)$, then the remarkable case can occur that any arbitrarily small interval $(\gamma \ldots \delta)$ in $(\alpha \ldots \beta)$ contains points of P. In such a case we will say that P is everywhere-dense in the interval $(\alpha \ldots \beta)$.[6]

Cantor's special designation of everywhere-dense sets transformed an earlier, undistinguished element of his research into a precisely defined concept. Consequently, he was able to consider the interrelations between everywhere-dense sets and derived sets, and he noted several straightforward ways in which the two were connected. A set P was everywhere-dense in an interval (α,β) whenever its first derived set P' contained (α,β) itself. Furthermore, everywhere-dense sets were necessarily sets of the second species, while first species sets could never be everywhere-dense. The question as to whether all sets of the second species were everywhere-dense was deferred for a later paper.[7]

The second major characteristic of point sets which Cantor elaborated in 1879 classified them according to their power [*Mächtigkeit*]. Once again, this idea had already been outlined in his earlier research. It had figured prominently in the paper on dimension of 1878, though he had given the first clear example of distinct powers in 1874 when he established the nondenumerability of the set of all real numbers. Though he was able to define what he meant by the *equality* of powers, he did not attempt to describe what the concept of power itself involved:

> *Definition:* Two sets are said to be of the same power if a one-to-one correspondence between their respective elements is possible.[8]

Cantor singled out two cases: denumerable sets, whose power was that of the natural numbers *N;* and continuous, or nondenumerable sets, whose power was that of the real numbers *R*. Examples of each were drawn directly from his 1874 paper: countably infinite sets included the natural numbers, the rational and the algebraic numbers. All sets of the first species were also of this first denumerable kind. But the rational and algebraic numbers showed that everywhere-dense sets, and hence sets of the second species, could also be denumerable. As an example of a nondenumerable set, Cantor turned to his 1878 paper and offered any continuous interval from which even a denumerably infinite set of points might be excluded.

Cantor was obviously working within the confines of a very limited number of specific examples. This may help explain the most striking omission of the entire paper: though it was clearly within reach, he did not choose to conjecture that the *only* two powers possible among point sets were those equivalent either to *N* or to *R*. Before he could offer what he felt was a satisfactory solution to this problem, he had to introduce his sets of the second species in a more systematic manner. In the meantime, he limited his discussion to the already familiar examples of sets of equivalent or of distinctly different powers.[9]

But working from examples, as Cantor had discovered, was often misleading. Intuitions were not to be trusted. For example, the surprising discovery of one-to-one mappings between lines and planes only attested to the richness of the real line. As to the profundity of its secrets, Cantor was understandably reticent to make any hasty generalizations. The interrelations between the natural, rational, and algebraic numbers were complicated and not easy to characterize. Transcendental numbers were even more elusive, and if they were subject to subgroupings as varied as those of the denumerable yet everywhere-dense sets, then the possibilities were by no means clear.

Leaving undecided the question of how many different powers of point sets there might be, Cantor went on to establish that the two sets *N* and *R* were indeed distinct. Though his argument was cast in the new language of everywhere-dense sets, its development was exactly the same as the one given in 1874, when he had first proven that the continuum was nondenumerable.

Except for the new terminology, Cantor's paper of 1879 offered nothing in the way of new results. But it was a beginning, and served as a fresh start for an entire series of investigations to follow. The concept of derived set, power, and everywhere-dense sets were about to assume major roles of their own. No longer were they tied to the language and requirements of trigonometric series or dimension.[10] Cantor was thus free to follow his new creations wherever they might lead.

THE PAPER OF 1880: DERIVED SETS AND TRANSFINITE SYMBOLS

Cantor's second paper of 1880 was brief.[11] It continued the bricklaying work of the article of 1879, and it too sought to reformulate old ideas in the context of linear point sets. It also introduced for the first time an embryonic form of Cantor's boldest and most original discovery: the transfinite numbers. As a preliminary to their description, however, Cantor introduced several definitions. He also pointed out that *first species sets* could be completely characterized by their derived sets.[12] However, sets of *second species* could not be entirely described in such terms. Thus additional properties had to be considered, new techniques developed, and more sophisticated means employed to deal adequately with sets of the second species.

To illustrate this point, Cantor considered a general second species point set P. He then observed that the first derived set P' of P could be given a disjoint decomposition: $P' \equiv \{Q, R\}$, where Q was the set of all points belonging to first species sets of P', and R the set of points contained in *every* derived set of P', thus making R a set of the *second* species. Cantor's notation for the intersection of sets, namely that set common to every set of a collection of sets P_n, was simply $\mathfrak{D}(P_1, P_2, P_3, \ldots)$; (the $\mathfrak{D}$ was taken from *Durchschnitt,* Cantor's term for "intersection"). Thus:

$$R \equiv \mathfrak{D}\ (P', P'', \ldots).$$

He went on to remark that since R was the set of points belonging to *every* derived set of P', it was equally true that:

$$R \equiv \mathfrak{D}\ (P'', P''', \ldots)$$
$$R \equiv \mathfrak{D}\ (P^{(n)}, P^{(n+1)}, P^{(n+2)}, \ldots).$$

Consequently he felt justified in denoting R, taken from P, as

$$R \equiv P^{(\infty)},$$

and thus he designated $P^{(\infty)}$ the derived set of P of order ∞.

Cantor hoped that such disjoint decompositions might later help to advance his study of the continuum. But in 1880, by specifically describing the basis for distinguishing between derived sets of the first and second species, Cantor was prepared to introduce in a rigorous form his definition of a radically new concept. Beginning with $R \equiv \mathfrak{D}(P', P'', \ldots, P^{(\nu)}, \ldots) = P^{(\infty)}$, and assuming $P^{(\infty)} \neq 0$, Cantor denoted the first derived set of $P^{(\infty)}$ by $P^{(\infty+1)}$, the nth by $P^{(\infty+n)}$. Continuing, as Cantor later described it, in this "dialectic fashion,"[13] it was possible to generate derived sets denoted generally as follows:

$$P^{(n_0\infty^{\nu} + n_1\infty^{\nu-1} + \ \ldots\ + n_{\nu})}.$$

In fact, allowing ν to be taken as a variable, one could produce an endless sequence of "concepts":

$$P^{(n\infty^{\infty})}, P^{(\infty^{\infty+1})}, P^{(\infty^{\infty+n})}, P^{(\infty^{n\infty})}, P^{(\infty^{\infty^{n}})}, P^{(\infty^{\infty^{\infty}})}, \ldots .$$

Cantor summarized the entire procedure in decisive terms: "Here we see a dialectic generation of concepts, which leads further and further, and thus remains in itself necessarily and consequently free of any arbitrariness."[14]

He might have added more specifically that the concepts being generated were, in actuality, transfinite numbers. But he did not take this step. Instead, he was reluctant to confer any reality or special significance upon his new creation. Until he came to terms with the metaphysical nature of the transfinites in writing his *Grundlagen*, they remained nothing more than "infinite symbols."[15] The derived sets remained the focus, and the transfinites, the infinite symbols, were taken only for useful tags by means of which derived sets themselves could be distinguished and identified.

The importance of the infinite symbols, however, was demonstrated without delay. The paper of the preceding year had asked (but left unanswered) the question of whether sets of the second species were necessarily everywhere-dense in any given interval.[16] The new infinite symbols made it easy to describe second species sets consisting of a single point. The method employed was similar to the one Cantor had used in 1872 to construct point sets of the first species, nth kind. His new point sets of the second species for which there were derived sets (indexed by one of the transfinite symbols) consisting of a single point were clearly *not* everywhere-dense. Consequently, these were losing candidates in his search for the complete characterization of the continuum, its properties, and structure.

Cantor made it clear that he expected his new symbols to be of major importance in helping characterize sets of the second species. First species sets had already been given complete characterization in terms of derived sets of finite index ν. By extending the idea of derived sets to $P^{(\infty)}$ and beyond, it seemed reasonable that complete characterization of continuous sets of the second species would likewise be possible. The new "infinite symbols" were perhaps the machinery necessary and sufficient for the task.

In explaining the status of his infinite symbols, mathematically, Cantor noted a feature of their creation which he later stressed with great emphasis in defending his transfinite numbers. They were neither arbitrary tags nor empty symbols, he said, but were dialectically generated in a very "concrete" way. Their reality stemmed from the reality of the derived sets themselves, and it was these sets, more than the symbols $n_0\infty^{\nu} + n_1\infty^{\nu-1} + \ldots + n_{\nu}$, that were the result, the actual *content* of the dialectic generation. In calling the new infinite symbols "concepts," rooted in the derived sets, but generated in a way free from any arbitrariness whatsoever, he had already begun preparation for the struggle to establish the legitimacy of his new numbers. Aware of the opposition aroused by his earlier work, he must have realized that it was best to advance cautiously. Rather than ascribe any reality to his infinite symbols, he chose to speak instead of derived sets alone. The new symbols were thus carried along with successive sets, little more than marks, notationally convenient. Without emphasis, he had indicated both the mathematical and the philosophical grounds upon which he would later have to argue the reasonableness of his new ideas. Cantor's position did not advance to a more positive and comprehensive statement of real, actually existent transfinite *numbers*, until he wrote the *Grundlagen* several years later. If anything, the 1880 paper best exemplified Cantor's caution and restraint.

THE PAPER OF 1882: MULTI-DIMENSIONAL DOMAINS

Cantor's third article went beyond the confines of linear point sets to show the valuable conclusions one could draw by extending results of his first two papers to domains of higher dimension.[17] Derived sets, power, everywhere-denseness, each had an analogous counterpart in the language of arbitrary dimension. By extending these to multiple-dimensional domains, he expected new insights, perhaps broader conclusions concerning point sets in general. He also hoped to gain a better understanding of the nature of continuity itself, linear or otherwise.

What *were* the essential differences between discrete and continuous domains? A decade earlier Cantor, as well as Dedekind, had suggested that many prevailing conceptions of continuity were largely the result of convention, convenience, or uncritical intuitions.[18] For example, in a letter to Dedekind of April 15, 1882, Cantor questioned those who accepted the continuity of space by virtue of perceived, apparently continuous motion. Before developing this idea further, however, Cantor needed to present some preliminaries. The ultimate goal of his latest paper, he claimed, was to extend the idea of powers:

> The concept of power which includes as a special case the concept of whole number, that foundation of number theory, and which ought to be regarded as the most general, genuine feature [*Moment*] of aggregates, is in no way limited to linear point sets. It can be considered much more as an attribute of every well-defined multitude, whatever conceptual characteristic its elements may have.[19]

Cantor admitted that the determination of well-defined sets sounded better than it actually was. In practice it could be exceedingly difficult, if not impossible. But he was quick to add a number of qualifications. He described a well-defined set in rather general terms:

> I call an aggregate (a collection, a set) of elements which belong to any domain of concepts *well-defined,* if it must be regarded as *internally determined* on the basis of its definition and in consequence of the logical principle of the excluded middle [*ausgeschlossenen Dritten*]. It must also be internally determined whether any object belonging to the same domain of concepts belongs to the aggregate in question as an element or not, and whether two objects belonging to the set, despite formal differences, are equal to one another or not.[20]

In order to further clarify his meaning, he elaborated upon the intent of the definition:

> In general the relevant distinctions cannot in practice be made with certainty and exactness by the capabilities or methods presently available. But here this is not of any concern. The *only* concern is the *internal determination* from which in concrete cases, where it is required, an actual (external) determination is to be developed by means of a perfection of resources.[21]

"Internal determination" and "perfection of resources" were phrases designed to cover the most unyielding cases. Cantor offered by way of example the set of all algebraic numbers. Given a real number x, it was not always clear (he suggested, for example, the familiar e and π), whether the number was algebraic or not. Hermite in 1873 had managed to show that e was *not* algebraic, but transcendental.[22] Ironically, the proof that π also failed to be algebraic was discovered by Lindemann[23] in 1882, the same year that Cantor published his ideas on well-defined sets. There could have been no better example of what Cantor meant by the "perfection of resources" than Lindemann's result.

Other cases also came to mind. Though at any given time available knowledge and resources might prove inadequate to characterize fully the set of all numbers satisfying Fermat's last theorem, Cantor not only believed that a definite solution must nevertheless exist, but that it would eventually be found. As an early example from Cantor's work, nothing quite rivals the sense of Platonism underlying his position concerning the determination of well-defined sets. Though his Platonist sympathies were to emerge even more conspicuously in his later writings, they nevertheless found early expression in such ontological conjectures as this one.

For well-defined sets, the notion of power followed directly. Like the derived sets $P^{(\nu)}$ of a given set P, the power of P was given intrinsically with the set P

itself. Following the one-to-one mapping described in 1878 between lines and planes, he was able to reiterate the fact that the powers of point sets in general, whatever their dimension, could be fully determined in terms of powers predicated of *linear* point sets alone.

This attempt to determine powers offered difficulties equal to those met in trying to determine well-defined sets. As problematic as it might be to identify the elements of such sets, the question of a set's power, once it was known to be well-defined, could be even more frustrating. Powers might be "internally determined," but Cantor had earlier met great difficulty in demonstrating the nondenumerability of the real numbers R. Nevertheless, the concept of power was a great unifier:

> If we only keep mathematics in mind and leave aside for now other conceptual areas, the theory of aggregates in the interpretation given here embraces arithmetic, function theory and geometry. It joins them in terms of the concept of power to a higher unity. Thus continuity and discontinuity find themselves considered from the same point of view, measured by the same standard.[24]

The suggestion was straightforward. Not only did Cantor seem to have applications of set theory already in mind, but he also hinted that the underlying secret of such applications was to be found in the concept of power.[25] He was clearly convinced that set theory somehow lay, necessarily, at the very foundations of mathematics.

In the attempt to separate further the essential characteristics of the discrete from the continuous, it was natural for Cantor to focus his attention upon the application of powers to infinite sets. The smallest sets, which were also infinite, constituted the class of denumerable sets. The study of denumerability, then, rested upon two easy theorems:

> *Theorem I:* Every infinite part of a denumerable set constitutes an infinitely denumerable set.[26]

> *Theorem II:* Given a finite or denumerably infinite set of sets $(E),(E'),(E''), \ldots$, each of which is denumerable, then the union of all elements of $(E), (E'), (E''), \ldots$ is likewise a denumerable set.[27]

Having presented these basic elements, Cantor was prepared to develop the major result of the article. His conclusions were more than surprising. They embodied the first major indications of his hopes and expectations for the future applications of his new theory. As a step in that direction, he offered a simple, extremely important theorem:

> *Theorem:* In an n-dimensional, infinite, continuous space A let an infinite number of n-dimensional, continuous sub-domains (a) be defined,

disjoint and contiguous at most on their boundaries; then the aggregate (a) of such sub-domains is always denumerable.[28]

Much of the theorem's simplicity was its forthrightness. Few assumptions were required. None was made concerning either the partition or the total space itself. The subsets (a) could be taken as arbitrarily small. The only necessary stipulation insisted that each subset (a) have definite volume and that two such subsets overlap, at most, only on boundaries.[29]

The proof was elegant in its line of reason. In mapping the space A^N by reciprocal radius vectors onto a unit n-sphere B^N, every subset (a) of A had a counterpart (b) in B. If the collection (b) were countable, then so was (a). The subsets (b) of B could be denumerably ordered in a direct way according to size. Given any number γ, the number of subsets (b) exceeding γ in volume was necessarily finite, since the volume[30] of B itself was necessarily less than $\frac{2\pi^{\frac{n+1}{2}}}{\Gamma\left(\frac{n+1}{2}\right)}$. In such a sequence, the smaller subsets (b) would follow the larger, diminishing without limit. Once proved in general, the specific cases were impressive.

For $n = 1$ the result was essential for any further development of set theory. Though no examples were immediately forthcoming, Cantor did stress the implications of his conclusion: for any collection of separate intervals (α,β) coinciding at most on their endpoints, defined on an infinite straight line, the collection was necessarily countable.[31] For $n = 2$, however, the implications were similarly far-reaching, implying the denumerability of any set of separate surfaces in two dimensions. For the plane he suggested that the theorem had important applications in the theory of complex variables.[32]

Cantor saved the most remarkable corollary for last. For $n = 3$, and the corresponding space of natural phenomena, he could advocate the outlines of an entirely new mechanics.[33] The ramifications were striking and represented the first direct extension of his set theory to considerations that were not strictly mathematical. He used the new result to question the standard interpretations and explanations of the continuity of space and the mechanics devised to describe phenomena of the physical world. Putting his theorem of 1874 to good use, Cantor was able to argue that continuous motion was possible in discontinuous spaces. He demonstrated this conclusion specifically for the case $n = 2$, the plane.

Considering any denumerable point set (M) everywhere-dense in a continuous domain A, Cantor introduced the domain $\mathfrak{A} = A - M$. $\mathfrak{A}$ for example might be the space of all points defined by n-tuples of transcendental real numbers, that is, the space of "real" n-tuples excepting those defined by algebraic numbers. For any two points N,N' in $\mathfrak{A}$, Cantor claimed that a continuous line l' could be given joining N,N', and lying entirely within $\mathfrak{A}$. To show this, he first assumed that N,N' was joined by any line l in A. There could then be, at most, countably many elements m in (M) on l. This meant that one could take an arbitrary, finite number of points N_n on l that were *not* elements of A. He then claimed that given such points NN_1, N_1N_2, . . . , N_nN', these could be connected by lines entirely in $\mathfrak{A}$. The proof, for NN_1, rested on the fact that given the set of all circular arcs passing through NN_1, their centers were given by points g_n falling on a single straight line. Since there were only countably many points of (M) that could fall on such arcs, by denoting each such arc by its center point p_ν, there had to be at least one center point η determining a circular arc joining NN_1 but encountering *no* point of (M). Not only did this arc give a line l' joining NN_1, one lying entirely in $\mathfrak{A}$, but it did even more: "After all, with the same resources it would be possible to connect the points N and N' by a continuously running line given by a *unique analytic rule* and completely contained in the domain $\mathfrak{A}$."[34]

The fact that such a continuous line could be defined in such a discontinuous space as $\mathfrak{A}$ by means of a single, unique analytic expression provoked some serious speculation. Cantor expressed his own opinion: belief in the continuity of space was generally based upon the evidence of continuous motion. Having shown that continuous motion was conceivable even in discontinuous spaces, he called for new research proceeding under alternative assumptions.

Indirectly, similar criticisms concerning the traditionally assumed continuity of space had already been made a decade earlier by both Cantor and Dedekind. Cantor's axiom correlating the arithmetic and geometric domains underscored his inability to *prove* that an arithmetically determined point of the real line identified a physically real point of geometric space.[35] The designation was purely arbitrary, as Cantor had argued in 1872. Ten years later he put it even more emphatically. In general there was *no* inner constraint for thinking of every point given by coordinates of real numbers x,y,z as actually belonging to three-dimensional space. The assumption, he insisted, "must be regarded as a free act of our mental constructive activity. The hypothesis of the continuity of space is therefore nothing but the assumption, arbitrary in itself, of the complete, one-to-one correspondence between the 3-dimensional purely arithmetic continuum (x,y,z) and the space underlying the world of phenomena [*Erscheinungswelt*]."[36]

Continuity thus seemed little more than a free act of the mind, constructed intellectually but by no means guaranteed as actually conforming to the reality of phenomenological space. Cantor was explicit on this point. The hypothesis of the continuity of space was entirely arbitrary.

To the sentences quoted above Cantor added a footnote stressing the importance of strictly arithmetic conceptualization in mathematics.[37] In particular, both Cantor and Dedekind, and subsequently Lipschitz, had striven for a theory of number wholly independent of any geometric content or intuition.[38] Such purely arithmetic treatments of both number and the theory of sets had been undeservedly misjudged, Cantor felt. In particular, authors like Paul du Bois-Reymond and Giulio Vivanti, who believed in the applicability of a theory of actual infinitesimals in mathematics, had failed to heed the evidence of purely arithmetic considerations, or so Cantor argued.[39] The infinitely small, as an existing, absolute quantity, was utterly inadmissible to Cantor's way of thinking. This entire issue became a subject for heated debate a decade later when he engaged the Italian mathematician Veronese in a polemical exchange of letters and articles.[40] But in 1882 the problem was relegated to a note, and Cantor's objections to infinitesimals were left unexplained, at least temporarily.

The closing paragraph of the paper of 1882 presented additional provocative conjectures. The mind, Cantor argued, could as easily abstract from individual points of space, even if their distribution were everywhere-dense, in order to deal conceptually with spaces like $\mathfrak{A}$. The possibility of continuous motion in such spaces immediately raised additional questions. He advocated that a revised mechanics be considered, under the hypothesis of a space with properties much like those of $\mathfrak{A}$. Such a study, he suggested, might possibly lead to new revelations and produce a better understanding of the true grounds upon which the assumption was made that space was everywhere continuous.[41]

Just as he had earlier faulted mathematicians for depending too heavily upon unreliable intuitions in the matter of dimension, he did the same in criticizing mathematical physicists for assumptions they had made concerning the continuity of space.[42] Intuition, as he had learned, was a poor guide in rigorous mathematics. Thus one could not hope to learn much about the continuum from geometry. Instead, reliance had to be placed upon purely arithmetic results. His article of 1882, despite its concentration upon point sets of various dimensions, nevertheless shared a common interest with the other papers in the series on linear point sets. It, too, was pressing on toward an imagined, expected solution to the problem uppermost in Cantor's mind: what was the identifying nature of continuity?

THE PAPER OF 1883: THE NEW SET THEORY AND ANALYSIS

Cantor published a fourth paper on linear point sets in 1883. This was the last to appear before publication of the *Grundlagen,* and it was devoted primarily to applications of his recent set theoretic conclusions to questions in the theory of functions.[43] The paper also added more examples to the growing stockpile of point set decompositions which later figured prominently in his attempts to provide a satisfactory characterization of continua. Though he continued to improve his understanding of discrete versus continuous sets, no real breakthrough had yet been made. None even seemed to be in sight.

To facilitate the treatment of set decompositions, he replaced his earlier notation $P \equiv \{P_1, P_2, P_3, \ldots\}$ for disjoint sums by $P \equiv P_1 + P_2 + P_3 + \ldots$. Thus, for $Q \subseteq P$, $R \equiv P - Q$ and $P \equiv Q + R$ could be handled simply in the context of addition and subtraction. Furthermore, he introduced explicit terminology for sets containing none of their limit points. Such sets, where $\mathfrak{D}(Q, Q') \equiv 0$, were denoted *isolated* sets.[44] Though it would certainly have been appropriate, he said nothing about the empty set, nor did he attempt to distinguish between a zero and a set containing no elements.

Given any set P, an isolated set Q resulted by simply removing $\mathfrak{D}(P,P')$ from P. Thus $Q \equiv P - \mathfrak{D}(P,P')$. Consequently $P \equiv Q + \mathfrak{D}(P,P')$. The formula offered some immediate insights. Clearly, any set P could be considered as the disjoint sum of an isolated set Q and any set that was a divisor of P'. Since $P^{(\nu)} \subseteq P'$, $P^{(\nu)} - P^{(\nu+1)}$ was *always* isolated. Two decompositions were of special importance:

$$P' \equiv (P' - P'') + (P'' - P''') + \ldots + (P^{(\nu-1)} - P^{(\nu)}) + P^{(\nu)}$$
$$P' \equiv (P' - P'') + \ldots + (P^{(\nu-1)} - P^{(\nu)}) + \ldots + P^{(\infty)}.$$

Cantor was then able to produce a number of related theorems which followed almost directly from the above decompositions:

> *Theorem I:* Every isolated point set is denumerable, and therefore belongs to the first class.[45]

Assuming Q to be isolated, Cantor picked $q \in Q$ and denoted the remaining elements of Q by $q', q'', q''', \ldots$. Considering the distances $\overline{qq'}, \overline{qq''}, \overline{qq'''}, \ldots$, it followed that there must be a least distance ρ among the $\overline{qq^{(\nu)}}$, different from zero, since Q was an isolated set. Then taking distances from q', ρ' corresponded to the least distance among $\overline{q'q}$, $\overline{q'q''}$, ... After such a least distance $\rho^{(\nu)}$ had been determined for each $q^{(\nu)}$ and its nearest point, Cantor supposed each q to be the center of a spherical neighborhood of radius $\frac{1}{2}\rho^{(\nu)}$. Each neighborhood was disjoint from every other, except perhaps for boundaries, and since Cantor's 1882 paper had shown that such a collection of neighborhoods was at most countably infinite, it followed that the isolated set Q must also be denumerable.[46]

Having shown that isolated sets were always countable, he went on to establish for the first time the important connections between power, species, and denumerability:

> *Theorem II:* If the derived set P' of a point set P is denumerable, then P is denumerable.[47]

The proof was almost immediate. If P' were countable, then so was $R \equiv \mathfrak{D}(P,P')$. Setting $P - R \equiv Q$, Q was isolated and therefore countable. Since $\mathfrak{D}(P,P') \equiv R \subseteq P'$, R was countable, and therefore $P \equiv Q + R$ was likewise countable. In a similar fashion Cantor argued the following:

Theorem III: Every point set of the first species and n^{th} kind is denumerable.[48]

Theorem IV: Every point set of the second species, for which $P^{(\infty)}$ is denumerable, is itself denumerable.[49]

Greater generality was given in the last of these theorems, in where Cantor appealed to any index α of the infinite symbols $\alpha > \infty$ defined in 1880.

Theorem V: Every point set P of the second species, for which $P^{(\alpha)}$ is denumerable, is itself denumerable.[50]

Though Cantor had made considerable gains with respect to his results for countable sets of both the first and the second species, nothing particularly revealing had been accomplished for continuous domains. So long as $P^{(\alpha)}$ was countable, then so was the original set P. Turning all this around, he was able to draw one conclusion concerning nondenumerable domains. Almost as a corollary to the preceding theorems, he offered a counter-result that did manage to say something about uncountable sets:

If P is a nondenumerable point set, then $P^{(\alpha)}$ is nondenumerable, not only when α is a finite whole number, but also when it is one of the infinite symbols.[51]

Though the status of his transfinite symbols remained unchanged, the application of point sets generally was not neglected in the paper of 1883. Cantor was prepared to establish one of the important results which his new theory made possible in functional analysis. In 1882 he had already stressed his theorem establishing the denumerability of any collection of disjoint intervals on the infinite line as an indispensable result for the advance of set theory. In 1883 he put that claim into practice. At the same time he took the opportunity to raise some doubts concerning generalizations that du Bois-Reymond and Harnack had offered in recent papers discussing theorems on integration.

Paul du Bois-Reymond and Axel Harnack had made noticeable use of set theoretic techniques to aid their studies of the integral.[52] Each had been particularly interested in point sets of negligible content.

Definition: A point set P is said to be of negligible content if its elements can be enclosed in intervals (c_ν, d_ν) such that the sum of the lengths of these intervals, $\Sigma(d_\nu - c_\nu)$ may be made arbitrarily small, in other words, $\lim \Sigma(d_\nu - c_\nu) = 0$.[53]

Despite early confusion between discrete sets, sets of the first species, nowhere-dense sets, and contentless sets, by 1882 enough examples had been given to make it clear that even sets of second species could be nowhere-dense, while nowhere-dense sets could be of positive content. Cantor managed to show that sets of zero content were necessarily nowhere-dense, but the converse was not true. Nowhere-dense sets were not necessarily of content zero. Above all, Cantor complained that his colleagues had failed to provide any indication of the conditions sufficient to ensure that a set *be* of negligible content:

> In research which du Bois-Reymond and Harnack have undertaken on certain generalizations of theorems of the integral calculus, linear point sets are used which have the property of being enclosable in a finite number of intervals *so that the sum of all intervals is less than any given number*.
>
> For a linear point set to possess this property it is clearly necessary that it be everywhere-dense in no arbitrarily small interval; but this last condition does not seem sufficient to confer upon a point set the property in question.[54]

It was one thing to insist that a set be of negligible content, but another to specify the structural requirements such sets could fulfill. Cantor offered a specific characterization of such sets, and presented conditions based upon the first part of his paper as alternatives to the incomplete results of Harnack and du Bois-Reymond:

> *Theorem:* If a point set P contained in an interval (a,b) is so constituted that its first derived set P' is denumerable, then it is always possible to enclose P in a finite number of intervals the sum of which is arbitrarily small.[55]

To prove his theorem, Cantor required three lemmas:

> *Lemma I:* A given function $\varphi(x)$, continuous in the continuous interval (c,d), which at the endpoints has unequal values $\varphi(c)$ and $\varphi(d)$, assumes every value y lying between the limits $\varphi(c)$ and $\varphi(d)$ at least once.[56]

> *Lemma II:* An infinite number of disjoint intervals lying in an infinite line, contiguous at most on their boundaries, is always denumerable.[57]

> *Lemma III:* If one has a denumerably infinite set of numbers ω_1, ω_2, $\omega_3, \ldots, \omega_\nu, \ldots$, then in any given interval a number η can be found which does not appear among those numbers.[58]

Narrowing his analysis of the theorem to the interval $(0,1)$, Cantor observed that for $P \subseteq (0,1)$, $P' \subseteq (0,1)$ and wrote their union:

$$Q \equiv \mathfrak{M}(P,P').$$

Points of $(0,1)$ not contained in Q constituted the set R, thus:

$$(0,1) \equiv Q + R.$$

Assuming P' to be countable, he could immediately conclude from his earlier results that P, and therefore $Q \equiv P + P'$, were also countable. Q could therefore be everywhere-dense in *no* interval of $(0,1)$, but its elements could be listed in a denumerable sequence:

$$u_1, u_2, u_3, \ldots, u_\nu, \ldots . \tag{4.1}$$

Consequently, the set R corresponded to all internal values of an infinite sequence of disjoint intervals contained in $(0,1)$:

$$(c_1, d_1),\ (c_2, d_2),\ \ldots,\ (c_\nu, d_\nu),\ \ldots .$$

Since only the *internal* values of the intervals (c_ν, d_ν) corresponded with points of R, then the points c_ν, d_ν corresponded with points of Q and therefore appeared in the sequence (4.1).

Taking $r \in R$, Cantor argued that no point of Q could lie arbitrarily close to r, otherwise r would be a limit point of P, and then belong to Q. Therefore, left of r there had to be a point c, right of r some point d, such that in (c,d) no points of Q were to be found. If c and d were not isolated points of Q, then in any neighborhood of c and d, outside (c,d), points of Q would occur. Since each limit point of Q belonged to Q, then c,d also belonged to Q. Therefore the intervals (c_ν, d_ν) could not overlap, and consequently they were necessarily countable by Lemma II.

Cantor's next concern was the matter of interval magnitude.[59] Assuming $c_\nu < d_\nu$, the sum of all the intervals (c_ν, d_ν) could be written

$$\sum_1^\infty (d_\nu - c_\nu) = \sigma.$$

Because the intervals (c_ν, d_ν) were non-overlapping, it was clear that $\sigma \leq 1$. Were $\sigma = 1$, the theorem would be established. Therefore, he assumed $\sigma < 1$, and worked for a contradiction.

Cantor began by defining $f(x)$ on $(0,1]$. Summing the length of all (c_ν, d_ν) falling in $(0,x)$, he denoted this sum by $f(x)$. Should any (c_ν, d_ν) fall only partially within $(0,x)$, he allowed only that part included in $(0,x)$ to be included in determining $f(x)$. It was thus clear, by definition, that $f(1) = \sigma$. If he could establish $f(0) = 0$, then $f(x)$ would be a continuous function of x on $[0,1]$. It was already clear by definition that $0 < f(x+h) - f(x) \leq h$, from which the continuity of $f(x)$ followed. Should distinct values x and $x + h$ fall in the same interval (c,d), then

$$f(x+h) - f(x) = h$$

$$(x+h) - f(x+h) = (x+h) - f(x) - h = x - f(x).$$

Introducing $\varphi(x) = x - f(x)$, $\varphi(x)$ was also a continuous function of x, varying from 0 to $1-\sigma$ as x increased from 0 to 1. It also followed by nature of $f(x)$ that $\varphi(x)$ was constant on the interior of any interval (c_ν, d_ν). Consequently, all values assumed by $\varphi(x)$ were necessarily enumerated by

$$\varphi(u_1),\ \varphi(u_2),\ \ldots,\ \varphi(u_\nu),\ \ldots . \tag{4.2}$$

If $x = u_\nu$, then $\varphi(x) = \varphi(u_\nu)$. If $x \in (c_\nu, d_\nu)$, $\varphi(x) = \varphi(c_\nu) = \varphi(d_\nu)$, but c_ν, d_ν were contained in the sequence (4.1), for some index λ: $c_\nu = u_\lambda$. Then $\varphi(x) = \varphi(u_\lambda)$. Therefore (4.2) contained all the values $\varphi(x)$ could assume, and the set itself was *denumerable*.

Assuming $\sigma > 1$, then $1 - \sigma > 0$, and by Lemma I the continuous function $\varphi(x)$ would assume every value between 0 and $1-\sigma$ at least once. Since (4.2) exhausted all values of $\varphi(x)$, it was necessary, by Lemma III, that there be at least *one* value $\eta \in (0, 1-\sigma)$ *not* enumerated by (4.2). By contradicting the continuity of $\varphi(x)$, Cantor's argument reached the desired conclusion: $\sigma = 1$.

This last paper published before the appearance of the *Grundlagen* was thus meant to cast doubt on the sufficiency of theorems which Harnack and du Bois-Reymond had produced, while it advanced Cantor's own proof that any set P whose derived set P' was denumerable was necessarily of negligible content. In his language of 1882, p could be enclosed in intervals of arbitrarily small total length.

As Weierstrass wrote later of Cantor's theory of content, it demonstrated the inadequacy of the Riemann integral and the need to produce an even more general theory of integration.[60] But at the same time, Cantor's work indicated the significant advances that had been made in functional analysis since Riemann's *Habilitationsschrift,* largely by the increased interest mathematicians had begun to take in the properties of point sets. While Cantor was producing his series of papers on infinite linear sets, others were busy with investigations of their own running along similar lines.

For example, by distributing intervals in an everywhere-dense manner, du Bois-Reymond[61] had described a function in 1879 which failed to be integrable (in Riemann's sense), though it was continuous, and thus everywhere-dense in parts of any arbitrarily small interval. By 1882 both Hankel and du Bois-Reymond had introduced specific examples of linear point sets that were also nowhere-dense.[62] Harnack too was enjoying significant progress with the theory of sets, and it was clear that a number of mathematicians were simultaneously making use of new results concerning point sets in promoting their work in analysis.

Cantor may easily have felt a certain amount of pressure owing to the advances his colleagues were making on fronts he had promised to develop. In fact, they had begun to deal with problems he had spent nearly a decade trying to resolve. His criticism of both Harnack and du Bois-Reymond was immediately followed in the paper of 1883 with his own conclusions concerning sets of negligible content. It was as though he were trying to show, with the success of one theorem, both the shortcomings of his competitors' results, and the superiority, the directness of his own.

There was an additional factor as well. Cantor's first paper of the series on linear point sets bore the date January 1879. It was the first explicit formulation of his terminology "everywhere-dense."[63] The following year du Bois-Reymond published a lengthy footnote at the end of a paper, also dated 1879, claiming his priority in the matter of designating "everywhere-dense sets" with his own terminology *pantachisch:*

> A distribution of points is called *pantachisch* if, for any arbitrarily small interval, points of the distribution in question occur. One is led to such distributions of points (of which I have various examples) if one studies accumulation points of infinite order, the existence of which I wrote to *Cantor* in Halle years ago. At another opportunity I will discuss these distributions, the accumulation points of finite and infinite order of increasingly diminishing *intervals,* and finally my choice of the expression pantachisch compared with the expression of everywhere-dense adopted later by Cantor.[64]

Du Bois-Reymond had alluded to such sets in 1875, and had named them explicitly in 1879.[65] Shortly thereafter, he made direct reference to Cantor's paper of 1879. He even suggested in not-so-oblique terms that Cantor was misappropriating ideas which were not entirely his own. Though the concept of everywhere-dense sets was certainly of long standing, du Bois-Reymond clearly felt some claim to priority in designating the property such sets exhibited. He must have hoped his terminology, *pantachisch,* would become accepted and widespread. Even more clearly, he felt Cantor posed a threat to his hope for undisputed recognition.

Similarly, Cantor must have felt challenged by Harnack and du Bois-Reymond. During the next few years, from 1880 to 1884, the pages of the *Mathematische Annalen* carried the results of a good deal of set theoretic work. Though Cantor dismissed the innovations of his rivals' products as meaningless, sometimes inaccurate, he could not ignore them. When asked for an opinion, he encouraged Klein (then editor of *Mathematische Annalen*) to publish Harnack's work, but he was also convinced that it was not very important: "Of course, I am not at all in agreement with Harnack's designation [*Bezeichnung*] of discrete, linear, etc., point sets; I do not regard them as suitable. But this is not saying much—even if he does not get far, I still take the work for a useful one, and in any case it is appropriate for acceptance by the *Annalen*."[66]

If the refinements du Bois-Reymond and Harnack had produced in their analysis of sets were of small impact upon Cantor's thinking, the same cannot be said of their possible significance in urging him to publish. Facing the rising tide of papers on point sets and analysis, the theory of integration and corresponding interest in the topology of domains of definition, Cantor must have sensed the direction in which research seemed to be developing. Anxious not to lose his rights of priority by failing to publish his ideas, he wrote the first four papers on infinite point sets within a space of two years. By the end of 1882, the first sections of the fifth paper were already finished for publication. This, in its separately published version (the *Grundlagen einer allgemeinen Mannigfaltigkeitslehre*), was to be a thorough statement of his results in set theory to date, including the development of his new transfinite numbers and a defense of his entire mathematical philosophy. He was unusually anxious that its publication proceed rapidly. Toward the end of the year, he was writing to Klein nearly every day, urging the greatest speed possible.[67] Cantor even visited the press in Leipzig himself,[68] hoping to expedite the appearance of his latest work, his defense, both mathematical and philosophical, of an entirely new mathematics. The *Grundlagen* was to establish Cantor's place as the founder of set theory and to overshadow the work of even his closest rivals. It was the beginning of something new, quite startling, profoundly original.

CHAPTER 5

The Mathematics of Cantor's *Grundlagen*

The *Grundlagen einer allgemeinen Mannigfaltigkeitslehre*[1] (1883) represented a milestone in Cantor's mathematical research. Not only did it go well beyond the results of his earlier papers on linear point sets (the *Punktmannigfaltigkeitslehre*), but it introduced an entirely new field of inquiry to mathematics, one which promised new gains, while carrying with it the hidden seeds of trouble yet to come. Above all, the *Grundlagen* heralded a new beginning.

When Max Simon, the Strassbourg mathematician, reviewed the *Grundlagen,* he praised Cantor highly for making an important contribution both to philosophy and to the foundations of arithmetic.[2] It was a friendly and laudatory appraisal. The significance of the work, Simon wrote, justified an extensive review. For Simon this meant a careful, though brief recapitulation of the mathematical sections, but only a single sentence concerning the extensive, more philosophical portions of the *Grundlagen*. No polemics were mentioned, no embarrassing points of detail or epistemology were made. An objective and uncritical summary of the *Grundlagen's* theoretical content demonstrated Simon's implicit acceptance of Cantor's new arithmetic. As for those who might be opposed to the appearance of the absolute infinite in mathematics, Simon remarked only that:

> The other parts [of the *Grundlagen*] are exclusively philosophical and are intended to obviate the quarrel with philosophers who have denied transfinite [*Überendlichen*] numbers since the time of Aristotle; the refutation of such objections seems entirely successful, to this reviewer, particularly in Sections 5 and 6.[3]

For a mathematician like Simon, the philosophy presented in the *Grundlagen* was of little consequence or importance, nothing more than an appendage. It was only the mathematics that counted. When Gösta Mittag-Leffler suggested that selected parts of Cantor's work should be translated for his new journal, the *Acta Mathematica,* he felt much the same, and with Cantor's consent the strictly philosophical sections were omitted from the French version.[4] But since the *Grundlagen* was treated in this way with Cantor's full approval, even supervision, it provides a useful dichotomy. While the present chapter concentrates on primarily mathematical portions of the *Grundlagen,* as well as Cantor's sixth and last paper in the *Punktmannigfaltigkeitslehre* series, the next chapter considers in detail the philosophical consequences of Cantor's transfinite set theory. However, despite the separation of these two very distinct aspects of Cantor's work, it must not be forgotten that in his own mind, the metaphysics and the mathematics went hand in hand.

THE TRANSFINITE NUMBERS: THEIR MODES OF GENERATION AND THE PRINCIPLE OF LIMITATION

The major achievement of the *Grundlagen* was its presentation of the transfinite numbers as an autonomous and systematic extension of the natural numbers. Until new means had been devised to penetrate beyond the limits of finite numbers, neither progress in set theory nor success in his study of the continuum was possible. He admitted that his new ideas might seem unorthodox, but he explained their simplicity in a straightforward way:

> As risky as this might seem, I can voice not only the hope, but my strong conviction, that in time this will have to be regarded as the simplest, most appropriate and natural extension [of the concept of number]. But I realize that in this undertaking I place myself in a certain opposition to views widely held concerning the mathematical infinite and to opinions frequently defended on the nature of numbers.[5]

Mathematicians were accustomed to using the infinite in the sense of a variable, either as one increasing beyond all limits, or decreasing to arbitrary smallness. In either case, *finite* magnitudes were retained, whether arbitrarily large or small. And while a great deal was said about eschewing any use of absolute or completed infinities in mathematics, Cantor pointed out that this was not always so. In the study of complex variables, for example, the exploration of the behavior of analytic functions at a point of infinity in the complex plane had proven of unmistakable advantage.[6] But instead of introducing a single point at infinity, Cantor was prepared to produce an unending succession of actually infinite numbers and to show that they possessed identifiable and determinate number-theoretic properties. In short, his transfinites possessed all the attributes expected of any number system and were thus as acceptable mathematically as were the irrational, real, or complex numbers.

Before introducing his transfinite numbers, Cantor explained the important distinction that had to be recognized between the potential and the actual infinite.[7] The former was used as the very roots of the calculus. Here the infinite as a variable, Cantor said, was prominent. Though limits involved the idea of a variable growing beyond any ascertainable bound, they were never regarded as completed or final. Such potential infinities were also referred to as *improper* infinities. In contrast to these, Cantor distinguished proper or *actual* infinities. The best examples of these, he suggested, were his new numbers, the transfinite numbers, first introduced in his paper of 1882, although with unfortunate ambiguity regarding their status as numbers. His realization now that the transfinities were equally as real in a mathematical sense as the finite whole numbers represented one of the most striking advances made in the *Grundlagen:* "A number of years ago I was led to the infinite real whole numbers [*die unendlichen realen ganzen Zahlen*] without having realized that they were concrete numbers of real significance."[8]

At last, with the help of his new transfinite numbers, Cantor was certain that he could define more precisely the concept of power [*Mächtigkeit*]. Until the appearance of the *Grundlagen,* he had been unable to make any simple, accurate determination of powers beyond the denumerably infinite. Though it was clear that the least power among infinite sets was that of the countable set of natural numbers, he had found no simple, natural definition of higher powers. The new transfinite numbers, however, would remedy this deficiency. Thus their definition was the first step toward a more comprehensive description of the powers of infinite sets generally.

Cantor explained how the sequence of natural numbers 1, 2, 3, . . . , had its origin in the repeated addition of *units*.[9] He called this process of defining finite ordinal numbers by the successive addition of units the *first principle of generation*. It was clear that the class of all finite whole numbers (I) had no largest element. Though it was incorrect to speak of a largest element for (I), Cantor believed there was nothing improper in thinking of a new number ω which expressed the natural, regular order of the *entire* set (I). This new number ω, the first transfinite number, was the *first* number following the entire sequence of natural numbers ν. It was then possible to apply the first principle of generation to ω, and to produce additional transfinite ordinal numbers:

$$\omega, \omega + 1, \omega + 2, \ldots, \omega + \nu, \ldots$$

Again, since there was no largest element, one could imagine another number representing the entirety, in order, of numbers $\omega + \nu$. Denoting this entirety by $2\,\omega$, it was possible to continue further:

$$2\omega, 2\omega+1, 2\omega+2, \ldots, 2\omega+\nu, \ldots.$$

In attempting to characterize this mode of generation, Cantor allowed that ω could be regarded as a limit which the natural numbers N (increasing monotonically) approached but never reached.[10] Lest the analogy seem entirely mistaken, he added that by this he meant only to emphasize the character of ω taken as the first whole number following next after *all* the numbers $n \in N$. The idea of ω as a limit served to satisfy its role as an ordinal, the smallest integer larger than any integer $n \in N$.

This then was the second principle of generation. Whenever a set of numbers could be considered as limitless in extent, new transfinite numbers could always be generated by positing the existence of some least number larger than any in the given class. Cantor expressed the essential feature of this second principle of generation in terms of its logical function:

> I call it the *second principle of generation* of real [*realen*] whole numbers and define them more precisely: if any definite succession of defined whole real numbers exists, for which there is no largest, then a new number is created by means of this second principle of generation which is thought of as the *limit* of those numbers, i.e., it is defined as the next number larger than all of them.[11]

By successive application of the two principles it was always possible to produce new numbers and always in a completely determined succession. In their most general formulation, such numbers could be written as follows:[12]

$$\nu_0 \, \omega^{\mu} + \nu_1 \, \omega^{\mu-1} + \, \ldots + \nu_{\mu}. \tag{5.1}$$

But by proceeding apparently without constraint, there seemed to be no end to the numbers of this second number class. Cantor was able to add, however, a third principle which he called the principle of limitation [*Hemmungsprinzip*] and which was designed to produce natural breaks in the sequence of transfinite numbers. This in turn made it possible to place definite bounds upon the second number class (II), and to distinguish it from the third and successively higher number classes.

> *Definition:* We define therefore the second number class (II) as the collection of all numbers α (increasing in definite succession) which can be formed by means of the two principles of generation:
>
> $$\omega, \, \omega + 1, \, \ldots, \, \nu_0 \, \omega^{\mu} + \nu_1 \, \omega^{\mu-1} + \, \ldots + \nu_{\mu}, \, \ldots, \, \omega^{\omega}, \, \ldots, \, \alpha, \, \ldots,$$
>
> which are subject to the condition that all numbers preceding α (from 1 on) constitute a set of power equivalent to the first number class (I).[13]

The difference between Cantor's definition of the transfinite numbers in the *Grundlagen* and his earlier introduction of "infinite symbols" was striking. Previously he had concentrated upon derived sets of the second species and used his transfinite "symbols" merely as indices to identify and to distinguish between derived sets themselves. But if the transfinites were to be of any help in determining the powers of sets and in solving the question of the power of the continuum, they could not be *defined* in terms of those same point sets. Instead, a completely independent formulation was required, one which would establish their legitimacy and character mathematically. This done, there would be no question as to the appropriateness of applying those numbers to the study of point sets and powers.

The significance of Cantor's new achievement was underscored by his choice of terminology and notation. At last his transfinites were *numbers,* as "real" mathematically as the real numbers R. The new symbol ω was introduced instead of the familiar ∞, Cantor said, in order to emphasize the fact that the transfinite ordinal numbers were completed, actual infinities.[14] The potential infinite, traditionally expressed by the symbol ∞, was wholly unsuited to his new purposes. The shift in notation from ∞ to ω thus reflected a significant transformation in his thinking, specifically the advance in status of the transfinites from symbols to numbers.

TRANSFINITE NUMBER CLASSES AND THE CONTINUUM HYPOTHESIS

Once the principle of limitation had been introduced, it was possible to consider the sequence of number classes and their comparative sizes in detail. Of major importance was the fact that the second number class was indeed distinct from the first. In Section 12 of the *Grundlagen* Cantor showed that the new number class (II) was of a greater power than that of the first number class (I).[15]

Beginning with a set of numbers $\{\alpha_\nu\}$ of the second number class, and by assuming that $\{\alpha_\nu\}$ was only denumerably infinite, that is, equivalent in power to the first number class (I), he argued that there must be numbers of the second class (II) which failed to be indexed in the sequence $\{\alpha_\nu\}$. Were there a largest element of $\{\alpha_\nu\}$, γ for example, then $\gamma+1$ (produced upon application of the *first* principle of generation) would produce the required element of (II) not indexed in $\{\alpha_\nu\}$, and the proof would follow immediately. The only interesting case assumed there was no largest element of $\{\alpha_\nu\}$. Cantor was still able to claim that there was an element $\beta \in$(II) larger than all elements of $\{\alpha_\nu\}$. Furthermore, for every integer $\beta'<\beta$, he claimed he could give elements of $\{\alpha_\nu\}$ such that they surpassed every such β' in magnitude. First he took α_{κ_2} as the first element of $\{\alpha_\nu\}$ greater than α_1, α_{κ_3} the number first surpassing α_{κ_2}, and so on, producing two sequences:

$$1 < \kappa_2 < \kappa_3 < \kappa_4 < \ldots,$$
$$\alpha_1 < \alpha_{\kappa_2} < \alpha_{\kappa_3} \ldots,$$
$$\text{and } \alpha_\nu < \alpha_{\kappa_\lambda} \text{ when } \nu < \kappa_\lambda.$$

Considering the set of all integers greater than 1 but less than α_1; the set of all integers greater than or equal to α_1 but less than α_{κ_2}; the set of all integers greater than or equal to α_{κ_2} but less than α_{κ_3}; and so on, Cantor claimed each set was countable, as was a countable union of such sets, and therefore the entire collection was of the first power. Thus, by definition of (II), there was always a definite number $\beta \in$ (II) that was larger than any α_{κ_λ}. β had to surpass any α_ν, since κ_λ could always be taken greater than any given ν, and therefore $\alpha_\nu < \alpha_{\kappa_\lambda} < \beta$. Similar manipulation of indices would ensure that for any $\beta' < \beta$, elements of $\{\alpha_\nu\}$ could be determined such that $\beta' < \alpha_{\kappa_\nu}$. The entire proof rested upon the consequences of having defined (II) with the help of the second principle of generation. This principle guaranteed the existence of the value $\beta \in$ (II) surpassing any value contained in only a denumerable collection of elements from (II). Thus, on the basis of the defining properties of (II), Cantor argued that in fact it was of a power greater than (I). But was it necessarily the set of next largest power? Was it possible that there might exist sets of power intermediate between the powers of (I) and (II)?

In Section 13 of the *Grundlagen* Cantor showed in fact that the power of the second number class was the *next* larger power following immediately after the smallest transfinite power represented by the denumerably infinite number class (I).[16] His argument again depended in large measure upon the definition he had already given of the second number class. As a first step toward the proof, he claimed that, given any set of elements $\{\alpha'\}$ of numbers of the second number class (II), then it was always possible to find a smallest number contained in such a set.[17] Assuming that $\{\alpha'\}$ be ordered so that its terms, sequentially, were monotonically decreasing, Cantor added that not only did the series necessarily end with a finite number of terms, but that the last term was also clearly the smallest. For finite integers the theorem was obvious. For infinite integers of the second number class (II) it was equally clear once those elements of $\{\alpha_\nu\}$ on a finite index were considered. Should the smallest of these numbers equal α_ρ, it then followed that, since $\alpha_\nu > \alpha_{\nu+1}$, the sequence $\{\alpha_\nu\}$ and consequently the entire sequence $\{\alpha_\beta\}$ must consist of exactly ρ terms.

> *Theorem:* If (α') is any set of numbers contained in the collection (II), then only the following three cases can occur: either (α') is a finite collection, i.e., it consists of a finite number of numbers, or (α') has the power of the first number class; or thirdly, (α') has the power of (II).[18]

The proof required the introduction of the initial number of the third number class (III), denoted Ω. Then clearly, by definition, every number of the second number class α' was less than Ω. Cantor imagined the set of elements from (II) ordered sequentially, monotonically increasing, on a transfinite index of numbers of the second number class. Specifically, $\{\alpha_\beta\}$, where $\beta = \omega, \omega+1, \ldots$. For all indices, $\beta < \alpha_\beta$. Because the indices β were drawn from the second number class (II), there were only three possibilities. Either $\beta < \omega + \nu$, in which case (α') would be finite by virtue of the value of ν; or β could assume *all* values of the sequence $\omega + \nu$, but remain less than an assignable number of the second number class (II), in which case (α') would clearly be a set of the first power; or lastly, β could run through *all* numbers of the class (II), in which case the set (α_β), in fact the set (α'), would be of the power (II). The proof rested upon the character of the index and its properties.

The theorem had two direct consequences. Given any well-defined set M of power of the second number class (II), any infinite subset M' of M was either denumerable, or of the same power as M. Furthermore, given any well-defined set M of the second power, a subset M' of M and a subset M'' of M', if M'' could be corresponded in a unique one-to-one way with M, then M' shared a similar correspondence with both M'' and M.

WELL-ORDERED SETS, *ANZAHLEN,* AND NEW DISTINCTIONS BETWEEN FINITE AND INFINITE SETS

An essential element in Cantor's definition of the transfinite numbers and their corresponding arithmetic was the concept of well-ordered sets. These had been an implicit part of his introduction of the transfinites from the start, since the natural numbers served as the prototype of well-ordered sets. Generation of the numbers of the first class (I) and, similarly, generation of the numbers of the second and higher number classes was accomplished in such a way that the entire sequence of transfinite numbers was automatically well-ordered:

> *Definition:* By a *well-ordered* set is to be understood any well-defined set in which the elements are arranged in a definite, given succession such that there is a first element of the set, and for every other element there is a definite *successor,* unless it is the last element of the succession.[19]

Well-ordered sets were extremely useful in drawing new distinctions between finite and infinite sets. For example, Cantor believed that a major advance made possible by his transfinite numbers was the creation of a new concept, that of the numbering [*Anzahl*] of the elements of a well-ordered, infinite set. Numberings expressed the order in which the elements of a given infinite set occurred. Though Cantor would later place much greater emphasis upon this point, he did note in the *Grundlagen* that this new concept conferred an immediate objectivity upon his transfinite numbers. In other words, the objective reality of the transfinites stemmed from the existence of well-ordered sets, whose order was expressed by numbers of Cantor's various transfinite number classes. Thus he set out to show how, given any infinite well-ordered set, there was always a number of the second number class (II) that would uniquely represent its order (or numbering). From a single, denumerable set (α_ν), different well-ordered sets could be given:

$$(\alpha_1, \alpha_2, \ldots, \alpha_\nu, \alpha_{\nu+1}, \ldots)$$

$$(\alpha_2, \alpha_3, \ldots, \alpha_{\nu+1}, \alpha_{\nu+2}, \ldots, \alpha_1)$$

$$(\alpha_3, \alpha_4, \ldots, \alpha_{\nu+2}, \alpha_{\nu+3}, \ldots, \alpha_1, \alpha_2)$$

$$(\alpha_1, \alpha_3, \ldots; \alpha_2, \alpha_4, \ldots).$$ [20]

Two sets, well-ordered, were defined to be similar, or of the same numbering, if they could be corresponded in a one-to-one fashion such that order was preserved between elements in each case. Thus, if $\alpha'_n < \alpha'_m$ in the one set, the corresponding elements of the second set, α'_n and α'_m would also have to be ordered such that $\alpha'_n < \alpha'_m$. As it turned out, such a stipulation was always uniquely determined. Given two such sets, Cantor showed that the natural succession of integers, including the transfinite numbers, could therefore be used to identify similar well-ordered sets. Given any number α of the first or second number class, it was clear that taken together with all the elements naturally preceding it, the numbering of all similar well-ordered sets was given by the number α in a unique way. For example, each of the three well-ordered sets:

$$(\alpha_1, \alpha_2, \alpha_3, \ldots, \alpha_\nu, \alpha_{\nu+1}, \ldots),$$

$$(\alpha_2, \alpha_1, \alpha_4, \ldots, \alpha_{\nu+1}, \alpha_\nu, \ldots),$$

$$(1, 2, 3, \ldots, \nu, \ldots)$$

had the same numbering.[21] By definition, it was ω. Similarly, the well-ordered sets:

$$(\alpha_2, \alpha_3, \ldots, \alpha_\nu, \ldots, \alpha_1),$$

$$(\alpha_3, \alpha_4, \ldots, \alpha_{\nu+1}, \ldots, \alpha_1, \alpha_2),$$

$$(\alpha_1, \alpha_3, \ldots; \alpha_2, \alpha_4, \ldots),$$

had, respectively, the numberings $\omega+1$, $\omega+2$, and 2ω.

Cantor's distinction between number and numbering brought new insights to the nature of the difference between finite and infinite sets. For a given finite set, regardless of ordering, the numbering of elements was always the same. Infinite sets were much more interesting because the different numberings one could find for sets of the same power were produced by different well-orderings. The numbering of sets, therefore, was a concept totally dependent upon the order in which the elements of the set were taken. Different orderings in general produced different numberings, though the number of elements in each case might be the same. And there was a correlation between the number of elements of a set and the numberings which its elements might produce depending upon their arrangement. Cantor considered sets of the first power given in any definite succession. As long as they were well-ordered, then their numberings were always numbers of the second number class (II), and *only* of the second number class. Conversely, given any number α of (II), any set of the first power could be ordered in such a way that its numbering coincided with α. He commented upon this connection between powers and numberings by noting that:

> Every set of the power of the *first* class is denumerable by numbers of the *second* number class and only by such. In fact, the set can always be arranged in such a way that it is enumerated by any given number of the second number class. This number gives the *numbering* [*Anzahl*] of elements of the set with respect to that succession.[22]

Analogous results held for sets of higher power. Thus every well-defined set with the power of the second class could be given a definite numbering in terms of numbers from the third number class (III), and so on.

As Cantor stressed, for finite sets, the concepts of power [*Mächtigkeit*] and numbering [*Anzahl*] coincided. Different orderings did not produce different numberings. No matter how one chose to count a set of n elements, its numbering remained the same. Since the concept of power was independent of the order in which elements were taken, the power of a finite set of n elements was equally expressed by the number n.

For infinite sets, however, the distinction between power and numbering was significant. Though every number α of the second number class distinguished a unique numbering of elements, the power of any set with a numbering α was always the same: denumerable. But there was still a connection between power and numbering for infinite sets, since for *any* well-ordered set of denumerable power, its numbering was always a uniquely determined number of the second number class. This means of identifying transfinite numbers, numberings, and number classes seemed a satisfying, harmonious feature to Cantor of his new set theory. That harmony arose basically from the properties of well-ordered sets. The concept of numbering, after all, was a generalization of the basic idea of counting. The numbering of an infinite set, given by one of Cantor's transfinite numbers, was *determined* by the "law of counting" [*Gesetz der Zählung*]. Once the significance of this interpretation of numbering had been appreciated, declared Cantor, the status of transfinite numbers was as secure as the finite numbers: "The existence of the infinite will never again be deniable while that of the finites is nevertheless upheld. If one permits either to fall, one must do away as well with the other."[23]

TRANSFINITE ARITHMETIC

Well-ordered sets were also indispensable because they provided the foundation for defining arithmetic operations for transfinite numbers. It is worth mentioning that Cantor did not believe the laws of arithmetic for his transfinites to be arbitrarily defined; they were to be inferred "immediately from the mind's inner intuition with absolute certainty."[24] The epistemological significance of this statement, however, is considered in greater detail in the following chapter.

The addition of two transfinite numbers α and β was defined in terms of two infinite well-ordered sets M and M_1, such that α and β represented the *numbering* of each set, respectively.[25] Then the sum $\alpha+\beta$ was defined in terms of the well-ordered set $M+M_1$ produced by the sequence of elements from M followed by the sequence of elements of M_1. Cantor cautioned that for infinite sets the operation was not necessarily commutative, though on finite sets the order could be reversed without consequence. A simple example illustrates this noncommutativity of transfinite addition:

$$1+\omega=1,\ 1,\ 2,\ 3,\ \ldots \neq 1,\ 2,\ 3,\ \ldots,1=\omega+1.$$

Associativity, however, was unaffected, and $\alpha+(\beta+\gamma)=(\alpha+\beta)=\gamma$.

Multiplication was similarly defined. Given a well-ordered set of numbering β, by replacing each of its elements by elements of numbering α, the product $\beta\cdot\alpha$ was defined with β the multiplier and α the multiplicand. For example, if $\beta=\omega$, $\alpha=2$, then:

$$\beta\cdot\alpha=(1,\ 2,\ \ldots,\ \nu,\ \ldots)\cdot\alpha=(a_1,\ b_1,\ a_2,\ b_2,\ \ldots,\ a_\nu,\ b_\nu,\ \ldots)=\omega.$$

On the other hand:

$$\alpha\cdot\beta=(a_1,\ b_1)\cdot\beta=(1,\ 2,\ \ldots,\ \nu,\ \ldots;\ 1,\ 2,\ \ldots,\ \nu,\ \ldots)=\omega+\omega=2\omega.$$

Clearly, multiplication failed to be commutative for transfinite numbers, and in general $\alpha\cdot\beta\neq\beta\cdot\alpha$. Associativity, however, was not lost.

Prime numbers were identifiable among the transfinites, but they were of two varieties.[26] Given a transfinite number $\alpha=\beta\cdot\gamma$, α could be said to be prime if the only possible decomposition required that $\beta=1$, or that $\beta=\alpha$. But in the case of transfinite decompositions, the multiplicand was not necessarily unique. There was a certain range of possible values it might assume. However, before Cantor could deal adequately with this feature of the transfinite numbers, he had to investigate the nature of inverse operations in greater detail.

INVERSE OPERATIONS AND PRIME FACTORIZATIONS FOR TRANSFINITE NUMBERS

Acceptance of the new set theory and the companion transfinite numbers hinged upon more than a few definitions of elementary properties. If the transfinite numbers were to be regarded as consistent extensions of the finite reals, their algebraic behavior and arithmetic properties were significant questions to settle. Two subjects were of special interest. How were the inverse operations of subtraction and division to be understood among transfinites? In other words, did the transfinite numbers satisfy the conditions of a field? Second, was unique prime factorization possible? Both questions were made more complicated by the fact that the standard operations of addition and multiplication were in general noncommutative for transfinite numbers.

Subtraction could be considered in two ways.[27] Given any pair of transfinite numbers α and β, the equation $\alpha+\xi=\beta$ always admitted a unique solution for ξ, where ξ could be a number of either the first or second number class, and was defined as equal to $\beta-\alpha$. Alternatively, subtraction made clear the significance of the noncommutativity of transfinite operations. Unlike the former equation, $\xi+\alpha=\beta$ was often insoluble for ξ. Cantor considered the example $\xi+\omega=\omega+1$. Whenever the equation $\xi+\alpha=\beta$ was soluble, however, it was by no means necessary that the solution be unique. It was entirely possible that ξ could assume infinitely many values and still solve the given equation. Among all such solutions, however, there was always a smallest one. Assuming, then, that the equation $\xi+\alpha=\beta$ was soluble, the solution was denoted $\beta_{-\alpha}$, which in general was always different from $\beta-\alpha$.

A similar analysis then followed for the case of division, in which the same distinctions were drawn.[28] $\beta=\xi\cdot\alpha$ was given a solution, always unique, $\xi=\frac{\beta}{\alpha}$. On the other hand, considering the alternative possibility, $\beta=\alpha\cdot\xi$, whenever a solution was possible there could also be infinitely many, but always a smallest, which Cantor denoted by $\frac{\beta}{\underline{\underline{\alpha}}}$.

With the inverse operations thus explained, Cantor began to study the numbers of the second number class in greater detail. He distinguished two varieties, those of the first kind, for which there was always an immediate predecessor in the number series, and those of the second kind for which there was no such immediate predecessor. Numbers of the first kind were those produced by application of the *first* principle of generation, while numbers of the second kind were those produced by means of the *second* principle of generation. Of the second kind, Cantor offered ω, 2ω, $\omega^{\nu}+\omega$, ω^{ω} as examples.[29]

Prime numbers of the second kind took, in order, the following form:

$$\omega,\ \omega^{\omega},\ \omega^{\omega^2},\ \omega^{\omega^3},\ \ldots\ .$$

Thus, among numbers represented as

$$\varphi=\nu_0\omega^{\mu}+\nu_1\omega^{\mu-1}+\ \ldots+\nu_{\mu}, \tag{5.2}$$

there was only the single prime number ω of the second kind. This should not lead one to conclude, warned Cantor, that the number of primes of the second kind was very small, for it turned out that they were actually as numerous as the numbers of the second number class itself.[30] Thus the power of the set of prime transfinite numbers of the second kind was the same as the power of (II).

Prime numbers of the first kind included those of the form:

$$\omega+1,\ \omega^2+1,\ \ldots,\ \omega^{\mu}+1,\ \ldots\ .$$

Under numbers represented by φ, equation (5.2), these were the only primes of the first kind that occurred, and the power of the set of all such primes was also the same as that of the second number class (II).

There were important characteristics of the prime numbers of the second kind which warranted further comment. Letting η represent any prime number of the second kind, Cantor showed that it was always true that $\eta \cdot \alpha = \eta$, whenever α was smaller than η. Consequently, given any two transfinite numbers α, β both less than η, then the product $\alpha \cdot \beta$ was also less than η.

Given two numbers of the second number class:

$$\varphi = \nu_0 \,\omega^\mu + \nu_1 \,\omega^{\mu-1} + \ldots + \nu_\mu,$$

$$\psi = \rho_0 \,\omega^\lambda + \rho_1 \,\omega^{\lambda-1} + \ldots + \rho_\lambda,$$

Cantor explained the procedure for addition and multiplication.[31]

For addition:

1. If $\mu < \lambda$, then $\varphi + \psi = \psi$.
2. If $\mu > \lambda$, then $\varphi + \psi = \nu_0 \,\omega^\mu + \ldots + \nu_{\mu-\lambda-1} \,\omega^{\lambda+1} + (\nu_{\mu-\lambda} + \beta_0)\,\omega^\lambda + \rho_1 \,\omega^{\lambda-1} + \ldots + \rho_\lambda$.
3. If $\mu = \lambda$, then
$\varphi + \psi = (\nu_0 + \rho_0)\,\omega^\lambda + \rho_1 \,\omega^{\lambda-1} + \ldots + \rho_\lambda$.

For multiplication:

1. If $\nu_\mu \neq 0$, then

$$\varphi \cdot \psi = \nu_0 \,\omega^{\mu+\lambda} + \nu_1 \,\omega^{\mu+\lambda-1} + \ldots + \nu_{\mu-1}\omega^{\lambda+1} + \nu_\mu \,\rho_0 \,\omega^\lambda + \rho_1 \,\omega^{\lambda-1} + \ldots + \rho_\lambda.$$

If $\lambda = 0$, then the last term on the right equaled $\nu_\mu \,\rho_0$.

2. If $\nu_\mu = 0$, then

$$\varphi \cdot \psi = \nu_0 \,\omega^{\mu+\lambda} + \nu_1 \,\omega^{\mu+\lambda-1} + \ldots + \nu_{\mu-1} \,\omega^{\lambda+1} = \varphi \,\omega^\lambda.$$

With these rules in hand, he could then go on to produce the decomposition of a given number φ into its prime factors. Given

$$\varphi = c_0 \,\omega^\mu + c_1 \,\omega^{\mu_1} + c_2 \,\omega^{\mu_2} + \ldots + c_\sigma \,\omega^{\mu_\sigma},$$

where

$$\mu > \mu_1 > \mu_2 > \mu_3 > \ldots > \mu_\sigma,$$

and $c_0, c_1, c_2, \ldots, c_\sigma$ were positive finite numbers all different from zero, then the following product was possible:

$$\varphi = c_0 \,(\omega^{\mu-\mu_1}+1)c_1(\omega^{\mu_1-\mu_2}+1)c_2 \ldots c_{\sigma-1} \,(\omega^{\mu_{\sigma-1}-\mu_\sigma}+1)c_\sigma \,\omega^{\mu_\sigma}.$$

Taking the elements $c_0, c_1, \ldots, c_\sigma$ as given in their own unique factorizations, then φ was given in terms of its own unique prime factors of the first and second kinds. Though he did not pursue the study of transfinite prime numbers in much detail in the *Grundlagen,* Cantor promised to return to the subject at another time. It was even possible, he noted, to establish the unique prime factorization of any transfinite α of the second number class, but no proof of this assertion was published until 1897.[32]

Thus Cantor had justified his claim made at the beginning of the *Grundlagen.* He had demonstrated that his transfinite numbers were unambiguously defined and possessed regular number-theoretic properties. Though their arithmetic laws were somewhat different from those for finite numbers, particularly in terms of their noncommutativity, this posed no serious difficulty. What mattered most was the fact that he had shown the consistency with which the laws of transfinite arithmetic could be established. He had managed to demonstrate that not only were the basic operations of addition and multiplication definable, as well as their inverse operations, but that two forms of prime numbers could be identified, and correspondingly, even unique factorization was possible for the new transfinite numbers.

Had he not been so thorough in describing transfinite arithmetic, had the basic operations not been explained and their regularity demonstrated, then the acceptability of the transfinite numbers would surely have been a significant question to mathematicians. Certainly a full discussion would help make it clear that the transfinites were as mathematically consistent and well behaved as more familiar and established number systems. With a basic, transfinite arithmetic carefully presented, Cantor's theory, at least in outline, seemed to gain a stature and a presence of its own.

THE CONTINUUM

When Cantor edited his German version of the *Grundlagen* for translation in Mittag-Leffler's *Acta Mathematica,* he placed the original Section 10 at the very end, thereby emphasizing the important role of the continuum in directing and motivating his research.[33] Basic to the progress of all science, he felt, was an acceptable concept of continuity. Its nature and properties had always stimulated passionate controversy and great differences of opinion, though he was certain that the roots of the trouble were easy to identify. Underlying the concept of the continuum, different features had always been stressed, but no exact or complete definition had ever been given. He assigned original blame to the Greeks, who had been the first to study the problem but in such ambiguous terms that myriad interpretations were left open to later doxographers and commentators.[34] For example, Cantor believed that Aristotle and Epicurus represented polar opposites on the subject of the continuum and its consistency. On the one hand, Aristotle and his followers believed in a continuum composed of parts which were divisible without limit. Epicurus, by contrast, developed the atomist position and regarded the continuum as somehow synthesized from atoms which were imagined as finite entities. In retracing the history of divided opinion over these two basic positions, Cantor dismissed those who held intermediate views meant to find a compromise between the two extremes. For example, he cited Thomas Aquinas, who believed that the continuum must be composed neither of infinitely many nor of a finite number of parts but of no parts whatsoever. Cantor criticized Aquinas sharply, saying that his view of the continuum was no clarification at all but actually a silent confession that the fundamentals of the problem had yet to be identified.[35] For Aquinas, evasion was better than failure to resolve the dilemma. Thus Cantor accounted for the scholastic origins of a view that continued to have its adherents even in his own day, where the continuum was taken as an indivisible concept, an *a priori* intuition inaccessible to the processes and analysis of human minds. This view, wrote Cantor, would reject any attempt at an arithmetic analysis of continuity as an unlawful intervention. It would argue the impossibility of any analysis depending on the discrete, which would violate the essential "unity" of the continuum. Cantor refused to tolerate any view of the continuum as an *a priori* categorization, as a product of intuition not subject to analysis by thought, and hence by mathematics. This was, in his own words, equivalent to relegating the problem of the continuum to the level of religious dogma.[36]

Cantor suggested that, traditionally, continuous magnitudes provided the foundation for an intuitive conceptualization of the continuum.[37] Such intuitions then led to ideas of continuous functions upon which a theory of analytic functions was ultimately based. This, in turn, permitted recognition of even more general functions with peculiar properties, for instance, continuous but nondifferentiable functions. Nevertheless, in all cases Cantor claimed that mathematicians had assumed as given the independent continuum, although continuity as such had not been the subject of any further consideration.

Cantor found this reliance upon intuition all the more objectionable because time and space were generally used to explain the mind's intuition of continuity. But reference to the notion of time, specifically the continuity of time, was not, in his view, admissible.[38] Continuity of time depended upon an assumed continuum *independent* of time and could in no way be interpreted as objective in the sense of substance or as subjective in the sense of a necessary *a priori* form of intuition. Time, for Cantor, was nothing more than an auxiliary, related concept binding together movements in the natural world.[39] Having shown in the early stages of his *Punktmannigfaltigkeitslehre* that continuous motion was entirely possible in spaces that *failed* to be continuous in the accepted sense of the word,[40] he was reluctant to accept any arguments which appealed to the intuitions of space and time to analyze the continuum. Thus it was no surprise that he should deny the existence of objective or absolute time in nature, and that he consequently rejected these as a basis for measure in dynamics. (On the other hand, he did suggest that motion could be taken as a measure of time.) Space and time, as subjective *a priori* forms of intuition, could contribute nothing, he felt, to the thoughtful and rigorous analysis of continuity.

Drawing upon his theory of real numbers, Cantor wanted to develop a purely arithmetic analysis of the continuum.[41] Given an n-dimensional space G_n, he defined an arithmetic point of the space as any system of n-tuples defined from real numbers ranging from $-\infty$ to $+\infty$. In his notation, $(x_1/x_2/\ldots/x_n)$ was such an arithmetic point of G_n. Distance was defined between any two points in the space by the familiar formula:

$$\left| \sqrt{(x_1' - x_1)^2 + (x_2' - x_2)^2 + \ldots + (x_n' - x_n)^2} \right| .$$

The question uppermost in Cantor's mind: when was a given set P of arithmetic points in G_n to be considered a continuum?

Turning to his study of derived sets, he emphasized his discovery that for any set P whose first derived set P' was nondenumerable, P' could always be written as $P' \equiv R + S$, where S was a perfect set unaltered by derivation ($S' \equiv S$), and R was either finite or denumerable.[42] In applying this result to the example of continuous sets, he concluded that they must be perfect sets. But he had also discovered that this condition, by itself, was not sufficient to describe the continuum.

When Cantor first defined the real numbers in his paper on trigonometric series of 1872, using fundamental sequences of rational numbers, the reals could be characterized as the first derived set of the rational numbers. Clearly the continuum was a perfect set, but perfect sets could be constructed which were dense in no interval, however small. To illustrate this point, he offered the example of his famous ternary set. Considering all numbers z defined by

$$z = \frac{c_1}{3} + \frac{c_2}{3^2} + \ldots + \frac{c_\nu}{3^\nu} + \ldots ,$$

where the coefficients c_ν took the values 0 and 2, he produced a set with unusual but very suggestive properties. Not only was the ternary set dense in *no* interval of the continuum, it was closed, perfect, and it contained no interior points. Every point of the set was an accumulation point and the set itself was uncountably infinite.[43] This example clearly demonstrated that perfect sets need not be continuous, and thus something more was required before the continuum could be completely described. The missing concept, Cantor believed, was that of connectedness:

> *Definition:* We call T a connected point set if for any two points t and t' of T, there always exists a *finite* number of points $t_1, t_2, \ldots, t_\nu$ of T, such that the distances $\overline{tt_1}, \overline{t_1t_2}, \ldots, \overline{t_\nu t'}$ are all less than any given arbitrarily small number ϵ.[44]

Connectedness was essentially an equivalent form of everywhere-denseness. It accounted for the metric property of the continuum.[45] The concept of connectedness, used to replace the idea of everywhere-denseness, freed Cantor from having to introduce the auxiliary concept of the space with reference to which a given point set could be said to be everywhere-dense. With the idea of connectedness, no such additional considerations were required.

Thus the characteristic features of continua were specified: they were both perfect and connected. These, stated Cantor, were the necessary and sufficient conditions under which a point set could be considered continuous. He was proud to add that the designation was completely general and conceptually definite by virtue of the set theoretic ideas introduced in the *Grundlagen*. "Here 'perfect' and 'connected' are not simply words, but through the preceding definitions they are quite general predicates, conceptually characterized most sharply, of the continuum."[46]

Thus Cantor concluded that he had found the most precise means of defining the essence of continuity. He supported this judgment by turning to the definition Bolzano had advanced in his *Paradoxien des Unendlichen*.[47] Bolzano had managed to express only *one* property of the continuum, Cantor declared, that corresponding to the property of connectedness. By way of example, Cantor recalled the strange space $\mathfrak{A}$ (presented in his paper of 1882), the product of a continuous space from which an everywhere-dense set of points had been removed. Nevertheless, continuously connected lines were still possible within $\mathfrak{A}$. Though the space was clearly not continuous in the sense of Cantor's characterization, since it was not perfect, it did satisfy the conditions of Bolzano's definition. Similarly, Cantor criticized Dedekind's presentation of continuity as given in his *Stetigkeit und irrationale Zahlen,* for Cantor believed it emphasized, one-sidedly, the "perfect" aspect of the continuum.[48] Interestingly enough, Cantor chose to omit this criticism of Dedekind's work in the French version of the *Grundlagen* for *Acta Mathematica*. But he did choose to emphasize that his own characterization of the continuum was a satisfying combination of both elements stressed separately by Bolzano and by Dedekind.

Despite all that the *Grundlagen* had accomplished, there was a serious lacuna. Though Cantor had made it clear that his introduction of the transfinite numbers, particularly those of the second number class, was essential to sharpening the concept of power, the question of the power of the continuum was still unanswered. He hoped a proof would be forthcoming, establishing his continuum hypothesis that the power of the continuum was none other than that of the second number class (II). The benefits of such a proof would be numerous. It would immediately follow that all infinite point sets were either of the power of the first or second number class, something Cantor had long claimed. It would also establish that the set of all functions of one or more variables represented by infinite series was necessarily equal in power to the second number class. Likewise, the set of all analytic functions or that of all functions represented in terms of trigonometric series would also be shown to have the power of the class (II).

The *Grundlagen* went no further in settling any of these issues. Instead, Cantor published a sequel in the following year as a sixth in the series of papers on the *Punktmannigfaltigkeitslehre*.[49] Though it did not bear the title of its predecessor, its sections were continuously numbered, 15 through 19; it was clearly meant to be taken as a continuation of the earlier 14 sections of the *Grundlagen* itself. In searching for a still more comprehensive analysis of continuity, and in the hope of establishing his continuum hypothesis, he focused chiefly upon the properties of perfect sets and introduced as well an accompanying theory of content.

PERFECT SETS

Though the *Grundlagen* had established the significance of perfect sets in helping typify the structure of the continuum, Cantor had offered no hints as to how it might also help advance his analysis of powers. In his sixth of the *Punktmannigfaltigkeitslehre* papers, he began to introduce the most important and characteristic properties of perfect sets, preparing the way for their later application.

First he demonstrated that no *denumerable* point set P could be *perfect*.[50] Since perfect sets were defined as those identical with their first derived sets, $S = S'$, he only had to show that assuming P to be denumerable meant that there existed limit points p' that were not elements of P. P would then fail to be perfect. Assuming a denumerable sequence of elements $p_1, p_2, \ldots, p_\nu, \ldots$ in P, he constructed a sequence of nested neighborhoods and then recast his argument used in 1874 to prove the nondenumerability of the real numbers, thereby producing the required limit point p' not contained in P. Consequently, P was not a perfect set.

Next he began to develop the implications of the transfinite numbers for the study of point sets generally, adding specifically:

> The conceptual formation (which is so important for the study of the nature of a point set P) of derived sets of various orders through the derivates just mentioned, certainly does not come to an end with a *finite* ordinal number ν. In general it is quite necessary to take into account sets derived from P whose orders are characterized by definite *transfinite* numbers of the second, third, etc., number classes.[51]

Thus Cantor launched the program he had only mentioned obliquely at the beginning of the *Grundlagen*. In order to advance further with the theory of sets, he had insisted that exact determination of the transfinite numbers of higher order was imperative. With the various number classes finally at his disposal, he could begin to make good his earlier claim that they were in fact indispensable in the further development of his work. Though the early applications of the transfinite numbers were straightforward, they began to build toward increasingly more subtle and complex conclusions illustrating the diversity of special classes of point sets. For example:

> *Theorem B:* If α is any number of the *first* or *second* number class and P in G_n is a point set of such a character that $P^{(\alpha)} \equiv 0$, then P' as well as P is of the first power, unless P and P' are finite sets.[52]

In working out the proof Cantor mentioned that, since P was either zero or perfect for some value α of the second number class, it followed "that the derived sets of an order greater than α are all identical to the derived set $P^{(\alpha)}$ and therefore their consideration is superfluous."[53] Consequently, except in specified cases, he could limit himself to consideration of numbers from either the first or second number classes, certain all the while that the important cases were necessarily included among these two classes of numbers. In order to prove his Theorem B, he drew directly upon techniques developed for the paper of 1883.[54] Appealing to the decomposition $P' \equiv \sum_{\alpha'}(P^{(\alpha')} - P^{(\alpha'+1)})$, where $(P^{(\alpha')} - P^{(\alpha'+1)})$ was always an isolated set, and consequently countable, it followed that both the set P' and therefore P as well were countable.

Arguing on similar grounds, Cantor followed with four more theorems:

> *Theorem C:* If P is a point set of G_n such that its first derived set $P^{(1)}$ is of the *first* power, then there are always numbers γ of the first or second number classes for which $P^{(\gamma)} \equiv 0$, and of all such numbers γ, there is a smallest one α.[55]

> *Theorem D:* If P is a point set of G_n such that its first derived set $P^{(1)}$ has a power greater than the first, then there are always points which belong simultaneously to all derived sets $P^{(\alpha)}$, where α is any number of the first or second number classes, and the collection of all these points, which is nothing other than $P^{(\Omega)}$, is always a perfect set.[56]

> *Theorem E:* If P is a point set of G_n such that its first derived set $P^{(1)}$ has a power greater than the first, and if $S = P^{(\Omega)}$ is the perfect set whose existence is mentioned in Theorem D, then the difference $R \equiv P^{(1)} - S$ is always a point set of at most the first power, and we can thus always and uniquely separate $P^{(1)}$ into two components R and S, so that $P^{(1)} \equiv R + S$, where S is a perfect set and R is either finite or of the first power.[57]

> *Theorem F:* If P is a point set of G_n such that its first derived set $P^{(1)}$ has a power greater than the first, then there is always a least number α of the first or second number classes such that $P^{(\alpha)} \equiv P^{(\alpha+1)}$, and consequently it is already the αth derived set of P. In other words, $P^{(\alpha)}$ is equal to the perfect set $P^{(\Omega)} \equiv S$.[58]

All these results had been known, Cantor noted, at the time he wrote the *Grundlagen*. Section 10, in fact, had given what he labeled Theorem E above, but in the *Grundlagen,* he had overstated his conclusion, arguing there that the decomposition $P' \equiv R + S$ could always be given where S was a perfect set, and where "R is so constituted that it is subject to a continuing reduction until it is annihilated by the repeated process of derivation, so that there is always a first number γ of the number classes (I) or (II) for which $R^{(\gamma)} \equiv 0$; I call such point sets R," he added, "*reducible*."[59]

In general, however, this was not true.[60] Cantor had been led into this error by assuming that, since R was at most of the first power and simultaneously a part of P', it must then have followed that R was the derived set of a certain subset of P, and therefore he concluded, now correctly, that there must be an α such that $R^{(\alpha)} \equiv 0$. But in general R need not be the derived set of another set. Consequently, Cantor should not have concluded that, given any set P with derived set P' of the power of the second number class (II), it was always capable of decomposition into a *reducible* and a *perfect* set.

Ivar Bendixson, the Swedish mathematician, had written to Cantor in May 1883, pointing out the error Cantor had made and showing how the situation could be reformulated with complete accuracy.[61] Not only did Bendixson rework a number of Cantor's own results, in particular those corresponding to Theorems D, E, and F, but he also was able to characterize the properties of the set R in a completely general way. Since it was not necessarily true that R^{α} for transfinite α must be 0, it was important to determine those properties of the countable set R which distinguished it from other point sets of the power of the first number class (I). Bendixson settled the matter with a theorem Cantor included as his Theorem G:

> *Theorem G:* If R is the set of first power in Theorem E, then there exists a smallest number α belonging to the first or second number class, such that $\mathfrak{D}(R, R^{(\alpha)}) \equiv 0$.[62]

Cantor joined to this theorem of Bendixson's several explanations and additional remarks. Again he was able to stress properties of the transfinite numbers he had developed in connection with the first half of the *Grundlagen*. He noted that if $P^{(\omega)} \equiv 0$, then there was some $n \in N$ as well for which $P^{(n)} \equiv 0$. But this result was only a special case of a more general theorem, expressed as follows:

> *Theorem:* If β is any number of the second kind belonging to the second number class, and P a point set of such a character that $P^{(\beta)} \equiv 0$, then there are always smaller numbers $\beta' < \beta$ of the first or second number classes, so that for these as well $P^{(\beta')} \equiv 0$.[63]

Cantor had already explained in the *Grundlagen* that numbers of the first kind were those for which there was always an immediate predecessor. This enabled him to characterize a definite property of the transfinite numbers identified in Theorem C. It was clear that the smallest number α in that theorem had to be of the first kind, so that there was always a next-smaller number α_{-1}. Were α of the second kind, a transfinite for which there was *no* immediate predecessor, then α could not be the *smallest* number for which $P^{(\alpha)} \equiv 0$, contradicting the fact that α was taken as the *smallest*. Once more he was able to show the distinctions his transfinite numbers were capable of making in the theory of sets.

Before proceeding further, Cantor introduced several conventions of terminology. For the unions of sets, instead of reserving the notation

$$P_1 + P_2 + P_3 + \ldots + P_\nu + \ldots \equiv \mathfrak{M}(P_1, P_2, P_3, \ldots, P_\nu, \ldots)$$

for disjoint sets only, he chose to allow the notation to include the union of any collection of point sets.[64] But it was necessary to add that while commutativity and associativity remained operative, that for subtraction, if $P \equiv P_1 + P_2 + P_3 + \ldots$, and $P - P_1 \equiv P_2 + P_3 + \ldots$, then P_1 could have no points in common with any other P_ν. Whenever subtraction was considered, the elements subtracted had to be disjoint with respect to the remaining sets. For finite collections,

$$(P_1 + P_2 + P_3 + \ldots + P_\nu)^\alpha = P_1^\alpha + P_2^\alpha + P_3^\alpha + \ldots + P_\nu^\alpha.$$

For infinite collections of sets, this did not hold in general.

So much for operations. Whenever a set P contained all of its limit points, $P' \subseteq P$, equivalently expressed by $\mathfrak{D}(P,P') \equiv P'$, Cantor described the set P as a *closed* set.[65] By way of example, he noted that the sets of singular points of analytic functions of complex variables were always closed sets. Given any set P, there was a corresponding closed set $\mathfrak{M}(P,P') \equiv P + P'$. Every set that was the first derived set of another set was necessarily closed by definition. Cantor added that any closed set could be represented in an infinite number of ways as the first derived set of any other set. He went on to offer counterparts of his theorems C, D, E, and F in terms of *perfect* point sets specifically.[66] And then he turned to introduce a new and important concept: the idea of a set dense-in-itself.

Given a set P, if P was a divisor of its derived set P', then $P \subseteq P'$, or $\mathfrak{D}(P,P') \equiv P$, then P was said to be *dense-in-itself*.[67] Furthermore, it was clear that if P were dense-in-itself, then P' was a perfect set. By contrast, Cantor defined a set P to be *separated* if no part of the set P was dense-in-itself. Isolated sets formed a special class of separated sets. It was also true that all closed sets of the first power (I) were separated sets. Otherwise they would not have been reducible. Consequently it followed that the sets obtained earlier in Theorems E and G of Section 16 were reducible, and therefore separated. He argued these results on the basis of contradictions obtained by applying the results of Bendixson's work to reducible sets. Furthermore, no matter what the character of P might be, he could ensure that $P - \mathfrak{D}(P, P^{(\Omega)})$ was always a separated set. It was also a set of the first power, a denumerable set.[68] Again he was able to demonstrate the distinctions his new results were capable of making, much more advanced than anything the earlier papers of the linear point set series had been able to produce.

Sets dense-in-themselves seemed to share obvious affinities with the concept introduced much earlier of sets everywhere-dense in given intervals. But Cantor was careful to explain the very important differences between the two ideas.[69] The concept of "dense-in-itself" was an internal property of any set. Given a collection of elements P, it was immediately determined whether the set was dense-in-itself, or not. But the concept of "everywhere-dense" was quite different. Given a set P, it could be everywhere-dense only with respect to another set $H \subseteq G_n$, and depending upon the character of the set H, P could either be everywhere-dense in H, or it could not. Cantor made the differences even clearer. P could be dense-in-itself and not be everywhere-dense in $H \subseteq G_n$. On the other hand, P could be everywhere-dense in $H \subseteq G_n$ and fail to be dense-in-itself. Any P with points lying outside H would be an example. But, if P was an everywhere-dense set in H, such that $P \subseteq H$, then it was always true that P was dense-in-itself as well.

With the new idea of a set dense-in-itself thus formulated and explained, Cantor went on to make one of the most significant new advances in the development of his set theory. In Section 18 of the sixth paper on linear point sets, he presented his theory of content, or the volume of sets in general.

CANTOR'S THEORY OF CONTENT

To any point set $P \subseteq G_n$, continuous or not, Cantor claimed it was possible to correspond a definite and unique, non-negative number that represented its content or volume with respect to the n-dimensional space G_n containing P. He described this as the content of P "with respect to G_n."[70] In familiar cases, the concept coincided with the usually understood n-fold integral:

$$\int dx_1 \, dx_2 \, dx_3 \, \ldots \, dx_n,$$

where the integral was extended over all subspaces of the set P in question. But it could also be defined in more unusual cases, for which the standard Riemann integral failed to be defined.[71]

Cantor's generalized integral was introduced in terms of a function which he called the "characteristic function of P with respect to G_n."[72] The procedure was as follows. Given a bounded set $P \subseteq G_n$, Cantor considered the closed set $\mathfrak{M}(P, P') \equiv P + P'$. For every point $p \in \mathfrak{M}(P, P')$, it was possible to enclose each in an n-dimensional spherical neighborhood of radius ρ, where $K(\rho, p)$ denoted each such neighborhood of a point p. The set of all such neighborhoods $\sum_{\rho} K(\rho, p)$ possessed a least common multiple denoted

$$\Pi(\rho, P \text{ in } G_n), \Pi(\rho, P) \text{ or simply } \Pi(\rho).$$

Since P was assumed to be bounded, $\Pi(\rho)$ consisted of a finite number of parts of the space G_n, where each $K(\rho, p)$ represented a continuous n-dimensional part of G_n, including boundaries. Consequently, the integral

$$\int dx_1\, dx_2\, dx_3 \ldots dx_n$$

taken over the elements $\Pi(\rho)$ was uniquely defined, depending on ρ. These values Cantor denoted $F(\rho)$. It was in terms of this function that he defined the characteristic function of P with respect to G_n:

$$F(\rho, P \text{ in } G_n) = \int_{(\Pi(\rho, P \text{ in } G_n))} dx_1\, dx_2\, dx_3 \ldots dx_n.$$

$F(\rho)$ was not only a continuous function of (ρ), but the limiting value of $F(\rho)$ was used to define the content or volume of the set P. Specifically:

$$I(P \textit{ in } G_n) = \lim_{p=0} F(p, P \text{ in } G_n) \qquad \text{or} \qquad I(P) = \lim_{p=0} F(p).$$

For disjoint sets, he could immediately claim that:

$$I(P + Q) = I(P) + I(Q).$$

Of primary importance was a theorem which stated that "the content of a set P is always equal to the content of its derived set P' with respect to the same space G_n, or equivalently: $I(P \text{in } G_n) = I(P' \text{in } G_n)$."[73] This theorem led to an even more general one:

> *Theorem:* If γ is any finite or transfinite number of the second number class, and P any point set in G_n, then it is always true that $I(P \text{ in } G_n) = I(P^{(\gamma)} \text{ in } G_n)$.[74]

The proof that $I(P) = I(P^{(\gamma)})$ produced a number of direct consequences. Were P a reducible set, then $I(P) = 0$. This followed from Cantor's Theorem C', Section 17 of the same paper, where he had shown that under such circumstances there was necessarily some smallest transfinite number α such that $P^{(\alpha)} \equiv 0$. Consequently $I(P^{(\gamma)}) = I(P) = 0$.

If a given set P were not reducible, then there was always a perfect set S such that it had the same content as P, or in other words, $I(P) = I(S)$. By Cantor's earlier Theorem E', in fact, there was always a smallest α such that $P^{(\alpha)} \equiv S$, and since $I(P) = I(P^{(\alpha)}) = I(S)$, the claim was justified directly. Thus the content of any set could always be considered as the content of its perfect sets.[75]

Cantor intimated that there were significant consequences and extensions to be drawn from the results of his theory of content and that he would develop them in greater detail on some later occasion. Though this never happened, he did add that perfect sets could have zero content but only when the perfect set P was "everywhere-dense in no interval of the space G_n." He had already offered an example of such a point set in the *Grundlagen,* where he added an explanatory note in which he introduced his famous ternary set. Not only was it an uncountably infinite, closed set, containing no interior points, every point of which was an accumulation point, but it was also nowhere-dense, perfect, and of content zero.[76]

For perfect sets everywhere-dense in a set H, however, Cantor could confirm that the content of P would always be positive. It was even possible for a perfect set to be nowhere-dense in H and yet still be of positive content. Though he did not substantiate these claims, he did suggest the direction in which he expected his research to lead: "In the mathematical-physical applications of set theory," he wrote, "about which I shall soon publish my research, an even more general concept than the one designated here by $I(P)$ plays an essential part."[77] For example, it was possible to extend the idea of areas beyond the notion of $I(P \text{ in } G_n)$, and Cantor hinted at the sort of generalizations he had in mind.[78] Assuming that $\varphi(x_1, x_2, \ldots, x_n)$ was any integrable function in G_n, and $P \subseteq G_n$, P to be bounded and reduced to the form of its least common multiple $\Pi(\rho, P \text{ in } G_n)$, then the integral

$$\int\limits_{(\Pi(\rho, P \text{ in } G_n))} \varphi(x_1, x_2, x_3, \ldots, x_n) dx_1 \, dx_2 \, dx_3 \ldots dx_n$$

represented a continuous function of ρ, whose limit for $\rho=0$ produced a value dependent on P and $\varphi(x_1, x_2, \ldots, x_n)$, which he denoted by

$$I(\varphi(x_1, x_2, \ldots, x_n), P \text{ in } G_n),$$

or simply by

$$I(\varphi(x_1, x_2, \ldots, x_n), P).$$

He concluded that his previously defined $I(P)$ was equivalent, in the generalized context, to $I(1, P)$.

Though he said nothing about this more general form of his theory of content, it was apparent that he had many new ideas in reserve. Unfortunately, events of the following year were to discourage him greatly from mathematical research, and from preparing the work already in mind for publication. Details of these events, which occurred shortly after the appearance of Cantor's *Grundlagen* and which affected his life dramatically, both personally and professionally, are discussed in the next chapter. Before turning to these, however, it will be helpful to summarize the state of transfinite set theory as Cantor brought his sixth and last paper to be published in the *Punktmannigfaltigkeitslehre* series to a close.

CONCLUSION

Cantor's study of point sets had made one feature abundantly clear: content could be reduced to the consideration of perfect sets alone. Because such sets were closed, complete, contained in themselves, they possessed very special properties. Continuous sets were perfect sets, perfect sets that were also connected. But despite their incredible variety, Cantor was able to prove that perfect sets were all of the same power.[79] Since the continuum was a perfect set, it followed that the continuum hypothesis would be established if only it could be shown that the power of any perfect set was equivalent to that of the second number class (II). Though he was unable to resolve this question completely, he was able to draw some promising conclusions for closed sets in general.

Given any closed set P, not of the first power, one that was not reducible, he gave its decomposition as the sum of two sets by virtue of his Theorem E': $P \equiv R + S$, where R was a set of the first power, hence denumerable, $\{r_n\}$, and where S was a perfect set.[80] Thus: $P \equiv \{r_n\} + S$.

He then established that $P \sim (0, 1)$, concluding that if P were a closed set of power greater than (I), then it must have the power of the linear continuum itself. Cantor regarded this as a strong and encouraging conclusion. He underscored the final sentences of his last paper in the *Punktmannigfaltigkeitslehre* series as follows:

> In future paragraphs it will be proven that this remarkable theorem has a further validity even for linear point sets which are not closed, and just as much validity for all n-dimensional point sets. From these future paragraphs, with the help of the theorems proven in Nr.5, §13 (the *Grundlagen*), it will be concluded that *the linear continuum has the power of the second number class (II).*[81]

But the aims Cantor set were high, too high. Though he began the struggle to establish the validity of his continuum hypothesis in the summer of 1884 and tried to exploit directly the properties of closed sets and perfect sets in order to do so, he was unable to produce an acceptable proof. Unfortunately, the goal toward which the *Grundlagen* and the entire collection of six papers in the *Punktmannigfaltigkeitslehre* series had been directed was not to be attained so easily. In fact, Cantor was never able to do more than conjecture that his continuum hypothesis was true. But despite this failure, he had nevertheless succeeded in presenting the first outline of a new mathematical discipline, one that would eventually enrich and transform mathematics by forcing it to come to terms with his new theory of the infinite.

Whatever the disappointments Cantor was to suffer, his transfinite set theory represented a revolution in the history of mathematics. Not a revolution in the sense of returning to earlier starting points, but more a revolution in the sense of overthrowing older, established prejudices against the infinite in any actual, completed form. Consequently, Cantor's transfinite numbers were to prove no less revolutionary for philosophers and theologians who were concerned with the problem of infinity. As the next chapter demonstrates, Cantor was deeply committed to probing the metaphysical and religious significance of his work.

In fact, for him the mathematical, metaphysical, and theological aspects of his transfinite set theory were mutually reinforcing. Cantor was convinced that his discoveries were not only essential for the future development of pure mathematics but that set theory could even be used to refine philosophy and to support theology.

CHAPTER 6

Cantor's Philosophy of the Infinite

Cantor made philosophy an equal and intentional partner to mathematics in his *Grundlagen einer allgemeinen Mannigfaltigkeitslehre*. In the German version issued as a separate monograph by Teubner in 1883, a simple introduction was added in which he stressed that the mathematical and the philosophical sections were inextricably connected.[1] In Cantor's view, the *Grundlagen* was much more than a strictly mathematical presentation of his new transfinite set theory. It offered as well his first published defense of the actual infinite, a concept which most philosophers, theologians and mathematicians had traditionally opposed.

Philosophers had been wary of the paradoxical nature of the infinite since the Pre-Socratics first began to explore its many contradictory forms. Aristotle's solution was generally followed, which simply rejected the use of completed infinities. Christian theologians were also opposed to the actual infinite; for the most part they regarded the idea as a direct challenge to the unique and absolutely infinite nature of God. Finally, mathematicians generally followed the philosophers in avoiding any application of the actual infinite, and their reluctance stemmed from the apparent inconsistencies such concepts always seemed to introduce. Gauss, in a celebrated letter to Heinrich Schumacher, expressed in most authoritative terms his opposition to use of such infinities:

> But concerning your proof, I protest above all against the use of an infinite quantity [*Grösse*] as a *completed* one, which in mathematics is never allowed. The infinite is only a *façon de parler,* in which one properly speaks of limits.[2]

Cantor was well aware that his new theory of infinite sets and transfinite numbers faced opposition, much of it traditional and of long standing. One goal

A portrait of Cantor in middle life.
In the possession of Wilhelm Stahl.

of the *Grundlagen* was to demonstrate that there was no reason to accept the old objections to completed, actual infinities and that it was possible to answer mathematicians like Gauss, philosophers like Aristotle, and theologians like Thomas Aquinas in terms they would find impossible to reject. In the process Cantor was led to consider not only the epistemological questions which his new transfinite numbers raised but to formulate as well an accompanying metaphysics. Arguing the mathematical consistency of the new theory, and thereby asserting its legitimacy, was not enough. He felt compelled to defend his work from any form of attack, on whatever level, and therefore he was anxious to consider philosophical and theological objections that might be raised against the concept of actual infinities.

PHILOSOPHY OF THE INFINITE

Cantor believed that opposition to the use of actual infinities in mathematics, philosophy, and theology was based upon a common and pervasive error. Whatever mathematicians may have assumed in the past, finite properties could *not* be predicated in all cases of the infinite. Such attempts led inevitably to contradictions and to misunderstandings. In explaining this to Vivanti in a letter of 1886, he pointed to Aristotle as the source for the medieval dogma: "infinitum actu non datur,"[3] something Cantor described as a basic tenet of the scholastics. As the inspiration for centuries' worth of opposition to the actual infinite, Aristotle required explicit confrontation. His assumption that there were only finite numbers, which led to the conclusion that only enumerations of finite sets were possible, precluded any consideration of infinite numbers from the start.

A typical argument used by Aristotle and by the scholastics involved the "annihilation of number."[4] Were the infinite admitted, it was said that finite numbers would be swallowed up by any infinite number or magnitude. For example, given any two finite numbers a and b, both greater than zero, their sum $a+b>a$, $a+b>b$. However, if b were infinite, no matter what finite value a might assume, $a + \infty = \infty$, and this seemed contrary to a well-known and basic property which the addition of any two positive numbers ought to exhibit. It was in this sense that any infinite number was thought to "annihilate" any finite number, and because this appeared to violate the way in which numbers were understood to behave, infinite numbers were consequently rejected as being inconsistent.

Cantor condemned this kind of argument, however, on the grounds that it was fallacious to assume that infinite numbers must exhibit the same arithmetic characteristics as did finite numbers. Moreover, by direct appeal to his theory of transfinite ordinal numbers, Cantor could demonstrate that infinite numbers *were* susceptible of modification by finite numbers. In fact, the distinction between Cantor's ω and $\omega+1$ showed expressly that finite numbers *could* be added to infinite numbers without being "annihilated." Thus Cantor believed that Aristotle was quite mistaken in his analysis of the infinite, and that his authority was exceedingly detrimental.[5]

Having dealt with Aristotle and the scholastics, Cantor undertook an investigation of other works by some of the most impressive thinkers of the seventeenth century, a century that witnessed serious and often profound analysis of the nature of infinity. He suggested that anyone interested in such things would do well to consult Locke, Descartes, Spinoza, and Leibniz, while Hobbes and Berkeley were highly recommended as additional reading.[6] These writers had produced the most convincing criticisms known to Cantor of the actual infinite. If he could demonstrate their error in rejecting completed infinities, then he was certain that his transfinite numbers could easily withstand criticism of a similar or less penetrating sort. Since he believed that refutation of systematic and coherent knowledge (as he characterized the thought of the seventeenth century) could only be done, legitimately, in terms of such systems and since God was inevitably called upon in the judgments of the seventeenth century, Cantor felt obliged to make similar considerations.

With this in mind, he summarized the position most commonly encountered in the seventeenth century: that number could only be predicated of the finite. The infinite, or Absolute, in this view, belonged uniquely to God.[7] Uniquely predicated, it was also beyond determination, since once determined, the Absolute could no longer be regarded as infinite, but was necessarily finite by definition. Cantor's inquisitive "*how* infinite" was an impossible question. To minds like Spinoza and Leibniz, the infinite in this absolute sense was incomprehensible, as was God, and therefore any attempt to assign a basis for determining magnitudes other than merely potential ones was predestined to fail.

Moreover, following Aristotle, the "annihilation" of finite quantities by completed infinities (were they to exist) seemed completely contradictory, and consequently no concept of infinite numbers could be accepted in mathematics. Cantor criticized this refusal to predicate anything of the absolute in terms of number, just as he had in showing the inadequacy of Aristotle's position on the same point, by showing that such conclusions depended upon a circular argument, a *petitio principii*. It was improper to assume that infinite numbers should be constrained to follow principles established for finite numbers. As he explained, though $1+\omega=\omega$ might be said to annihilate the unit on the left, $(\omega)+1=(\omega+1)$ just as clearly did not.[8] Added on the right the unit did in fact, under the rules of transfinite arithmetic, modify the actually infinite number ω.

Although Cantor at times made it seem as if he were the first and only mathematician to have taken the absolute infinite seriously, he nevertheless drew some solace from two of his predecessors.[9] Both were important figures in the historical development of the concept of infinity, and both had written on the mathematical *and* philosophical consequences of the actual infinite. One of these was G.W. Leibniz, the other was Bernhard Bolzano.

Leibniz was particularly difficult because his opinions concerning the infinite seemed different depending upon occasion and context. As Cantor showed from various citations, Leibniz frequently denied any belief in the absolute infinite. However, in several instances Leibniz did advance a useful and significant distinction between the actual infinite and the absolute infinite, and Cantor was happy to claim him as a supporter of the former, at least in the following passage:

> I am so in favor of the actual infinite that instead of admitting that Nature abhors it, as is commonly said, I hold that Nature makes frequent use of it everywhere, in order to show more effectively the perfections of its Author. Thus I believe that there is no part of matter which is not—I do not say divisible—but actually divisible; and consequently the least particle ought to be considered as a world full of an infinity of different creatures.[10]

Cantor developed this idea to great advantage several years later, particularly in attempting to obviate the confrontation some had feared between a theological interpretation of the infinite and Cantor's new transfinite numbers.

Unlike Leibniz, Bolzano was an unequivocal champion of the absolute infinite.[11] Cantor particularly admired Bolzano's attempt to show that the paradoxes of the infinite could be explained, and that the idea of completed infinities could be introduced without contradiction into mathematics. In fact, Bolzano's *Paradoxien des Unendlichen,* published in 1821, received high praise from Cantor for having done an important service to mathematics and philosophy alike.

Even so, Cantor criticized Bolzano's treatment of the infinite for two reasons. Not only was Bolzano's concept of the actual infinite mathematically unclear, but the important ideas of power and the precise concept of numbering were never developed.[12] Though in certain instances one could find suggestions of both ideas, they were never accorded lucid and independent development. These were essential concepts, Cantor stressed, for a proper understanding of infinite number, and without them, no completely successful theory of the actual infinite was possible. Nevertheless, despite such reservations and disagreements with certain particulars of Bolzano's program, Cantor was nevertheless impressed by the boldness and the audacity with which he defended the actual infinite in mathematics.

One feature of Bolzano's work which particularly impressed Cantor was the distinction he made between *categorematic* (actual) and *syncategorematic* (potential) infinities. The *Grundlagen* placed much stress on this point, and in his more philosophical papers (published several years later), Cantor went even further in exposing the faulty assumptions of those who failed to admit the distinction. For example, among contemporary German philosophers opposed to the idea of completed infinities, Cantor singled out John Frederick Herbart and Wilhelm Wundt as prime offenders. Their preoccupation with potential infinities precluded any satisfactory discussion of the actual infinite. In a letter to the Swedish mathematician and historian Gustav Eneström, Cantor summarized his opposition as follows:

> All so-called proofs against the possibility of actually infinite numbers are faulty, as can be demonstrated in every particular case, and as can be concluded on general grounds as well. It is their *πρῶτον ψεῦδος* that from the outset they expect or even impose all the properties of finite numbers upon the numbers in question, while on the other hand the infinite numbers, if they are to be considered in any form at all, must (in their contrast to the finite numbers) constitute an entirely new kind of number, whose nature is entirely dependent upon the nature of things and is an object of research, but not of our arbitrariness or prejudices.[13]

Herbart was particularly open to Cantor's criticism. By defining the infinite in terms which admitted only potential forms of infinity, there was no way Herbart could have consistently allowed the idea of a completed or an actual infinity. Cantor believed that the actual infinite had to be studied without such arbitrariness or prejudice. Cantor once said that the nature of things had to be taken as given, and he was certain that the nature of things both abstractly in mathematical terms and concretely in physical terms affirmed the existence of his transfinite numbers.[14] Moreover, the connections between the two, the abstract and the concrete, provided yet another level upon which Cantor hoped to justify his new theory.

METAPHYSICS AND CANTOR'S MATHEMATICS

Whenever Cantor spoke of metaphysics he meant the philosophical study of the relations between the constructs of mind and the objects of the external world. Thus the study of the abstract theory of the transfinite numbers was the business of mathematics, but the study of the realization or embodiment of the transfinite numbers in terms of the objects of the phenomenological world was the concern of metaphysics. And so metaphysics assumed its place in Cantor's continuing program to establish the legitimacy of his new theory, particularly in the years following publication of the *Grundlagen*.

In the *Grundlagen* itself, Cantor was careful to distinguish between real (*reellen*) numbers and *real* (*realen*) numbers, though it was a distinction lost in both the original French and in the recent English translations of the work.[15] Considering the French version which appeared in *Acta Mathematica*, this is not surprising, for Mittag-Leffler had specifically requested that Cantor prepare a version of his monograph eliminating entirely all philosophical aspects of his arguments in favor of the transfinite numbers. Since the contrast between the real and *real* numbers was essentially one of metaphysical significance to

Cantor, the French version appeared without any emphasis upon the reality which he attributed to the new concepts being advanced.

Even so, the distinction was of major importance to Cantor for epistemological reasons. It was essential to differentiate between the *reellen Zahlen*, real numbers as opposed to the complex numbers in a formal mathematical sense, and *realen Zahlen*, *real*, actual numbers which enjoyed more than a purely formal existence. In fact, he insisted that his transfinite numbers were *real* in the same sense that the finite whole numbers might be regarded as real.[16] Since the positive integers were taken to have an objective existence in terms of actual sets of finitely many objects, exactly the same was true for his transfinite numbers, because they too were derived from actual sets of infinitely many objects.

Thus transfinite numbers reflected, in terms of their origins, the same character as did finite numbers, although Cantor's new numbers were grounded in infinite rather than in finite sets. And, just as the finite whole numbers possessed an objective reality, so did the transfinite numbers. Their existence was naturally reflected in the matter and space of the physical world, and in the infinities of concrete objects. There was a particularly interesting passage elaborating this idea at the very end of a paper he published in the *Acta Mathematica* in 1885. Mittag-Leffler had suggested that Cantor demonstrate the utility of his new concepts by suggesting some possible applications of transfinite set theory to other branches of science.[17] Cantor did so by introducing two hypotheses concerning the nature of matter and the aether. Utilizing a consciously Leibnizian terminology, he raised two questions: what was the power [*Mächtigkeit*] of the set of all monads that were *material?* Similarly, what was the power of all monads comprising the *aether?* Cantor replied by saying that for years he had held the hypothesis that the set of all material monads was of the first power, while the set of all aetherial monads was of the second power.[18] He claimed there were many reasons to support this view, but he did not produce any at the time. Instead, he went on to suggest that transfinite set theory, applied in this way, would be of great benefit to mathematical physics and could help solve problems of natural phenomena including the chemical properties of matter, light, heat, electricity, and magnetism.

Just as the irrational numbers were in a sense concretized by virtue of their geometric representation (the $\sqrt{2}$ by the diagonal of a square, for example), Cantor found a similar objectivity for his transfinite numbers in the material world. This material verification of the transfinite numbers was achieved by virtue of the actually infinite sets of monads, corporeal or aetherial, in which the transfinites were reflected. Though he never placed primary emphasis upon this concrete, metaphysical component of his thinking in trying to justify his theory, he did believe it to be a reasonable auxiliary. In his own mind, the applications of transfinite numbers in physical terms was direct evidence of their very real existence.[19]

There were other approaches, too, by which he argued the objective reality of his transfinites. He offered a particularly interesting argument whereby he turned the position of finitists like Kronecker to his own ends. If the integers were granted real and objective status in mathematics, Cantor reasoned that the same was then true for his transfinite numbers. Finitists, who only allowed arguments of the sort: "For any arbitrarily large number N there exists a number $n > N$," necessarily presupposed (said Cantor) the existence of *all* such numbers $n > N$, taken as an entire, completed collection which he called the *Transfinitum*.

In a lengthy footnote to remarks which he had made at a meeting of the *Gesellschaft Deutscher Naturforscher und Aerzte* in Freiburg during September 1883, he elaborated his metaphor of the road, which illustrated in a graphic and charming way his argument for the very real and necessary existence of the actual infinite:

> Apart from the journey which strives to be carried out in the imagination [*Phantasie*] or in dreams, I say that a solid ground and base as well as a smooth path are absolutely necessary for secure traveling or wandering, a path which never breaks off, but one which must be and remain passable wherever the journey leads.[20]

In strictly mathematical terms, Cantor translated this as follows: "Thus every potential infinity (the wandering limit) leads to a *Transfinitum* (the sure path for wandering), and cannot be thought of without the latter."[21] Cantor, in all seriousness, believed that his theory of the transfinite numbers was the "Highroad of the Transfinites," and that it was absolutely necessary for the existence and applicability of the potential infinite. In a letter to the Italian mathematician Vivanti, he asserted the existence of the transfinite numbers even more emphatically.[22] In the more general language of domains, be they in algebra, number theory, or analysis, the domain of values had to be taken, he insisted, as actually infinite.

This position, however, cleverly led Cantor to press another reason for the justification of his transfinite numbers. Once the existence of absolutely infinite sets was established, then the transfinite numbers were a direct consequence. He explained his reasons for this assertion in the above-mentioned lecture he presented in Freiburg, in 1883. Since finishing the *Grundlagen* he had considerably sharpened both his concept of power and that of *Anzahl,* or ordinal number, by defining them as general concepts abstracted from given, existing sets.[23] This development (one of great significance for Cantor's evolution of transfinite set theory in the years following the *Grundlagen*) is discussed in greater detail in the following chapter, but for now it is enough to appreciate how he had come to regard the transfinite numbers as naturally produced, by abstraction, from the existence of actually infinite sets. For example, "By the numbering [*Anzahl*] or the ordinal number of a well-ordered set $\mathfrak{M}$ I mean the general concept or universal [*Allgemeinbegriff, Gattungsbegriff*] which one obtains by abstracting the character of its elements and by reflecting upon nothing but the order in which they occur."[24]

Thus, if one admitted the existence of infinite sets, the transfinite numbers followed as little more than a direct consequence. To support further the reasonableness of such thinking, Cantor reexamined the grounds for granting the irrational numbers mathematical legitimacy. He emphasized the similarities which bound the established, acceptable irrationals with his new, unorthodox transfinites:

> The transfinite numbers themselves are in a certain sense *new irrationals,* and in fact I think the best way to define the *finite* irrational numbers is entirely similar; I might even say in principle it is the same as my method described above for introducing transfinite numbers. One can absolutely assert: the transfinite numbers *stand or fall* with the finite irrational numbers; they are alike in their most intrinsic nature [*inneresten Wesen*]; for the former like these latter (numbers) are definite, delineated [*abgegrenzte*] forms or modifications (ἀφωρισμένα) of the actual infinite.[25]

To define the irrational numbers, infinite collections of rational numbers had been required. Believing that any potentially infinite collection necessarily presupposed the existence of infinite collections, while asserting that the concept of number proceeded directly by abstraction from the existence of a given set, finite or infinite, Cantor concluded that the existence and reality of the transfinite numbers was immediate. The only further requirement was consistency. As long as new concepts were not inconsistent, there was no reason why they should not find acceptance and application in mathematics.

CONSISTENCY AND THE IMPORTANCE OF CANTOR'S FORMALISM: IRRATIONALS, TRANSFINITES, AND INFINITESIMALS

Though Cantor believed in the objective reality of his transfinite numbers and argued that their existence was confirmed directly by abstraction from the existence of infinite sets, he did not believe that mathematicians needed either to consider or to accept such arguments. If one wished to do so, such lines of reasoning might be regarded as compelling, but they were not essential. For mathematicians, only one test was necessary: once the elements of any mathematical theory were seen to be consistent, then they were mathematically acceptable. Nothing more was required:

> In particular, in introducing new numbers, mathematics is only obliged to give definitions of them, by which such a definiteness and, circumstances permitting, such a relation to the older numbers are conferred upon them

> that in given cases they can definitely be distinguished from one another. As soon as a number satisfies all these conditions, it can and must be regarded as existent and real in mathematics. Here I perceive the reason why one has to regard the rational, irrational, and complex numbers as being just as thoroughly existent as the finite positive integers.[26]

Logical consistency was the touchstone which Cantor applied to any new theory before declaring it existent and a legitimate part of mathematics. Though he had argued the theoretical consistency of his transfinite numbers, he was not content to leave that as their only justification. He had cleverly devised a series of arguments, leading directly from the more generally accepted irrational numbers to the corresponding mathematical reality of the transfinite numbers as well. Realizing that more mathematicians might find it initially easier to admit the consistency and reality of the irrational numbers, he suggested that it was then only a short step to accepting his new transfinite numbers without reservation.

As he had said before, one could not accept the irrationals and reject the transfinites. Ontologically, their mathematical status was the same. Both were defined in terms of infinite sets, by similar procedures. Since he took his transfinite numbers to be consistently defined, with regular number-theoretic properties established in the *Grundlagen,* there were no grounds to deny his new theory. This kind of formalism, stressing the internal conceptual consistency of his new numbers, was all mathematicians needed to consider before accepting the validity of the transfinite numbers.

Cantor's formalism was nowhere better represented than by his lifelong opposition to infinitesimals. Contrary to the efforts of mathematicians like du Bois-Reymond, Stolz, and Vivanti who had seriously attempted to establish theories of actual infinitesimals, Cantor believed that his transfinite numbers could be called upon to demonstrate the inadmissibility of such ideas.[27] He used the results of his theory of ordinal numbers, the numberings of well-ordered sets, to show the contradictory nature of any assumption of the existence of infinitesimally small linear magnitudes.

There is even some irony in Cantor's position. To many mathematicians, his theory of actually infinite transfinite numbers seemed to justify intrinsically the infinitely small as well as the infinitely large. Benno Kerry, in a review of the *Grundlagen,* expressed his own faith in a formal definition of infinitesimals in a footnote:

> In my opinion a *formal definition* of definite, infinitely small numbers is indeed given in fixing the greatest of such numbers as one which produces the sum 1 by adding itself to itself ω times; the next smaller is then the one which produces 1 by adding itself to itself $\omega-1$ times, etc. The definite, infinitely small numbers would accordingly be denoted as:

$$\frac{1}{\omega}, \frac{1}{\omega+1}, \ldots, \frac{1}{2\omega}, \ldots, \frac{1}{\omega^2}$$

> etc. Of course whether numbers so defined have any *empirical applicability* is not decided here.[28]

Taken as reciprocals, it seemed that the transfinite numbers ought to provide an immediate basis for a theory of actual infinitesimals. But to Cantor's mind, such a step was irresponsible and under no circumstances could it be rigorously justified. In the *Grundlagen,* he tactfully expressed his skepticism of infinitesimals, saying only that attempts to consider the infinitely small in absolute terms were unsound and without purpose. Later he remarked, however, that he had been too reserved in the *Grundlagen* by failing to deny explicitly the existence of actually infinitesimal quantities.[29]

In 1887 Cantor explained his rejection of any consideration of infinitesimals on purely formal grounds in a letter to his old teacher in Berlin, Karl Weierstrass.[30] Cantor presented his contention directly in the form of a theorem:

> Non-zero linear numbers ζ (in short, numbers which may be thought of as bounded, continuous lengths of a straight line) which would be smaller than any arbitrarily small finite number do not exist, that is, they contradict the concept of linear numbers.[31]

Cantor intended to refute attempts (made by Stolz and du Bois-Reymond) to produce infinitesimals by discounting the Archimedean Axiom. This axiom asserted that given any real numbers $a<b$, it was possible, for some arbitrarily large $n \in N$, to ensure that $na>b$. Cantor based his proof of the impossibility of infinitesimals upon the concept of linear numbers. Numbers were *linear* if a finite or infinite number of them could be added together to produce yet another linear magnitude. If ζ were an infinitesimal, Cantor argued that $\zeta \cdot \nu$, even if ν were transfinite, would remain arbitrarily small, less than *any* linear quantity. Thus he asserted: "that ζ cannot be made *finite* by any multiplication, however great, not even an actually infinite one, and thus it certainly cannot be an element of finite numbers."[32]

On the strength of this conclusion, Cantor went on to demonstrate that the Axiom of Archimedes was no axiom, but a proposition about the system of real numbers that could be proven. Non-Archimedean systems could not be generated, he argued, simply by discarding the axiom, since it was not an axiom at all. As long as he believed it was actually a theorem following directly from the concept of number itself, infinitesimals were necessarily contradictory.

In 1893 Vivanti, the Italian mathematician, wrote to Cantor suggesting that his rejection of infinitesimals was unjustified. As du Bois-Reymond had shown, "the orders of infinity of functions constitute a class of one-dimensional magnitudes, which include infinitely small and infinitely large elements. Thus there is no doubt that your [Cantor's] assertions cannot apply to the most general concept of number."[33]

In defending his point of view, Cantor wrote back to Vivanti with the vitriolic remark that to the best of his knowledge, Johannes Thomae was the first to "infect mathematics with the Cholera-Bacillus of infinitesimals."[34] Paul du Bois-Reymond, however, was soon to follow. Cantor claimed that in systematically extending Thomae's ideas, du Bois Reymond found "excellent nourishment for the satisfaction of his own burning ambition and conceit."[35] Cantor went on to discredit du Bois-Reymond's infinitesimals because they were self-contradictory, since he rejected without compromise the existence of linear numbers which were non-zero yet smaller than any arbitrarily small real number. Infinitesimals, Cantor replied, were complete nonsense. He reserved his strongest words for du Bois-Reymond's *Infinitäre Pantarchie* and his orders of infinitesimals: "Can one still call such things numbers? You see, therefore, that the 'Infinitäre Pantarchie,' of du Bois-Reymond, belongs in the wastebasket *as nothing but paper numbers!*"[36] Cantor placed the theory of actual infinitesimals on a par with attempts to square the circle, as impossible, sheer folly, belonging in the scrap heap rather than in print. Ironically, much of Cantor's criticism could have been turned as effectively against the transfinite numbers as against infinitesimals. And his rejection of infinitesimals was certainly fallacious in *its* reliance upon a *petitio principii*. Having assumed that all numbers *must* be linear, this was equivalent to the Archimedean property, and thus it was no wonder that Cantor could "prove" the axiom. The infinitesimals were excluded by his original assumptions, and his proof of their impossibility was consequently flawed by its own circularity.

In addition to the arguments Cantor made explicit, there were other persuasive reasons for his rejection of infinitesimals. Mittag-Leffler had once asked him about the possible existence of infinitesimals as numbers interpolated between the rational and irrational numbers.[37] But Cantor believed in the *completeness* of the finite real numbers, in terms of rational and irrational numbers alone. His work in analysis and set theory had always assumed the reciprocal and unique correspondence between the real numbers and points on the real line. His continuum hypothesis, too, would only have been complicated by admitting infinitesimals as an entirely new sort of number, sandwiched between the rational and irrational numbers. Were there infinitesimals, somehow, in addition to the standard real numbers, then the continuum hypothesis might even prove to be false. There would have been many more numbers to contend with than the familiar set of reals, and this would have made the likelihood of his hypothesis more doubtful and certainly more complex.

Nevertheless, the overriding basis for Cantor's rejection of infinitesimals rested upon the question of consistency. On the basis of his own definition of linear quantity, the infinitesimals were contradictory. Hence they were unacceptable. The argument was the clearest expression of his formalist justification for ideas in mathematics. Ultimately, appeals to consistency were sufficient to advance the reality and legitimacy of any sort of real number—rational, irrational, complex, or transfinite. The true nature of Cantor's position concerning the nature of mathematical ontology in general, and the legitimacy of his transfinite numbers in particular, was only vaguely discernible in the *Grundlagen* itself. But in the succeeding years, as Cantor's interests became more philosophical, this kind of formalism became increasingly apparent.

THE NATURE OF MATHEMATICS

Cantor reinforced his study of the philosophical status of his new numbers with a simple analysis of the familiar and accepted natural whole numbers. For both finite and infinite integers, they could be considered in essentially two ways. Insofar as they were well-defined in the mind, distinct and different from all other components of thought, they served "in a connectional or relational sense to modify the substance of thought itself." This reality, which the whole numbers consequently assumed, he described as their *intrasubjective* or *immanent* reality.[38] In contradistinction to this immanent reality was the reality numbers could assume concretely, manifest in objects of the physical world. He explained further that this second sort of reality proceeded from whole numbers as expressions or images of processes in the world of physical phenomena. This aspect of the whole numbers, be they finite or infinite, he termed *transsubjective* or *transient*.[39]

Cantor asserted the reality of both the physical and ideal aspects of the number concept. These dual realities, in fact, were always found in a joined sense, insofar as a concept possessing any immanent reality always had a transient reality as well.[40] It was one of the most difficult problems of metaphysics to determine the nature of the connection between the two.

Cantor ascribed the necessary coincidence of these two aspects of number to the unity of the universe itself.[41] This meant that it was possible to study only the immanent realities, without having to confirm or conform to any objective content. This set mathematics apart from all other sciences and gave it an independence that was to imply great freedom for mathematicians in the creation of mathematical concepts. It was on these grounds that Cantor offered his now famous dictum that the essence of mathematics is its freedom. As he put it in the *Grundlagen:*

> Because of this extraordinary position which distinguishes mathematics from all other sciences, and which produces an explanation for the relatively free and easy way of pursuing it, it especially deserves the name of *free mathematics,* a designation which I, if I had the choice, would prefer to the now customary "pure" mathematics.[42]

Cantor thus asserted the freedom of mathematics to accept the creation and application of new ideas solely on the grounds of intellectual consistency.

Though there were counterparts to the immanent reality of number in the phenomenological world, that did not matter. Instead, the formal consistency of mathematical ideas in the mind provided the ultimate criterion for Cantor in determining the advance of mathematics. Its application to physical phenomena of the external world was of considerable but subsidiary importance.

Mathematics was therefore absolutely free in its development, and bound only to the requirement that its concepts permit no internal contradiction, but that they follow in definite relation to previously given definitions, axioms, and theorems. On these grounds, what were the criteria for introducing new numbers? The matter rested entirely in terms of definition. So long as new numbers were distinct and could be distinguished from other kinds of numbers, as well as from each other, then a new number was defined and must be taken as existing.

The only possible objection to this doctrine of freedom in mathematics, said Cantor, might be the arbitrary creation of new ideas without correctives. But there were, he insisted, correctives nonetheless. If an idea was fruitless or unnecessary, this quickly became evident, and by reason of failure, was abandoned or forgotten.[43] Alternatives, like the safe but limiting restriction of mathematics to the realm of finite numbers advocated by Kronecker and his followers, were very dangerous from Cantor's point of view. Any restriction or narrowness in mathematics would have direct and obvious consequences. Controls and artificial philosophical presuppositions retarded or even prevented any growth of mathematical knowledge. Cantor appealed to the great and legendary figures of the history of mathematics to lend support to his contention.

Without the freedom to construct new ideas and connections in mathematics, Gauss, Cauchy, Abel, Jacobi, Dirichlet, Weierstrass, Hermite, and Riemann would never have made the significant advances they did. Kummer would never have been able to formulate his ideal numbers, and consequently the world would be in no position, Cantor added with a note of cunning, to appreciate the work of Kronecker and Dedekind.[44]

Mathematics, Cantor believed, was the one science justified in freeing itself from any metaphysical fetters. Applied mathematics and theoretical mechanics, on the contrary, were metaphysical in both their content and goals.[45] Any attempt to free them from metaphysical substrata resulted in nothing but idle speculation describing worlds without substance. But mathematics, by virtue of its independence from any constraints imposed by the external reality of the spatial-temporal world, was quite free. Its freedom, as Cantor insisted was its essence.

THE CRISIS OF THE CONTINUUM AND CANTOR'S THEORY OF ORDER TYPES

Cantor once described the terrain of set theory, as he regarded it shortly after the first part of the *Grundlagen* had been published, as a *terra incognita*.[46]

Unquestionably, the most perplexing lacuna was the continuum hypothesis. Though he had tried repeatedly to show that the real numbers were equal in power to the second number class (II), he had never succeeded in proving the conjecture. Every time a new solution seemed promising or near perfect, something always turned up to dash his hopes. In May 1883 he wrote to Mittag-Leffler, saying that he had worked on the problem for a long time and yet every proof he had found eventually revealed mistakes or errors that required new beginnings.[47]

While Cantor was continually frustrated by his failure to resolve the continuum hypothesis, he was also having to suffer Kronecker's campaign to discredit transfinite set theory. Cantor complained, in correspondence with the French mathematician Hermite, that Kronecker was attacking his work and calling it "humbug."[48] Shortly before Christmas 1883 Cantor was so angered by these attacks that he wrote directly to the Ministry of Education, hoping to annoy Kronecker by applying for a position in Berlin available the following spring.[49] Cantor was convinced that the time might at last have come when Berlin needed him. In his opinion he was the only person in Germany who could be said to know all of mathematics, both old and new. He was convinced that his historical interests incorporated explicitly as part of the *Grundlagen* represented special preparation and gave him a unique appreciation for mathematics. He also felt strongly that he deserved the honor of a position at a German university known for its great mathematicians: Göttingen, or preferably, Berlin. But in a letter to Mittag-Leffler of December 30 he admitted that the application in Berlin would come to nothing.[50] He had heard from Weierstrass that the obstacles were largely financial, owing to Kronecker's large salary. Nonetheless, the letter to Mittag-Leffler also provided Cantor with another occasion to reiterate the bitterness he felt toward his position at Halle and to emphasize how poorly he was compensated there in comparison with Schwarz, Fuchs, Kronecker, and other of Cantor's least favorite colleagues. The following short remarks reveal how vehement Cantor could be in his judgment of Kronecker and of Kronecker's allies in Berlin:

> You understand quite rightly the meaning of my application; I never thought in the least I would actually come to Berlin. But since I plan to do so eventually and since I know that for years Schwarz and Kronecker have intrigued terribly against me, in fear that one day I would come to Berlin, I regarded it as my duty to take the initiative and turn to the Minister myself. I knew precisely the immediate effect this would have: that in fact Kronecker would flare up as if stung by a scorpion, and with his reserve troops would strike up such a howl that Berlin would think it had been transported to the sandy deserts of Africa, with its lions, tigers, and hyenas. It seems that I have actually achieved this goal![51]

If Cantor had sought to annoy Kronecker, Kronecker returned the challenge masterfully. There can be no doubt that Kronecker's subsequent maneuver, following in early January 1884, certainly had an overwhelming effect upon Cantor. Kronecker wrote to Mittag-Leffler asking to publish in the *Acta Mathematica* a short paper on his position with regard to certain mathematical conceptions in which he would show "that the results of modern function theory and set theory are of no real significance."[52] At first Cantor was mildly receptive to the idea, believing that at last it would bring Kronecker's opposition into the open, where it could be directly countered, and presumably rejected. But Cantor began to have second thoughts, and he feared that Kronecker might reduce their differences to personal polemics. To Cantor it seemed that Kronecker, by wanting to publish in *Acta Mathematica,* was trying to drive him out of the one journal in which he had found a sympathetic editor, just as Kronecker years earlier had tried to prevent Cantor from publishing any of his work in Crelle's *Journal*. The analysis Cantor made of the matter to Mittag-Leffler was highly skeptical of Kronecker's intentions. He warned that all was surely not what it seemed, and added for good measure the maxim: "timeo Danaos et dona ferentes."[53] Should any polemical writings appear in the *Acta Mathematica* under Kronecker's signature, Cantor threatened that he would withdraw his support from the journal.[54] Though he was soon to make good this threat for other reasons, the matter did not come to such an end because of Kronecker's paper. Apparently, Kronecker never sent anything for Mittag-Leffler's perusal, and certainly no provocative confrontation materialized in the *Acta Mathematica*. But the threats Cantor was willing to make, even to his friend Mittag-Leffler, showed how vindictive he could be in response to the conspiracy he felt was brewing against his work under the auspices of a single man: Leopold Kronecker.

By the early part of 1884 Cantor was again at work on the continuum hypothesis, but he found it as intransigent as ever. On April 4, 1884, he wrote to Mittag-Leffler and announced that while the problem had been reduced to showing that perfect sets were equivalent in power to the second number class (II), the continuum hypothesis still had not been solved.[55] Though he did not seem to register despondency over his prolonged inability to solve the continuum conjecture, it must nevertheless have been an ever-present source of frustration. Having said in the last paragraph of the sixth paper of the *Punktmannigfaltigkeitslehre* series that he expected a solution in the near future, he was committed to producing one.[56]

However, the strain of such anxieties was apparently more than Cantor could bear, and in the spring of 1884 he suffered the first of his serious nervous breakdowns. The illness came upon him swiftly and unexpectedly.[57] Evidence in a letter to Felix Klein suggests that it must have occurred shortly after May 20, 1884, and apparently lasted somewhat more than a month. By the end of June he was sufficiently recovered to write to Mittag-Leffler but complained that he lacked the energy and interest to return to rigorous mathematical thinking and was content to take care of trifling administrative matters at the university. He felt capable of little more. But it is significant that Cantor also added that he was anxious to return to work and would prefer his research to confining himself to the preparation of his lectures:

> I thank you heartily for your kind letter of May 15; I would have answered it sooner, but recently I have not felt so fresh as I should, and consequently I don't know when I shall return to the continuation of my scientific work. At the moment I can do absolutely nothing with it, and limit myself to the most necessary duty of my lectures; how much happier I would be to be scientifically active, if only I had the necessary mental freshness![58]

In fact, as soon as he had recovered sufficient strength, Cantor set off for his favorite vacation resort in the Harz mountains and returned to his analysis of perfect sets. He also undertook the bold step of writing (on August 18, 1884) directly to Kronecker and attempted to put their differences aside.

Kronecker's response was as positive as anyone might have expected, but the attempted peace was not to last.[59] Though Cantor tried to explain several technical details of his new theory in a second letter following the first by less than a week, he failed to convince Kronecker of the legitimacy of his transfinite numbers. Cantor assumed that Kronecker had been preoccupied with the philosophical parts of the *Grundlagen,* thus overlooking the more concrete grounds upon which the transfinite numbers could be justified:

> I am of the opinion that the greatest part of what I have done scientifically in the last few years, which I include under the rubric of set theory, is not so very much opposed to the demands which you place upon "concrete" mathematics as you seem to believe. It may be the fault of the presentation (which may not be entirely clear), that you have given less attention to the concrete mathematics in my research than to its other, namely, philosophical content.[60]

Cantor realized that overcoming the breach with Kronecker was not really a question of persuasion but that only time would reveal the strength and rightness of his new work. During an evening's conversation at Kronecker's home in Berlin, early in October 1884, Cantor was further convinced that there was little hope in winning Kronecker away from the narrowness of his preconceptions. Cantor described the meeting, which ran into the early hours of the following morning, in a letter to Mittag-Leffler:

> It seems to me of no small account that he and his preconceptions have been turned from the offensive *to the defensive* by the success of my work. As he told me, he wants to publish soon his opinions concerning arithmetic and the theory of functions. I wish it luck![61]

The attempted though abortive reconciliation with Kronecker was but one of several events which were equally significant in contributing to Cantor's growing disillusionment with mathematics during the year following his first nervous breakdown. In the fall of 1884 he again took up the intricate problem of the continuum hypothesis. On August 26, 1884, little more than a week after his letter of reconciliation to Kronecker, Cantor had written to Mittag-Leffler announcing, at last, an extraordinarily simple proof that the continuum was equal in power to the second number class (II). The proof attempted to show that there were closed sets of the second power. Based upon straightforward decompositions and the fact that every perfect set was of power equal to that of the continuum, Cantor was certain that he had triumphed. He summarized the heart of his proposed proof in a single sentence: "Thus you see that everything comes down to defining a single *closed* set of the second power. When I've put it all in order, I will send you the details."[62]

But on October 20 Cantor sent a lengthy letter to Mittag-Leffler followed three weeks later by another announcing the complete failure of the continuum hypothesis.[63] On November 14 he wrote saying he had found a rigorous proof that the continuum did *not* have the power of the second number class or of any number class. He consoled himself by saying that "the eventual elimination of so fatal an error, which one has held for so long, ought to be all the greater an advance."[64] Perhaps he was thinking back to the similar difficulties he had encountered in trying to decide whether or not the real numbers were denumerable, or how lines and planes might be corresponded. Cantor had come to learn that one should never be entirely surprised by the unexpected. Nevertheless, within twenty-four hours he had decided that his latest proof was wrong and that the continuum hypothesis was again an open question. It must have been embarrassing for him to have been compelled to reverse himself so often within such a short period of time in his correspondence with Mittag-Leffler. But even more discouraging must have been the realization that the simplicity of the continuum hypothesis concealed difficulties of a high order, ones that, despite all his efforts and increasingly sophisticated methods, he seemed no better able to resolve.

In attempting to find a solution for his continuum hypothesis, Cantor was led to introduce a number of new concepts enabling more sophisticated decompositions of point sets. These, he hoped, would eventually lead to a means of determining the power of the continuum. His attempt to publish these new methods and results marked the final and most devastating episode responsible for his disillusionment with mathematics and his discontent with colleagues both in Germany and abroad.

MITTAG-LEFFLER AND THE WITHDRAWAL OF CANTOR'S "PRINCIPIEN EINER THEORIE DER ORDNUNGSTYPEN" FROM PRESS

Early in 1885 Cantor chose to present a number of new ideas in two brief articles which were to be published in *Acta Mathematica* as letters to the editor and designated as "first" and "second" communications. The first of these, the

"Erste Mittheilung" of his "Principien einer Theorie der Ordnungstypen"[65] was set in type and dated February 21, 1885 (the mathematical significance of the *Principien,* as well as Cantor's innovation of order types, is discussed in Chapter 7). On February 25 he added another four paragraphs and sent these off as well. But to his dismay, on March 9, Mittag-Leffler wrote back suggesting that perhaps the "Erste Mittheilung" on order types should be withdrawn from press:

> I am convinced that the publication of your new work, before you have been able to explain new positive results, will greatly damage your reputation among mathematicians. I know very well that basically this is all the same to you. But if your theory is once discredited in this way, it will be a long time before it will again command the attention of the mathematical world. It may well be that you and your theory will never be given the justice you deserve in our lifetime. Then the theory will be rediscovered in a hundred years or so by someone else, and then it will subsequently be found that you already had it all. Then, at least, you will be given justice. But in this way [by publishing the *Principien*], you will exercise no significant influence, which you naturally desire as does everyone who carries out scientific research.[66]

Mittag-Leffler likened Cantor's work to Gauss' research on non-Euclidean geometry, saying that Gauss had always been reluctant to publish on the subject. He added that Cantor's work was no less revolutionary and needed to be treated with similar care. Mittag-Leffler's caution that few mathematicians were prepared for Cantor's new terminology and increasingly philosophical manner was probably genuine enough. But to Cantor it seemed that Mittag-Leffler was only concerned about the reputation of his journal. In recalling the event more than a decade later, he confided to Poincaré his real feelings:

> Suddenly it was clear to me that he must wish to see my work withdrawn in the interests of his *Acta Mathematica.* The connection is as follows! Even my earlier work published since 1870 has not enjoyed the approval of the powers in Berlin: Weierstrass, Kummer, Borchardt, Kronecker. Had Mittag-Leffler ever published my theory of transfinite order types (which was even more audacious and went much further) in the *Acta Mathematica,* he would have endangered in the highest degree the existence of his undertaking, which is still very young and which depends principally upon the good will of the Berlin academics.[67]

Cantor was deeply hurt by Mittag-Leffler's rejection of his latest research. More than his polemic with Kronecker, more than his nervous breakdown or the trouble he was having in finding a proof for his continuum hypothesis, Mittag-Leffler's suggestion that Cantor not print his new article in the *Acta Mathematica* was the most devastating. Though he never admitted that the incident affected his personal regard and friendship for Mittag-Leffler, thereafter he wrote less frequently and only seldom did he mention matters concerning his research. Cantor apparently believed that he had been abandoned by the last mathematician at all sympathetic with his struggle to establish the transfinite numbers. Consequently, he never published again in the pages of the *Acta Mathematica*.

Following this incident with Mittag-Leffler, Cantor was led to abandon mathematics almost entirely. Writing to the Italian mathematician Gerbaldi, he was candid about why he had decided to give up mathematics:

> The real reasons why publication was interrupted then remains a *mystery* to me, and even today I do not know. I suddenly received a letter from M.L. [Mittag-Leffler] in which he wrote (to my great astonishment) that after serious consideration he regarded this publication as "about a hundred years too soon." Had M.L. had his way, I should have to wait until the year 1984, which to me seemed too great a demand!
>
> Since I was therefore disgusted, as you will understand, with the mathematical journals, I began to publish my work in the *Zeitschrift für Philosophie und philosophische Kritik*. I did not decide until nine months ago to present the mathematical side of my theory once again in mathematical journals. But of course I never want to know anything again about Acta Mathematica.[68]

By the end of 1885 Cantor was in many respects a disillusioned man. Largely but not entirely the result of Mittag-Leffler's suggestion that he not publish his article on order types, Cantor chose to renounce *Acta Mathematica* just as he had decided in 1878 never to publish again in Crelle's *Journal*. In both cases Cantor was piqued by what he considered to have been personal affronts, and in his characteristically hasty, even angry, way he would have no more to do with either journal. It was an unfortunate feature of his personality, but Cantor always took criticism of his work deeply and personally. Thus in turning his back on *Acta Mathematica,* it was inevitable that his friendship with Mittag-Leffler would suffer. Without Mittag-Leffler, whom Cantor regarded as one of the few professional sources of encouragement and understanding upon which he could rely, there seemed little reason to continue a frustrating battle against both German and foreign mathematicians. With little hope of advancing to a position in either Berlin or Göttingen, Cantor found his prospects as a mathematician hopelessly bleak, certainly demoralizing.

It was ironic that Schoenflies, in writing his account of Cantor's illness and several crises of 1884, could end by saying that he owed so much to Mittag-Leffler, "who was then his only friend to remain scientifically faithful, despite the many attempts to dissuade him."[69] In part what Schoenflies said was true. Mittag-Leffler had been one of the first mathematicians to use Cantor's work in establishing one of his own most important theorems. Mittag-Leffler had also been responsible for suggesting that Cantor's early work and parts of the *Grundlagen* be translated into French and thus be made available to a wide readership through the *Acta Mathematica*. Even so, one still must remember the consequences (though doubtless unintended) of Mittag-Leffler's suggestion that Cantor withdraw his paper on the theory of order types. Mittag-Leffler's suggestion came at an unfortunately critical moment in Cantor's life, at the end of a long and upsetting series of events which he must have felt himself powerless to control. His disenchantment with mathematics dates, not from his first serious breakdown nor from the escalating series of polemics with Kronecker, but from Mittag-Leffler's letter of March 1885. Until then, Cantor had continued to work on the continuum hypothesis. Even after his first nervous breakdown, he continued to make new and significant advances with his theory of order types. But in response to Mittag-Leffler's letter, Cantor immediately sent a telegram asking for the return of any of his manuscripts still in Mittag-Leffler's possession. Thereafter he apparently put his mathematics aside and began to devote more and more of his time to certain historical interests that were entirely nonmathematical, or to problems of philosophy and theology. Above all, he soon discovered among theologians of the Roman Catholic Church a level of interest and encouragement that he had never found among mathematicians.

However, before it is possible to evaluate the significance of these interests, and above all Cantor's lengthy and often detailed correspondence with Catholic intellectuals and theologians over his transfinite set theory, it is necessary to consider the new attitude toward science proclaimed by Pope Leo XIII in 1879. Above all, something must be said of the new climate of interest in scientific ideas, a climate greatly stimulated by Pope Leo's attempt to promote a revival of Thomistic philosophy as prescribed in his encyclical *Aeterni Patris*.

POPE LEO XIII AND THE ENCYCLICAL *AETERNI PATRIS*

Pope Leo XIII's attempt to reconcile the new and often perplexing discoveries of science with scripture and with the authorities of the Church encouraged a number of Catholic intellectuals to study the various branches of natural science in detail.[70] By virtue of one of his most influential encyclicals, *Aeterni Patris,* Leo XIII was to foster interest in scientific ideas generally and from one quarter at least to arouse interest in Cantor's mathematics in a very direct, if somewhat surprising, way.[71]

Gioacchino Pecci (1810–1903), later Pope Leo XIII, was given a classical education at the Jesuit school in Viterbo. After receiving the doctorate of theology from the Collegio Romano in 1832, he entered the Accademia dei Nobili in Rome. At the same time he studied both canon and civil law at the Sapienza University. Five years later he decided to enter the priesthood and was ordained in December 1837. Following papal assignments in Naples and Perugia, he was appointed nuncio to Brussels by Pope Gregory XVI, who called Pecci back to Italy in 1844 to serve as the Archbishop of Perugia. Ten years later, in 1853, he was made a cardinal.

At this time Gioacchino's brother, Giuseppe Pecci, was also in Perugia serving as a Jesuit professor at the local seminary, and it was under his influence that Gioacchino became interested in Thomistic philosophy.[72] He even founded an Academy of St. Thomas in 1859. As Archbishop of Perugia, in the years 1876–1877, Gioacchino Pecci issued an important series of pastoral letters devoted to matters of the Church and civilization, in which he argued that the Church had to enter the current of modern times or be left behind.[73] To Pecci science was one of the most immediate representatives of modernity. But his pastoral letters were only a presentiment of the papal encyclicals to follow.

In 1878, having been elected by the College of Cardinals to succeed Gregory XVI, Gioacchino Pecci took the name Pope Leo XIII. He was meant to be a transitional Pope, but his pontificate lasted for twenty-five years. Among the first of his pronouncements was the encyclical *Aeterni Patris,* delivered on August 4, 1879. Urged by his brother, by then Cardinal Giuseppe Pecci, Leo XIII sought a renewal of philosophical thought along the lines of a revived Thomism.[74] Seen as an opposing force to political and social liberalism, Leo XIII's new program included several attempts to revitalize and to modernize the thought of the Church by reorganizing the Academy of St. Thomas, and by nominating Désiré Mercier (later to become one of the most prominent figures of the neo-Thomist movement), to a chair of Thomism at the University of Louvain. In fact, the Institut Supérieur de Philosophie at Louvain was organized by Mercier under the auspices of the Pope himself.[75]

The basic position of the neo-Thomists may be characterized succinctly, if somewhat oversimply, by their conviction that contemporary evils were the result of false philosophy. Improper or incorrect views of nature, they held, resulted in two consequences: atheism and materialism. Since the time of Thomas Hobbes, science had been repeatedly charged with having spawned these two undesirable offspring, but rarely was science then called upon to fill the gap and to rebuke these dreaded Leviathans on their own grounds. The encyclical *Aeterni Patris* was clear on one point in particular: that science could profit from scholastic philosophy and in the process could further the ideals and goals of the Church itself:

> For, the investigation of facts and the contemplation of nature is not alone sufficient for their profitable exercise and advance; but, when facts have been established, it is necessary to rise and apply ourselves to the study of the nature of corporeal things, to inquire into the laws which govern them and the principles whence their order and varied unity and mutual attrac-

> tion in diversity arise. To such investigations it is wonderful what force and light and aid the Scholastic philosophy, if judiciously taught, would bring.[76]

In dealing with the metaphysics of science, neo-Thomism was expected to guide the human understanding of the natural and spiritual world in the proper and unobjectionable direction prescribed by the Church, specifically along the lines defined implicitly in the encyclical *Aeterni Patris*. But the program which Leo XIII envisioned was less a subordination of philosophy and Thomistic teachings to science than an attempt to show how science ought to proceed and how it ought to be reconciled with the true principles of Christian philosophy. On April 21, 1878, in his first papal encyclical *Inscrutabili,* Leo wrote:

> Above all [education] must be wholly in harmony with the Catholic faith in its literature and system of training, and chiefly in philosophy, upon which the foundation of other sciences in great measure depends.[77]

But there was opposition to Leo XIII's efforts to commend the study of St. Thomas as the epitome of true philosophy. In fact, there was strong resistance among the clergy in Rome. As a result, it was suggested that a free course on Thomism should be established; and thus Father Giovanni Cornoldi (1822–1892), a Jesuit with a high reputation as both a scientist and a philosopher, was imported from Bologna to establish Thomism in the papal city.[78] Cornoldi went so far as to promise that "in the *Summa Theologica* was to be found the key to all the difficulties of modern science."[79] In stressing the harmony between Thomism and science, Cornoldi paved the way for renewed interest among churchmen in the affairs of science. Shortly after Cornoldi had begun his lectures in Rome, the encyclical *Aeterni Patris* was issued.

The impetus Leo XIII gave to scholarly and scientific study among intellectuals of the Church cannot be overestimated. Of special significance was the interest his encyclical generated among Germans who were to become interested in reconciling Cantor's work on absolute infinities with doctrines of Catholicism. Of particular importance in this respect was Constantin Gutberlet.

GUTBERLET AND THE APPRECIATION OF CANTOR'S WORK IN GERMANY

Constantin Gutberlet studied philosophy and theology at the Collegio Romano from 1856 to 1862, attending simultaneously the German University in Rome.[80] From 1862 to 1865 he lectured on the natural sciences at the seminary in Fulda, where he was made a professor of philosophy, apologetics, and dogma in 1866. In 1888 he founded the review journal, *Philosophisches Jahrbuch der Görres-Gesellschaft,* which proved to be a leading advocate of neo-Scholastic thought. But of special interest here was an article he published in 1886 drawing upon Cantor's set theory in a defense of his own views on the theological and philosophical nature of the infinite.[81]

Gutberlet realized that the study of infinity had entered a new phase with the appearance of Cantor's mathematical and philosophical studies. The question uppermost in Gutberlet's mind concerned the challenge of mathematical infinity to the unique, absolute infinity of God's existence. It was not long before correspondence over this question began to deepen Cantor's interest in the theological aspects of his theory of the transfinite numbers. It was Cantor's claim that instead of diminishing the extent of God's nature and dominion, the transfinite numbers actually made it all the greater.[82]

It is not important to follow the details of Gutberlet's article. It is important to know only that it was, in its broadest outline, an attempt to support arguments he had advanced several years earlier on the existence of the actual infinite. Subsequently, Gutberlet's ideas had been attacked in severe terms by another German, Caspar Isenkrahe.[83] Isenkrahe argued that the actual infinite was self-contradictory, and thus any attempts to support it were necessarily hopeless. The discussion was couched in terms of the infinite duration and eternity of the world, and it raised objections to theories of actual infinities made by thinkers as diverse in time as Aquinas and Herbart.[84] Gutberlet presented Cantor's work to support his own claims against opposition from theologians like Isenkrahe. By describing Cantor's mathematics, Gutberlet hoped that any reader could then decide:

> if they were correct, when they supposed they could dispose of my theory of actually infinite magnitude [*Grösse*] so easily. Above all we [Gutberlet] now want to explain the Cantorian theory and then to defend our conception against criticism, which this journal published, with Cantor's corresponding interpretation of infinite magnitude.[85]

The use which Gutberlet made of Cantor's ideas is of some interest. Particularly when it came to defending the existence of the absolute infinite, Gutberlet used a ploy reminiscent of Berkeley's use of God as a guarantor of the reality of the external world. In short, Gutberlet argued that God Himself ensured the existence of Cantor's transfinite numbers:

> But in the absolute mind the entire sequence is always in actual consciousness, without any possibility of increase in the knowledge or contemplation of a new member of the sequence.[86]

God was similarly called upon to ensure the ideal existence of infinite decimals, the irrational numbers, the true and exact value of π, and so on. God was not only capable of resolving the problem of the continuum hypothesis, but he also ensured the concreteness and objectivity of the cardinal number representing the collection of all real numbers.[87] Gutberlet even argued that, since the mind of God was believed to be unchanging, then the collection of divine thoughts must comprise an absolute, infinite, complete, closed set. Again, Gutberlet offered this as direct evidence for the reality of concepts like Cantor's transfinite numbers. Either one assumed the existence and reality of the actual infinite, or one was obliged to give up the infinite intellect and eternity of the absolute mind of God.

Gutberlet thus called upon Cantor's analysis of the infinite to defend his own use of actually infinite numbers. In the process he encouraged Cantor's interest in the philosophical and theological aspects of his work. That Gutberlet was prepared to argue the objective possibility of the transfinite numbers on the basis of the infinite intellect of God must have appealed to the mind of a religious man like Cantor.[88] It was also a complementary approach to Cantor's own Platonism, in which the legitimacy of the actual infinite was established in the immanent world of the mind by virtue of the consistent forms of reason alone.[89]

INFINITY, PANTHEISM, AND NEO-THOMIST RESPONSE TO TRANSFINITE SET THEORY

Though Gutberlet was one of Germany's leading neo-Thomists,[90] he was by no means the only philosopher of the Catholic Church who was interested in Cantor's mathematics. Cantor counted among his correspondents Tillman Pesch, Thomas Esser, Joseph Hontheim, and Ignatius Jeiler. All of these were closely involved with the revival of scholastic philosophy in the spirit of *Aeterni Partis*. Pesch[91] and Hontheim[92] were associated with a group of Jesuits at the Abbey of Maria-Laach, (on Lake Laach, near Andemach in the Rheinland, Germany), where a series of important contributions to neo-Thomism were published under the title *Philosophia Lacensis*. Pesch, in a work which attempted to describe the foundations of a Thomistic cosmology, the *Institutiones Philosophiae Naturalis,* was chiefly concerned with the progress of science in the nineteenth century.[93] Hontheim, writing for the same series, published a work on mathematics and logic in 1895: *Der logische Algorithmus*. Jeiler was a prominent figure among another group of neo-Thomists at Quaracchi, near Florence, and at Leo XIII's commission, he undertook a new edition of the works of St. Bonaventura in direct response to Leo's encyclical *Aeterni Patris*.[94]

Equally representative of the Church's interest in Cantor's mathematics was Thomas Esser, a Dominican in Rome. Apparently, Esser represented a group of Dominicans who were engaged in careful study of the theological implications of Cantor's work. In 1896 Cantor described their project in a letter to Jeiler:

> Now everything concerning this question (I tell you this in confidence) will of course be examined by the Dominicans in Rome, who are conducting a scholarly [*wissenschaftliche*] correspondence with me about it which will be directed by Father Thomas Esser, O.Pr.[95]

An important concern of the Catholic intellectuals who knew of Cantor's work involved the question of whether the transfinite numbers could be said to exist *in concreto*. Gutberlet was always clear that he differed fundamentally with Cantor on the matter, admitting the actual infinite as a "possible," and even real in the immanent, nonphysical dimensions of God's mind.[96] But Gutberlet, like his teacher Cardinal Franzelin, denied the possibility of a concrete, objective *Transfinitum* for reasons that make the Church's concern for Cantor's work understandable.

Cardinal Johannes Franzelin (1816–1886) was a leading Jesuit philosopher and papal theologian to the Vatican Council, the twentieth ecumenical council (1869–1870) called by Pope Pius IX, and famous for its promulgation of the dogma of papal infallability.[97] Cardinal Franzelin responded to Cantor's belief that the Transfinitum existed in *natura naturata* by explaining that it was a dangerous position to hold. Franzelin held that any belief in a concrete Transfinitum "could not be defended and in a certain sense would involve the error of Pantheism."[98] Pantheism, of course, had long been anathema to the Christian Church but was not condemned formally until 1861 by decree of Pius IX.[99] Spinoza, a philosopher Cantor had studied carefully, used the *natura naturans/natura naturata* distinction in a form similar to that of his heretical forerunner Giordano Bruno.[100] Both had been led to advocate a monistic philosophy of substance identifying God with the natural world. The question of the infinite was an easy touchstone identifying pantheistic doctrines. Any attempt to correlate God's infinity with a concrete, temporal infinity suggested Pantheism. Thus infinite space and infinite duration were both instances where the infinite, predicated of objects in the natural world, were held to be inadmissible on theological grounds. Any actual infinite *in concreto,* in *natura naturata,* was presumably identifiable with God's infinity, in *natura naturans.* Cantor, by arguing his actually infinite transfinite numbers *in concreto,* seemed to some to be aiding the cause of Pantheism.

These were the grounds upon which Gutberlet and Franzelin had been led to object to the admissibility of the actual, concrete infinite as more than a "possible" infinite. Cantor however, believing in the "actual," concrete infinite, was able to add a distinction between two sorts in infinity that was to satisfy at least some theologians, and in particular Cardinal Franzelin. On January 22, 1886, Cantor wrote to the cardinal explaining that in addition to differentiating between the infinite in *natura naturans* and in *natura naturata,* he further distinguished between an "Infinitum aeternum increatum sive Absolutum," reserved for God and his attributes, and an "Infinitum creatum sive Transfinitum," evidenced throughout created nature and exemplified in the actually infinite number of objects in the universe.[101] Cantor added that the important difference between absolute *infinitum* and actual *transfinitum* should not be forgotten. Cantor's clarifications turned Franzelin's reluctance into an imprimatur of sorts, when Franzelin chose to endorse Cantor's distinctions as follows:

> Thus the two concepts of the Absolute-Infinite and the Actual-Infinite in the created world or in the *Transfinitum* are essentially different, so that in comparing the two one must only describe the former as *properly infinite,* the latter as improperly and equivocally infinite. When conceived in this way, so far as I can see at present, there is no danger to religious truths in your concept of the *Transfinitum.*[102]

Cantor was always proud of the acceptance his new theory had found in the estimation of Cardinal Franzelin and would frequently remind his friends in the Church, through correspondence, that he had been assured on the cardinal's authority that the theory of transfinite numbers posed no theological threats to religion.[103] In fact, Cantor believed that the real existence of the *Transfinitum* further reflected the infinite nature of God's existence. Cantor even devised a pair of arguments from which the existence of transfinite numbers *in concreto* could be deduced on both *a priori* and *a posteriori* grounds. *A priori,* the concept of God led directly on the basis of the perfection of His being to the possibility and necessity of the creation of a *Transfinitum*.[104] Approaching the same conclusion with *a posteriori* arguments, Cantor believed that the assumption of a *Transfinitum* in *natura naturata* followed because the complete explanation of natural phenomena was impossible on exclusively finite assumptions.[105] Either way, Cantor felt he had demonstrated the necessity of accepting the *Transfinitum in concreto,* and he was not reluctant to call upon the nature and attributes of God in order to do so.

CHRISTIAN PHILOSOPHY AND ITS IMPORTANCE FOR CANTORIAN SET THEORY

The interest generated by Leo XIII's encyclical *Aeterni Patris* was a tonic for Cantor's own declining spirits. Following his nervous breakdown in the late spring of 1884, Cantor returned to mathematics only to be doubly discouraged. Not only was the continuum hypothesis a hopeless tangle, but early in 1885, Mittag-Leffler seemed to have closed the last door on Cantor's hopes for encouragement and understanding among mathematicians. Isolated in Halle, Cantor began to teach philosophy and to correspond with theologians who provided a natural outlet for Cantor's need to communicate the importance and implications of his work.[106]

In turn, Cantor's contact with Catholic theologians may have made his own religious sympathies all the stronger. By the early part of 1884, he could write to Mittag-Leffler that he was not the creator of his new work, but merely a reporter. *God* had provided the inspiration, leaving Cantor responsible only for the way in which his articles were written, for their style and organization, but not for their content.[107] Apparently hoping to disassociate himself as much as possible from having to assume responsibility for his controversial research,

Cantor was trying to shield himself from the criticism his transfinite numbers were destined to generate. Psychologically, the letter to Mittag-Leffler is revealing because it clearly demonstrates that *before* his first nervous breakdown, he was no longer anxious to take credit for his work but was willing to place the burden of responsibility for the provocative new ideas elsewhere.

It is also significant that Cantor believed in the absolute truth of his set theory because it had been *revealed* to him. Thus he may have seen himself not only as God's messenger, accurately recording, reporting, and transmitting the newly revealed theory of the transfinite numbers but as God's ambassador as well. If so, Cantor would not only have felt it appropriate, but more accurately, his duty, to use the knowledge which was his by the grace of God to prevent the Church from committing any grave errors with respect to doctrines concerning the nature of infinity. In writing to Jeiler during Whitsuntide 1888, Cantor declared:

> I entertain no doubts as to the truth of the transfinites, which I have recognized with God's help and which, in their diversity, I have studied for more than twenty years; every year, and almost every day brings me further in this science.[108]

Cantor was even more direct in a letter written to Hermite during the first month of 1894, in which he claimed that it was God's doing that led him away from serious mathematics to concerns of theology and philosophy:

> But now I thank God, the all-wise and all-good, that He always denied me the fulfillment of this wish [for a position at university in either Göttingen or Berlin], for He thereby constrained me, through a deeper penetration into theology, to serve Him and His Holy Roman Catholic Church better than I have been able with my exclusive preoccupation with mathematics.[109]

At one stroke, Cantor signaled the many disappointments and doubts accumulated over more than two decades. His lack of confidence in himself and his mathematical powers reflected the frustration he must have felt at being unable to solve the continuum hypothesis, compounded by the disastrous effects of Kronecker's attacks and of Mittag-Leffler's seemingly negative response to his recent work on order types. Realizing that no positions were ever going to be offered him in either Göttingen or Berlin,[110] Cantor turned to other interests less demanding than his mathematics, and more positively reinforcing. Later he interpreted his disaffection from mathematics and his deepening interest in philosophy and theology as the work of God. By the end of his life, in the spirit of *Aeterni Patris,* Cantor saw himself as the servant of God, a messenger or reporter who could use the mathematics he had been given to serve the Church. As he told Esser in early February 1896: "From me, Christian Philosophy will be offered for the first time the true theory of the infinite."[111]

Cantor had given up his mathematical colleagues and had found both consolation and inspiration among theologians and philosophers of the Church. Religion renewed his confidence and sustained his belief in the truth and significance of his research. Inspired and helped by God, Cantor was sure that his work *was* of consequence, despite the failure of mathematicians to understand the importance of his discoveries.

CHAPTER 7

From the *Grundlagen* to the *Beiträge*, 1883–1895

Before Cantor severed relations completely with Mittag-Leffler's *Acta Mathematica* in the spring of 1885, he had begun to develop new ideas and to write a fresh series of papers on the subject of his transfinite set theory. Despite his anxieties and nervous breakdown of the previous year, he was still confident that an entirely new approach to solving the continuum hypothesis might be found. On several occasions he had promised to produce a seventh in the series of six papers on the *Punktmannigfaltigkeitslehre*. In fact, in the sixth article he twice referred explicitly to this intended successor, and at the very end added, quite optimistically, "Continuation follows."[1] None did. Rather than produce yet another paper in the series on linear point sets, he began to publish separate articles in the *Acta Mathematica*. In 1884 an extract from a letter he had sent to Mittag-Leffler appeared in French. It was devoted to theorems on the power of perfect sets, and again, toward the end of this paper, he announced a forthcoming article in which he would produce a solution of the continuum problem.[2]

In the following year, Cantor did publish a lengthy analysis, the "Zweite Mittheilung," which fulfilled a number of earlier promises. In many respects it contained features that might easily have constituted a seventh paper. Not only was the "Second Communication"[3] a continuation of results from Cantor's sixth paper in the *Punktmannigfaltigkeitslehre* series, but it also offered a solution which he had promised in Number 6: "According to a known theorem proved in Crelle's *Journal,* volume 84, it follows that the perfect set S also has the same power as $(0 \ldots 1)$, and thus *all* linear perfect sets have the same power. In a later paragraph I want to prove the same theorem for perfect sets in any n-dimensional space."[4] But the central question concerning the power of the continuum was not even mentioned.

Nevertheless, the "Zweite Mittheilung" did introduce some new ideas, including three varieties of sets related to any given point set P.[5] The *adherence Pa* of any set P identified the set of all *isolated points* of P; the *coherence Pc* of P identified the set of all its limit points; and the *total inherence Pi* of P denoted higher orders of coherence Pc^{Ω} of P. Sets were described, depending upon their properties, as being homogeneous, separated, dense-in-themselves, or perfect. All these distinctions were designed to identify a large variety of subtle properties among point sets, and were introduced in hopes of promoting a proof of the continuum hypothesis. But no solution was forthcoming, and instead, Cantor concluded the paper by declaring: "I originally undertook this research on *point sets,* here brought to a kind of close, not only out of speculative interest, but also in hopes of possible applications to *mathematical physics* and other sciences."[6]

These lines clearly implied that the "Zweite Mittheilung" was indeed the Number 7 and now the acknowledged end of the *Punktmannigfaltigkeitslehre* papers. Though Cantor went on with a brief commentary concerning the application of all these new ideas to the study of matter and the aether, the mathematical content of the "Zweite Mittheilung" seemed hollow in the face of a single major omission. The continuum hypothesis remained unsolved.

By 1885 Cantor seems to have realized that exclusive concentration upon point sets might prove to be a mathematical *cul-de-sac*. They might never lead to any means of resolving the continuum problem. Perhaps his brief investigation of the fascinating properties of perfect sets and the question of measure had suggested another approach. The continuum problem was really not one that depended upon metric properties at all, but upon the nature of *order*. Persistent concentration upon point sets was unnecessarily specific, perhaps unnecessarily limiting as well. Cantor saw that he should try to reduce the problem of the continuum to still more essential features, and this he now set out to do.

It was as if the foundation for Cantor's new approach had already been laid in the "Zweite Mittheilung" of 1885. Cantor began by translating the ideas of adherence, inherence, and coherence (first introduced for point sets in the "Zweite Mittheilung") into the language of well-ordered sets. This new interest in a theory of order types was significant for its attempt to give an entirely new emphasis, a wholly new direction to Cantor's study of the continuum and its structure and power. The subject of order types was also much more interesting mathematically than most of the comparatively straightforward characteristics he had been able to identify among either point sets or powers.

ADVANCES IN THE THEORY OF ORDER TYPES: CANTOR'S UNPUBLISHED WORK

The "Principien einer Theorie der Ordnungstypen: Erste Mittheilung," written for volume seven of Mittag-Leffler's *Acta Mathematica* (but withdrawn), was Cantor's first concerted effort to suggest a new and independent theory of ordered sets in general.[7] The *Grundlagen* had only considered well-ordered sets in defining the transfinite ordinal numbers and the higher number classes, but Cantor had discovered that the idea of *simply*-ordered sets was also a very rich one conceptually. To illustrate this point, he noted that the well-ordered sequence of natural numbers 1, 2, 3, . . . , represented but one type of order, an important but nevertheless limited case. The set of rational numbers taken in their natural order represented a distinctly different type of order, since between any two rationals there was always another, and this property was not one shared by well-ordered sets. The collection of all real numbers taken in their natural order provided yet another example of a simply-ordered set distinct in type from either the natural or rational numbers. It was Cantor's hope that further study of the types of order of simply-ordered sets might provide new insights of a general nature to help advance his study of the continuum.

Cantor launched the *Principien* with a definition of power, described in terms of equivalence but then explained directly as a general concept (or "category") which included all sets equivalent to a set of given power. This broader, more comprehensive definition represented an important advance in Cantor's thinking:

> The *power* of a set M is determined as the *concept* [*Vorstellung*] of that which is common to *all* sets *equivalent* to the set M and *only these*, and thus also *common* to the set M itself. It is the *representatio generalis*, the τὸ ἓν παρὰ τὰ πολλά for all sets *of the same class as M*. Thus I take it to be the *most original* (both *psychologically* and *methodologically*) and the *simplest basic concept* arising by *abstraction* from all *particulars* which can represent a set of a *definite class*, both with respect to the *character* of its *elements* and with respect to the *connections* and *orderings* between the *elements*, be it with respect to *one another* or to *objects lying outside the set*. Insofar as one reflects *only upon that which is common to all sets belonging to one and the same class*, the concept of *power* or *valence* arises.[8]

At last Cantor had described power for what it was, not simply in terms of the equivalence of sets similar to known sets like the natural or real numbers but as a concept abstracted from all particulars of a given set, including order. This method of defining set-theoretic concepts by abstraction later proved to be of great importance in Cantor's most mature formulation of the transfinite numbers and his defense of the theory both mathematically and philosophically. For the *Principien*, however, the major benefit of this new approach was the definition Cantor gave of simply-ordered sets.

A set was said to be simply-ordered if, given any two elements, it was possible to determine exactly one of the relations $e < e'$, $e = e'$, $e > e'$. Cantor described what he meant by the type of such a simply-ordered set as follows:

"Every *simply-ordered set* has a definite order type, or, expressed more briefly, a definite *type;* by this I mean that general concept under which all sets similar to the given set fall, and only those (including the given ordered set itself)."[9]

The order type of the natural numbers 1, 2, 3, . . . , ν, . . . was written as ω. Similarly, the order type of the set of rational numbers $1-\frac{1}{\nu}$ was also ω. But, taken in their natural order on (0,1), the rational numbers were not of type ω, but represented a completely different type of order which Cantor denoted by η. Above all, taken as a simply-ordered set, he expressed the order type of all *real* numbers on (0,1) by θ.

Nowhere did the importance of Cantor's new theory of order types make itself more plainly felt than here. The question of continuity could not rely, in general, upon the question of power alone. The rationals, depending upon how they were ordered, though always denumerable, might be of type ω or of type η, but it was only in the natural form of type η that they exhibited a property of interest to the study of continuity. Taken in their natural order, the rational numbers were everywhere-dense. Similarly, though there were numerous examples of sets of the power of the second number class (II), and though Cantor believed that the continuum was also of this power, there was another and equally crucial question which concerned the nature of continuity characteristic of the continuum (0,1) taken in its natural order. Thus Cantor's attention focused more and more specifically upon the theory of simply-ordered sets, and order types generally. With their help he hoped to develop subtle, highly refined techniques, ones that might eventually help characterize completely the order type θ of the linear continuum.

As Cantor set out to chart the new territory, he did not want to rely upon examples of given number sets like the rationals or reals. Earlier he had disengaged the theory of transfinite numbers from point sets and their derivations and thereafter preferred to refer their construction to the more general principles of well-ordered sets. Similarly, he now sought to give an entirely general characterization of the order type η, without specific reference to the rational numbers at all. Instead, he chose to rely upon characteristics of abstract sets which would exhibit all the properties of the rationals taken in their natural order:

> One can also prove by the same means the following theorem, in which nothing is assumed about the character of the elements constituting the sets: if M is a simply-ordered set of the first power which has neither a least nor a greatest element, and which is so constituted that between any two elements e and e' there is always an infinite number of other elements, then M has the order type η.[10]

As if to demonstrate the power of the new approach, Cantor added the observation that if α denoted the order of a given simply ordered set, and $*\alpha$

the *reverse* order, then $*\alpha=\alpha$ was *always* true for finite sets, though not always true for infinite sets. And if the infinite sets in question were well-ordered, then the corresponding types α and $*\alpha$ were *never* the same.[11] This suggested that well-ordered sets represented a very special sort of simply-ordered set, and as a consequence Cantor went on to give them special treatment.

The eventual significance of this interest in well-ordered sets can be measured by the fact that in the *Grundlagen* such sets were only accorded a single section (Number 5), whereas in the *Beiträge* (Cantor's best known, most complete, and last major publication on set theory) the entire second half (1897) was devoted to study of their properties. Well-ordered sets were of such importance in Cantorian set theory because their order-types characterized the natural numbers; on finite sets they produced the familiar finite numbers, on infinite sets they produced the transfinite numbers of the second and higher number classes. Above all, they defined the transfinite order types of well-ordered sets of the first power, which comprised the second number class (II) and therefore assumed great importance in terms of the continuum hypothesis.

As in the *Grundlagen,* Cantor went on in the *Principien* to define addition and multiplication of order types but now for simply-ordered sets, adding the standard caveats concerning the generally noncommutative character of operations for transfinite types.[12] Of considerable interest were operations other than addition and multiplication which were uniquely characteristic of infinite order types, though they had no obvious counterparts for finite order types. First Cantor defined *principal elements* (*Hauptelements*):

> If e is an element of A, then the following can arise: if e' denotes *any* element of A preceding e, setting $e'=e$ if A has no element less than e; and if e' is *any* element of A which comes *later* than e, setting $e'=e$ if there is no element in A which is greater than e; then between e' and e' there are always infinitely many elements of A; if e fulfills these conditions, then we shall call e a *principal element* of A.[13]

If one then considered the set of all principal elements of a set A, allowing them to retain their order as originally given in A, a new simply-ordered set was created which Cantor designated the coherence of A, denoted Ac.[14] The order type of Ac was denoted αc, and called the coherence of α. Corresponding roughly to the idea of derived sets, the coherences of successive sets could be written:

$$Ac,\ Ac^2,\ \ldots,\ Ac^\nu,\ \ldots,\ \text{where } Ac^{\nu+1} \subseteq Ac^\nu.$$

Consequently the ordered sets $\mathfrak{D}\,(A, Ac, Ac^2, \ldots, Ac^\nu, \ldots) = Ac^\omega$ gave rise to the type αc^ω.

It was then possible to interpret directly the conditions under which an order type might be said to be dense-in-itself. If a given set A was so constituted that $Ac = A$, then the order type α of A was called a dense-in-itself type. As examples of such types, Cantor listed:

$$\eta,\ 1+\eta,\ \eta+1,\ 1+\eta+1,\ \theta,\ 1+\theta,\ \theta+1,\ 1+\theta+1.$$

He then went on to offer a number of types that were *not* dense-in-themselves, including:

$$\omega c = 0,\ (\omega + \nu)c = 1,\ (\omega 2)c = 1,\ (\omega 2 + \nu)c = 2,\ (\omega\mu)c = \mu - 1,$$
$$(\omega\mu + \nu)c = \mu,\ \omega^2 c = \omega,\ (\omega^2 + 1)\ c = \omega + 1,\ \omega^\omega c = \omega^\omega.$$

Particular emphasis was placed upon this last example, since one might otherwise have supposed that if $Ac{=}A$ was a condition (necessary and sufficient) for the *set* A to be dense-in-itself, then $\alpha c{=}\alpha$ should have implied that α was a dense-in-itself *type*. But since $(\omega^\omega)c{=}\omega^\omega$ was *not* dense-in-itself as a type, it was not enough to require that $\alpha c{=}\alpha$. Before α could be described as dense-in-itself, it was also necessary that $Ac{=}A$. This was a fact Cantor emphasized time and again.[15] In general, conclusions made with respect to simply-ordered *sets* were not automatically transferrable to similar conclusions with respect to the behavior of their corresponding order *types*.

In addition to types dense-in-thenselves, defined in terms of the coherence of ordered sets and their types, Cantor also introduced the notions of isolated ordered sets and of isolated types. For a simply-ordered set A, all elements which were not principal elements were said to be isolated. Taken in their order as given in A, they formed an ordered set of their own which Cantor called the *adherence* of A and denoted Aa. The type corresponding to Aa Cantor denoted αa. Since Aa contained no principal elements of A, Cantor described it as isolated and thus αa could be taken as an isolated type. Any set A could thus be given a unique and disjoint decomposition: $A = Aa + Ac$, where the order of elements in A had to be strictly observed. More generally:

$$A = \sum_{\rho' = 1,2,\ldots<\rho} Ac^{\rho'}a + Ac^{\rho}.$$

This offered Cantor yet another opportunity to stress the fact that formulae for ordered sets did not always parallel corresponding formulae for order types themselves. Thus he warned that, in general, $\alpha \neq \alpha a{+}\alpha c$.[16]

In consequence of the definitions of sets and of types dense-in-themselves, Cantor added that every part of the set A that *was* dense-in-itself must then belong to the coherence of A, thus a part of Ac^{ρ}. As a result, the set $\sum Ac^{\rho'}a$ could have no dense-in-itself part and was therefore, in Cantor's terminology, separated. The corresponding order type was then said to be a separated type. This led to the definition of closed types and then to the definition of perfect types. These were central to probing more exactly the nature of the continuum as a simply ordered set on (0,1) with its corresponding order type θ. Before he could define closed types, however, Cantor had first to make clear what he meant by a special kind of principal element in A:

> If we focus upon a simply-ordered set A which satisfies the following condition: if $e, e', e'', \ldots, e^{(\nu)}, \ldots$ is any simply infinite sequence of

> elements (of the simply-ordered set A) which either increases or decreases monotonically as ν increases, then there is a definite element f of A, which in the first case has a higher place than all $e^{(\nu)}$, whereby every lower element f' of the $e^{(\nu)}$ will be exceeded for sufficiently large values of ν; and in the second case (there is a definite element f of A) which has a lower place than all $e^{(\nu)}$, whereby any element f' greater than f has a higher order than the $e^{(\nu)}$ for a certain ν on; then f is clearly a principal element of A.[17]

If A was a set of elements f, Cantor defined it as a closed set and the corresponding order type α was then said to be a closed type. Closed types included $\omega+1$, $\omega^{\nu}+1$, $1+\theta+1$. On the other hand, ω, ω^{ν}, η, $1+\eta$, $\eta+1$, and $1+\eta+1$ were not closed types. Assuming, however, that A was a closed set, then it was clear that all coherences Ac of A also had to be closed types. Consequently, if α was a closed type, then so was αc^{ρ}. Moreover, if a simply-ordered set was dense-in-itself and closed, then it was said to be a perfect set with perfect type. Above all, $1+\theta+1$ was a perfect type.

In a concluding section written in February 1885, nearly two months after the rest of the *Principien* was finished, Cantor tried to make clear the differences between the approach he had taken in the *Principien* and that of the earlier "Zweite Mittheilung." In so doing, he also made clear the chronology of the two papers:

> Here [in the *Principien*] a generalization is made of concepts which we first encountered in the research on point set theory and particularly in the "Zweite Mittheilung über verschiedene Theoreme etc." (*Acta mathematica, VII,* page 105); thus it does not seem to me superfluous to indicate expressly a definite difference [between the two papers] which is related to the same idea in that more special [the "Zweite Mittheilung"] and in our more general area [the *Principien*].[18]

Cantor was concerned that some confusion might arise because of the similarity of terminology used in both papers. But in the "Zweite Mittheilung," concepts like coherence and adherence always retained a very special, direct connection to the point sets from which they were derived, while in the *Principien* Cantor was working in a much more general context. Point sets were necessarily bound with limitations that the order types could, in general, avoid. Point sets were first of all collections of distinct objects, points, and intrinsically part of their nature as point sets was the notion of distance. Moreover, Cantor had shown that there were only two possible powers among point sets. Consequently, they lacked sufficient generality to establish a comprehensive theory of transfinite numbers. What Cantor wanted to describe in its purest form was the idea of order alone. The theory of order types was designed to do so in a direct and general way. To stress the nature of that generality, Cantor reminded his readers that the concept of principal element defined for simply-ordered sets, when applied to point sets, was not exactly the same as the concept of limit point. Though it was true that every limit point of a set P was always a principal element of P, the converse was by no means always the case. Thus Cantor's theory of order types sufficed to deal with point sets specifically but surpassed the results obtained for point sets both in extent and in generality.

TOWARD A GENERAL THEORY OF ORDER TYPES

Though Cantor had resolved not to publish in the mathematical journals following his withdrawal of the *Principien* from *Acta Mathematica* in 1885, he nevertheless continued to press on with his research and even published several articles in the *Zeitschrift für Philosophie und philosophische Kritik*. These, of course, were predominantly philosophical in character, but he did publish one article which discussed, in addition to matters of philosophy and theology, strictly mathematical problems as well. This was Cantor's *Theorie der Ordnungstypen,* published as one section of a longer article: "Mittheilungen zur Lehre vom Transfiniten," which appeared in two parts in 1887 and 1888.[19]

Adopting the same general and abstract attitude which characterized the *Principien,* Cantor explained what he meant whenever he referred to sets. Given a set, it was to be thought of as a "thing in itself" [*Ding für sich*],[20] whether of distinct objects or of concepts, and these were to be taken as universals either with or without respect to their order. Cantor emphasized the process of such abstraction in defining the power of a set M by introducing the symbol $\overline{\overline{M}}$ with two bars across the top, each bar symbolizing a different level of abstraction.[21] Power involved a double abstraction, first from the *nature* of the elements of the set in question, and second from their given *order*. If one abstracted only the nature of the elements of a set, but not their order, then the convenient symbol $\overline{M}$ with a single bar could be used to represent its order type. Epistemologically, these distinctions were of great importance.[22]

Throughout all his writings Cantor took great pains to justify at every step the introduction of transfinite numbers and to explain the nature of their real mathematical existence. In the *Theorie der Ordnungstypen,* he was particularly careful to stress that the theorems concerning transfinite numbers followed, as he put it, "from the logical power of proofs, based upon definitions which are neither arbitrary nor artificial, but which arise naturally and regularly through the process of abstraction."[23] Of major concern was the relation between ordinals and cardinals generally.

Part of Cantor's struggle to establish his transfinite set theory and to silence opposition involved the dispute over *which* numbers were to be regarded as of original and primary importance in mathematics: the ordinals or the cardinals? If the ordinal numbers were given primacy, then the introduction of any number, finite or infinite, would have to be capable of formulation, originally, in terms of the ordinal process involving only the successive addition of units.

But the process was never regarded as complete, new units could always be added, and the bounds of finite number could never be surpassed. Thus Cantor's first transfinite number would never be reached. Moreover, since they could not be produced from finitely inductive considerations, they could not even be said to exist if such ordinal procedures were regarded as the only legitimate foundations upon which any concept of number had to be established. Thus Cantor's entire theory of transfinite numbers could easily be dismissed as an elaborate fantasy by such mathematicians as Kronecker. It is of little wonder, therefore, that Cantor insisted upon the cardinal concept of number as being separate and totally independent from any concept of ordinal number. It was the *cardinal* form that concerned the character of sets taken as completed wholes, without regard to order, and this was essential to production of the first and of all subsequent transfinite numbers.[24]

Thus the *Theorie der Ordnungstypen* opened by briefly outlining the major features of his theory of cardinal numbers. Cantor then went on to probe the nature of ordinal numbers in greater detail. As for the general theory of order types, he began to sketch its major features in the eighth section of the article. He began by defining n-dimensional order types, which he took to be nothing but units differentiated only by their order in a given n-dimensional set. Again the role of abstraction was emphasized in introducing the definition of pure order types, for which given any two units of the ordered set M, say e and e', the two elements were differently ordered in at least one direction n.[25] Were this not the case, then groups of units would be indistinguishable, since they would all have the same order in every direction; these Cantor called mixed types.

Where previously Cantor had only introduced the concept of the inverse type $*\alpha$ of a type α in the *Principien*, he now offered a second "conjugate type" as well, which he defined as the permutations transformation [*Vertauschungs-transformation*] with respect to two directions (μ,ν):

> One can change the order of elements of the type α so that the order of units $e, e', e'', \ldots$ is the same except for the elements at the μth and νth places, which are exchanged. Two units e and e' thus have the same relation to the μth and νth coordinates in the transformed type as they did to the νth and μth coordinates in α, while no change in the order of units on the other $n-2$ coordinates occurs. We call this transformation the *permutations transformation* with respect to the μth and νth coordinates.[26]

As if to reassure his readers that all these elaborate distinctions were to some avail, he added that:

> The *pure* simple types ($n = 1$) coincide with the finite ordinal numbers $1,2,3,4, \ldots$, because finite simply-ordered sets of *pure* type are

> always well-ordered sets [compare this concept with the *Grundlagen*, page 4], whose types I generally call ordinal numbers. To any finite cardinal number *m* there is only one *pure* simple type, i.e., *only one* ordinal number; this is intimately connected with the theorem proved in Number 7, where [it is established that] a finite set is never equivalent to one of its subsets.[27]

Once again Cantor had discovered a means of directly distinguishing finite from infinite, based upon the behavior of cardinal and ordinal numbers, viewed not separately, but in conjunction with one another. To every finite cardinal number *m* there was only one pure type, or only one ordinal number, because no finite set was ever equivalent to a proper subset. Rearrangement of elements did not affect the type of order. But for infinite sets, say of the first power, there might occur any number of pure types, since order did affect the type, though power remained unaffected. This was all reflected in the difference with respect to type between the natural numbers and the rational numbers taken in their natural order; the former was ω, the latter η, though both were countable types.

Cantor brought the *Theorie der Ordnungstypen* to a close with an elaborate discussion of formulae by which the actual number of order types, both pure and mixed, produced by given numbers of elements might be calculated.[28] He not only offered explicit diagrammatical guides for the case of pure types of two dimensions arising from the elements of a given finite cardinal number *m* but went on to derive recursive formulae for determining the exact number of types, pure and mixed, for a given *n*-dimensional ordering and a given cardinal number *m*. But the paper ended without offering any new clues concerning the power of the continuum or the order type θ of all real numbers on $(0,1)$.

The major advance that Cantor could claim in his published writing between 1885 and 1891 was the considerable degree of generality he had brought to defining the concepts of power and of order type, namely his transfinite cardinal and ordinal numbers. He had also amassed an impressive collection of new ideas, ones he hoped might bring him closer to solving the continuum hypothesis and to revealing the most general features of the continuous order-type θ. The whole spirit of set theory had undergone considerable transformation following the original appearance of the *Grundlagen* in 1883. He had broken away from the context of point sets and created in its place a new and abstract realm, where units were disentangled from their physical realities and even from the order in which they might be given, to produce new mathematical entities: transfinite ordinal and cardinal numbers. But it all seemed rather artificial and hollow until Cantor could show that his most recent developments were actually useful. Not all of them, in fact, survived to find a place in the last great summation of his set theory published between 1895 and 1897. By then, virtually no trace remained of the most general theory of *n*-dimensional order types. No mention was made of the cumbersome attempts undertaken to calculate the numbers of pure and mixed types a given cardinal number might engender. Such inspirations were like dinosaurs of his mental creation, fantastic creatures whose design was interesting, overwhelming, but impractical to the demands of mathematicians in general.

On the other hand, in addition to the advances made in the "Zweite Mittheilung" and in the *Principien,* an extremely important event took place in Cantor's life in 1891. It reversed the tide of those years immediately following the *Grundlagen* which were disappointing and at times intensely depressing ones for Cantor, both personally and professionally. In 1891 he was elected president of the newly founded Deutsche Mathematiker-Vereinigung, and on the occasion of the Union's first meeting, he read another of his papers destined to open a new chapter (and the last) in his continuing effort to establish transfinite set theory and to resolve the continuum hypothesis.

CANTOR'S DISAFFECTION FROM MATHEMATICS

In the years immediately following publication of the *Grundlagen* Cantor felt that his career had failed to produce the sort of recognition that he thought he deserved. As a result, following the critical years 1884–1885, he turned more and more to areas of research lying totally apart from the traditional German mathematics which he had come to find so disappointing, even confining. His publications of the late 1880s give the impression of a complete shift in interests. Between 1885 and 1891 what material he did publish appeared in journals devoted to philosophy,[29] and what survives from this period of his personal correspondence reveals a sustained preoccupation with the Bacon-Shakespeare controversy, Rosicrucianism, Freemasonry, and various historical-literary pursuits.[30] Finally, nothing could indicate more dramatically his apparent disenchantment than the desire he expressed as early as 1884 to teach philosophy at Halle instead of mathematics, thus abandoning his "erster Flamme" altogether.[31]

Toward some German mathematicians Cantor felt resentment, even bitter hostility. He also felt persecuted, threatened by the prospect of new attacks upon his set theory with every paper he might publish. Resenting the lack of appreciation for his work in Germany and elsewhere, he gave up the traditional forums of scientific journals, preferring to broaden his interests and to concentrate in print upon justifying the philosophical foundations of his work.

Cantor confined his public pronouncements on mathematics to an occasional lecture at the annual meeting of the *Gesellschaft Deutscher Naturforscher und Aerzte* (GDNA) and to students in his seminars at Halle. Though his interest in theoretical problems persisted, it was all sequestered in notes or in correspondence and carefully kept from the mathematical journals.[32] Mittag-Leffler might request a paper for his *Acta Mathematica,* but Cantor was in no hurry to provide one.[33] To Eneström, Mittag-Leffler's assistant, he was surprisingly candid: "As for your inquiry concerning my work, I am increasingly convinced that there is no reason to rush in publishing it, since for now there is no great interest in the transfinite numbers."[34]

Worse yet, Cantor's own enthusiasm was at low ebb. By 1887 he could write, quite openly, "At present of course I observe little interest in my mathematical research. My own inclinations do not urge me to publish and I gladly leave this activity to others."[35] But this did not mean that he had entirely abandoned his theory of sets and transfinite numbers. Actually, he had only given up the most superficial and socially expected part of research: publication. His commitment to mathematics was never really at question. Nothing could be better proof of this than his untiring efforts to reorganize the haphazard, uncoordinated meetings of the section for mathematics and astronomy of the Society of German Scientists and Physicians (GDNA).

The society, founded in 1822 by Lorenz Oken, had neither a permanent directorship nor a permanent home; both changed annually. Special sections were first established in 1828 at the meeting of the society in Berlin, under the direction of Alexander von Humboldt. Originally, no special provision was made for mathematics, although a joint section was designated for astronomy and geography.[36] Germany, as a union of free states, seemed unwilling to create a more centralized organization for the GDNA. Only in 1893 did the Society finally decide to establish its headquarters in Leipzig, and to adopt a more permanent form of leadership.[37] The fact that such relative instability of administration persisted for so long, coupled with the lack of a separate section for mathematics, contributed greatly to the eventual success of Cantor's efforts to provide an independent organization for mathematicians in Germany.

THE GERMAN MATHEMATICIANS' UNION

Deservedly enough, Cantor is almost universally credited with having been the driving force behind the successful creation of the Deutsche Mathematiker-Vereinigung (DMV).[38] Nevertheless, he was not the first to advocate an independent organization for German mathematicians. As early as 1867 at the GDNA meeting in Frankfurt, Alfred Clebsch had suggested that the mathematicians ought to form a separate group. The idea was informally discussed in the spring of 1868, and though no agreement was reached, it was decided that a new journal should be established, and by year's end the first issue of *Mathematische Annalen* made its appearance.[39] Not until 1872 was any serious attempt made to form a new society. Clebsch felt strongly that the younger mathematicians should promote the idea, and with this in mind Klein and Kiepert set off during Easter of 1872 for Berlin, where Klein appeared before a gathering of mathematicians to outline their plans. The concept of a new and independent organization of German mathematicians was well received, and a committee met in September to plan the first meeting for the following year in Göttingen. With much inspiration, encouragement, and financial support from Clebsch, it seemed the new union could not fail. But Clebsch died suddenly in November of the same year. Despite the loss, however, plans were carried out to meet in Göttingen as scheduled. All signs seemed to indicate a positive conclusion to Clebsch's dream, but the beginning, which seemed so promising, led nowhere. Owing largely to personal misunderstandings, ill feelings, and bad tempers, the union faded from existence. Without Clebsch, there was a void no one seemed willing or able to fill.[40]

Cantor eventually took the stage where Clebsch had left it. He hoped to work within the framework of the GDNA, in particular within its Section I for astronomy and mathematics, in order to promote a new and independent organization for mathematicians alone. By the end of the nineteenth century the traditional identification of astronomy and mathematics, so apparent in their respective histories and reflected in their joint section within the GDNA, was no longer indicative of the interests of the two groups. In fact, by the time Cantor decided to urge the creation of a new union for mathematicians, the astronomers had already taken their own step toward increased specialization by deciding to meet separately from the GDNA.[41] Thus there was already a precedent for the formation of a separate group for mathematicians. In addition, it had become clear by 1889 that the GDNA was no longer equipped to give mathematicians the type of forum they deserved. Mathematics had undergone a startling transformation since the founding of the GDNA more than fifty years earlier. It was now much more professional and specialized. For meetings to be efficient, and above all constructive, something separate and more stable than the GDNA was required. This could easily be provided by an independent union of German mathematicians, one which would establish simultaneously the desired combination of organization, stability, and professional identity.[42]

Compelling as such arguments were, why should Cantor have taken such an active interest in creating the new union? Why should he, almost single-handedly, have been so anxious to promote the DMV? In part, he had reasons that were quite personal, reasons which were deeply involved with the nature of German mathematics at the time, for Cantor saw the union as a new professional forum where mathematicians might be free to present their ideas openly without fear of prejudiced censure or unfair criticism from a small group of the mathematical elite.

As a permanent, professional organization, the DMV was thus designed to allow more candid and impartial discussion of new ideas than was possible in the authority-ridden universities, led by Berlin and Göttingen. Arthur Schoenflies, in writing a short account of Cantor's life for the *Mitteldeutscher Lebensbilder,* put it most simply by characterizing the DMV as the product of a struggle undertaken in the name of scientific freedom, one waged against the overpowering positions of a few German mathematicians.[43] Here Cantor's famous aphorism bears repeating: "the essence of mathematics is its freedom."[44] Freedom, it might be added, to work and to publish without the constant and prejudiced interference of an opinionated enemy like Leopold Kronecker, who in a single lecture, to believe Cantor's opinion, might discourage an entire generation of young mathematicians from ever reading a word of the "humbug" called Cantorian set theory.[45]

Cantor believed that the unfortunate fate of his own work was the product of a system in which a single individual could ruin any chances that a young, controversial mathematician might have of gaining recognition. Time and again, Cantor bitterly remarked that his salary in Halle was half that of other Professors Ordinarius because he had cut himself off from the two Brahmans of German mathematics: Kronecker and Weierstrass.[46] The price of "freedom" was isolation and discrimination. According to Cantor, being radical and unorthodox meant that one suffered poverty and recrimination.[47]

Consequently, for younger mathematicians he sought the kind of protection he had never enjoyed. In doing so, he was not being entirely unselfish. After all, it was the younger mathematicians who seemed to be taking an interest in his work. If set theory were to find its disciples, it was imperative to shelter them from the destructive power of opposition like that of Kronecker.[48] Through the open and diverse membership of a healthy society, Cantor hoped to ensure an impartial forum for new ideas and freedom for the younger mathematicians to pursue their own work, however heterodox it might seem to the older generation. The DMV could break down the established lines of authority associated with the prestige of university position by offering a new locus of evaluation centered in a group of peers. To Cantor, professionalism carried with it a sense of impartiality, at least ideally.

Surely the scars of Kronecker's polemic opposition to Cantor's work, coupled with a distasteful mix of academic jealousies and professional rivalries, provided strong motives for Cantor's hopes for the DMV. But despite the long history of their hostility, Kronecker was to figure quite prominently in the union's first meeting, and at Cantor's invitation. As a sign of respect for Kronecker's age and position among German mathematicians, Cantor had felt obliged to ask him to address the inaugural meeting in Halle. But Cantor's intentions were not entirely generous. In fact, he secretly wrote to Leo Königsberg in June 1891 and remarked that the invitation had been issued in hopes that Kronecker would take the opportunity to polemicize openly against Cantor and specifically against his set theory.[49] It is doubtful, however, that Kronecker would ever have been so tactless, or so direct, particularly since he had successfully avoided direct confrontation with Cantor in print all along. Moreover, an occasion like the opening of the DMV was not appropriate for the venting of personal vendettas. Nevertheless, Cantor fully expected Kronecker to repeat the same attacks he had made in his lectures that summer. If he did so again at the first meeting of the DMV, then Cantor predicted that "many who were previously blinded would have their eyes opened (to the *real* Kronecker) for the first time."[50] Cantor had always believed that once Kronecker's position was known publicly, then the merits of the confrontation would be equally clear, and both set theory and Cantor's position concerning the infinite would be directly and persuasively vindicated. But the opening session of the DMV was not to be the scene for acrimonious debate. Quite unexpectedly, Kronecker's wife was injured while mountain climbing late in the summer, and she died shortly afterwards. Consequently, Kronecker did not appear in Halle but instead sent a friendly, straightforward letter praising the objectives of the union and wishing it every success.[51] At the union's inaugural meeting, Cantor was elected to be its first president, while Kronecker was voted to a position on the board of directors.[52] Cantor later confided to Mittag-Leffler (as he had to Wilhelm Thomé) that it was just as well Kronecker had been unable to appear, since his presence would surely have lead to unpleasantness and would have provided Kronecker with yet another chance to intrigue against Cantor, and in his own university at that.[53]

The outcome of the 1891 meeting was successful. Finally the effort to create an independent Deutsche Mathematiker-Vereinigung had taken root, thanks largely to the boundless energy and enthusiasm Cantor had invested in the project. Many years later Schoenflies recalled that at one conference he had been asleep when Cantor came by to shout in his window: "*he* had more important things to do than sleep"—Cantor was out to convince everyone of the importance of the DMV.[54]

THE FIRST INTERNATIONAL CONGRESS OF MATHEMATICIANS

Though it is easy to understand Cantor's interest in forming a professional organization in which the encouragement of new ideas and the protection of individual freedom might be fully guaranteed, there was also another, less obvious reason for his energetic promotion of the DMV, one that seems to have been operative from the beginning. Ultimately, there were more far-reaching motives involved than a simple desire on Cantor's part to promote independence and freedom for German mathematicians.

By 1890 Cantor had come to resent the bureaucratic structure and institutionalized workings of professional mathematics in German universities. He had always wanted to leave Halle for a position of honor and respect, especially in Göttingen or Berlin. Opposition to his work made that impossible. Thus he was forced to remain in Halle where he had to suffer additional discrimination, financially, and he complained that his salary made it difficult, if not impossible, to travel. This in turn prevented him from maintaining the direct sort of professional contacts he might otherwise have enjoyed.[55] Intrigue had delayed his work from being published. He had given up Crelle's *Journal, Acta Mathematica,* and finally, his interest in publishing anything of mathematical significance seemed to have disappeared completely. At last, Cantor came to explain the fate he had been forced to suffer in an interesting way. He was not

German. As time went on, he was at greater and greater pains to renounce his ties to the nation that had failed to appreciate his work. Instead, he preferred to stress his Russian origins and he claimed that he felt no loyalties or sympathies whatsoever to Germany or to German mathematics.[56] He pointed out that by insisting upon freedom to pursue his own mathematical interests, he had effectively cut himself off from the powerful circle of mathematicians in Berlin. At its worst the list included Kummer, Kronecker, Borchardt, Weierstrass, Schwarz, and Fuchs.[57] Together such a grand conspiracy made it impossible for Cantor to expect aid or sympathy from any quarter in Germany. Consequently, he began to turn more and more to international contacts. Above all, he placed himself at the forefront of his own movement to promote the first international congress of mathematicians.

The idea for an international congress had been in Cantor's mind since the founding of the DMV itself, and he was as energetic in the pursuit of the congress as he had ever been in support of the DMV.[58] By promoting the idea of an international congress he could turn away from the closed circle of German mathematicians and renounce strictly German interests.[59] Cantor might well have expected fairer treatment for new mathematical ideas in less provincial, less national circles than even the DMV could provide. An international forum would be a freer organization by virtue of its diversity of membership. In part, Cantor must have regarded the DMV as a necessary stepping-stone. Without a central organization on the national level, it was difficult, if not impossible, to organize effectively meetings on an international scale. Political antagonism between nations was frustrating enough, but division within Germany itself would have made the question of any official international organization immensely complicated. As it was, the DMV refused to take the initiative in sponsoring any international meeting.[60] Few seemed opposed to the idea (in principle) in Germany, but no one was willing to assume responsibility for the organizational work. As it turned out, the self-appointed director, working behind the scenes to provide continuous inspiration, maintaining support, even drafting circulars and statutes, was Georg Cantor. By 1895 a steady stream of letters was going out to Hermite, Poincaré, above all to Charles Laisant and Emile Lemoine, working out details of the operation and suggesting the proper diplomacy to use against Germans like Schwarz (or Frenchmen like Picard) who seemed opposed to the idea of international congresses.[61]

The interest Cantor demonstrated in trying to encourage the basic principle of an international congress can be interpreted as a product, in spirit, of his efforts to establish the DMV. Both were similar attempts, on differing scales, to provide mathematicians with broader forums in which their work could be presented, discussed, evaluated. The hallmark of both enterprises was the recognition that mathematics had come of age, in Germany and in Europe. Professionalization was a direct consequence of specialization and of the growing number of mathematicians who made a life's work of theoretical, abstract research. But underneath it all lay the more obscure and personal motives of the one man who gave conscious energy and specific direction to both movements.

It is significant, and not a little touching, that Cantor used the occasion of the first meeting of the DMV to present the first piece of mathematical research he had completed for publication in over five years. The appearance of his paper in the first issue of the Union's *Jahresbericht* was direct evidence that he had decided to place his work once again before his colleagues, perhaps expecting a more positive judgment, fairer treatment, and freer appraisal.

THE DIAGONALIZATION PROOF OF 1891

The paper Cantor read to his colleagues in 1891, at the first meeting of the new Deutsche Mathematiker-Vereinigung, presented an entirely new proof for the existence of nondenumerable sets.[62] In 1874 he had demonstrated that the real numbers were nondenumerable, but he was no longer satisfied with that earlier proof, and in 1891 he sought to establish a much more general result. The techniques he devised in order to do so were extremely powerful, and this led him to produce an ascending hierarchy, in fact a limitless sequence, of transfinite powers.

Cantor began by exposing what he had come to regard as the most undesirable flaw of the proof of 1874: "It is possible," he wrote, "to produce a much simpler proof of that theorem, one which is independent of the consideration of irrational numbers."[63] In Chapter 3 Kronecker's objections to Cantor's work and specifically to his theory of real numbers has already been discussed. Not only did this opposition prompt Kronecker to cause the delay in publication of Cantor's *Beitrag* of 1878, but Cantor's earlier proof of the nondenumerability of the real numbers was also affected. In trying to deemphasize his significant discovery of that proof, the existence of sets that were larger in cardinal terms than countable sets, Cantor had entitled his paper "On a Property of the Set of All Algebraic Numbers." His reasons for doing so, and for making no reference to the much deeper and significant property he had discovered concerning the set of all real numbers, are closely related to his declaration in 1891 that he could at last avoid any reliance upon the irrational numbers in showing the existence of nondenumerable sets. Quite simply, irrational numbers were controversial. Coupled with dispute over the use and nature of the infinite in mathematics, they represented a major issue in his disputes with Kronecker.

But in divorcing set theory from point sets, irrational numbers, or specific objects of any kind, Cantor was also striving for complete and compelling generality. If his earlier conclusions might be criticized, even rejected because they incorporated elements that were questionable, then he would find alternative means. His paper of 1891 was an important step in this direction. Though it reconfirmed his earlier results and established the existence of nondenumerable sets, his conclusion was no longer related exclusively to the continuum of real numbers. In fact, his interest was not really the nondenumerability of the real numbers at all but the impressive observation that:

> This proof seems remarkable not only because of its great simplicity, but also because the principle which it follows can be extended directly to the general theorem, that the powers of well-defined sets have no maximum or, what is the same, that in place of any given set L another set M can be placed which is of greater power than L.[64]

Central to the proofs offered in the paper of 1891 was his famous method of diagonalization, relying upon only two elements, m and w. With these he considered the collection M of elements $E=(x_1,x_2, \ldots, x_n, \ldots)$, where each x_n was either m or w. As examples he suggested:

$$\begin{aligned} E^I &= (m,m,m,m,\ldots), \\ E^{II} &= (w,w,w,w,\ldots), \\ E^{III} &= (m,w,m,w,\ldots). \end{aligned}$$

Cantor then asserted that the collection of all such elements M was nondenumerable and offered a proof to that effect for the following theorem:

> *Theorem:* If $E_1, E_2, \ldots, E_\nu, \ldots$ is any simply infinite sequence of elements of the set M, then there is always an element E_0 of M which corresponds to no E_ν.[65]

First Cantor produced a countable listing of Elements E_μ in terms of the corresponding array (7.1), where each $a_{\mu,\nu}$ was either m or w:

$$\begin{aligned} E_1 &= (a_{11}, a_{12}, \ldots, a_{1\nu}, \ldots) \\ E_2 &= (a_{21}, a_{22}, \ldots, a_{2\nu}, \ldots) \\ &\vdots \\ E_\mu &= (a_{\mu 1}, a_{\mu 2}, \ldots, a_{\mu\nu}, \ldots) \\ &\vdots \end{aligned} \tag{7.1}$$

Then he defined a new sequence $b_1, b_2, \ldots, b_\nu, \ldots$. Each b_ν was either m or w, determined so that $b_\nu \neq a_{\nu\nu}$. By formulating from this sequence of b_ν the element $E_0=(b_1, b_2, \ldots, b_\nu, \ldots)$, it was immediately clear that $E_0 \neq E_\nu$ for any value of the index ν. Whatever elements E_ν one might choose to consider, E_0 was always different on the νth coordinate. What Cantor had shown (using only two elements m and w) was the fact that there were distinct powers among sets of infinitely many elements. In fact, the method of diagonalization provided an easy means of establishing directly that the ascending sequence of powers of well-defined sets had no maximum. In other words, given any set L, it was always possible to produce from elements of L another set M which was necessarily of higher power than L itself.[66]

As an example, Cantor considered the linear continuum L, the set of all real numbers on $(0,1)$, and demonstrated that the collection M of single-valued functions $f(x)$ which assumed only the values 0 and 1 for any value of $x \in (0,1)$ was greater in power than was L. First he argued that M had to be at least as large as L by showing that it contained a subset equal in power to L: namely the set N of all functions $f(x)$ which were equal to 0 throughout $(0,1)$ with the single exception of one point x_0 for which $f(x_0) = 1$. The set N of such functions was clearly equal in power to L, since there was a direct and natural one-to-one correspondence between N and L. What Cantor had yet to show was that no such correspondence was possible between L and M.

He did so by appealing to a suitably constructed form of the diagonalization argument. Were M and L of equal power, then it would be possible to give a one-to-one correspondence between the two through an intermediate function $\varphi(x,z)$, where every value z produced an element $f(x) = \varphi(x,z)$ in M, and every element $f(x)$ in M was produced from $\varphi(x,z)$ by exactly one value z. But this was impossible. Cantor turned to the case of a function $g(x)$ which assumed only the values 0 and 1 but in such a fashion that $g(x) \neq \varphi(x, x)$ for any given value of x. Thus $g(x)$ was clearly an element of M, but *no* specification $z = z_0$ would give rise to $g(x)$ from $\varphi(x,z)$, since $\varphi(z_0,z_0)$ was always different from $g(z_0)$.

Thus Cantor had introduced a means of showing that given any set, the set of all its subsets was always of a power greater than the parent set itself. By means of this new procedure he could claim the existence of his distinct, forever increasing succession of transfinite cardinal numbers, the powers of transfinite sets. These powers, Cantor added, represented the sole and necessary generalization of the concept of *finite* cardinal number whereby the *actually infinite* cardinal numbers were produced.[67] Thus regarded as powers, he argued that they should enjoy the same reality and definiteness as did the finite cardinals. In the course of history, the rational, irrational, and complex numbers had come to be accepted mathematically by virtue of their utility and consistency. Each form of number was merely a consistent generalization, in Cantor's view, of less comprehensive concepts. And he regarded the concept of power as the most comprehensive and natural of all.

Powers served a unifying and comprehensive function in Cantor's revised program to advance his theory of sets and transfinite numbers. Accordingly, stress was placed upon the cardinal numbers, finite or infinite, as being nothing other than the powers of sets, finite or infinite. This new emphasis was incorporated directly in his last and greatest statement concerning his life's work: the *Beiträge* of 1895 and 1897. The paper of 1891 hinted directly at the changes in Cantor's own mind concerning the nature of his set theory and the way its legitimacy, mathematically, could be most effectively argued.

Consequently, though Cantor's paper published in the first volume of the *Jahresbericht* of the DMV looked back to the results of research completed a decade earlier, it also looked to the future. If he invited Kronecker to address the union's first meeting in hopes of drawing him into direct confrontation, then his paper might easily have been calculated to provoke debate directly. Having removed all reference to irrational numbers, or to limits of infinite sequences, he may have believed that Kronecker would be forced to make his more fundamental objections to the set theory known. In doing so, he was convinced that the tenuousness of Kronecker's criticisms would be made apparent. But the confrontation never occurred. Kronecker died in 1891, and shortly thereafter, Cantor published a new, detailed explanation of his transfinite set theory. At last he seemed ready to offer mathematicians a clear account of his ideas by explaining the general and abstract principles upon which the transfinite numbers could be justified directly. By 1891 the transition from the *Grundlagen* to the *Beiträge* was in large measure complete and would soon be realized in a systematic and carefully written pair of articles within the next five years.

CHAPTER 8

The *Beiträge,* Part I: The Study of Simply-Ordered Sets

The *Beiträge zur Begründung der transfiniten Mengenlehre* was Cantor's last major mathematical publication.[1] When confronted with the task of presenting his work for the close scrutiny of a skeptical if not hostile forum of his contemporaries, he at last chose to present his revolutionary new ideas in a direct, straightforward way. After more than twenty years of development and painstaking consideration, he was prepared to summarize the strictly mathematical elements of his theory of transfinite numbers.

Many of the *Beiträge's* advances were methodological and philosophical in nature, though these were not, perhaps, readily apparent from its almost exclusively mathematical character. Nonetheless, the conclusions Cantor had reached by 1895 concerning the nature of mathematics were reflected in the philosophical assumptions of the entire work. Such considerations, which ran as an undercurrent throughout the pages of the *Beiträge,* are discussed in Chapter 10. For the present, only the specifically mathematical aspects of the work will be considered.

Unlike the *Grundlagen,* which began with a lengthy justification of unfamiliar and radical concept of actually infinite numbers, the *Beiträge* omitted such discussion entirely. Philosophy was absent in the belabored forms it had assumed in the *Grundlagen* and in the *Mittheilung zur Lehre vom Transfiniten*. Instead, in 1895, Cantor opened the *Beiträge* with the most basic elements of his set theory. There was no attempt at apology or explanation. Clear, forthright presentation would be sufficient, Cantor hoped, for mathematicians to realize the value of his new creations. Perhaps he had come to believe that the last ten years spent in trying to justify the transfinite set theory in a combined appeal to philosophy, theology, and mathematics had been to little avail. The merits of his work would ultimately be judged by mathematicians in terms of his mathematical ideas alone. Thus he set out to provide as clear an outline of the fundamental principles of his theory of transfinite sets as he possibly could. He hoped that his presentation might help create new interest in set theory, and to this end, as much as to any other, the work was directed.

SETS, POWERS, AND CARDINAL NUMBERS

The *Beiträge's* first sentence was a classic. It set the tone for all that was to follow:

> *Definition:* By a "set" we mean any collection M into a whole of definite, distinct objects m (which are called the "elements" of M) of our perception [*Anschauung*] or of our thought.[2]

It is significant that Cantor even bothered to define the concept of set at all. His doing so indicates how much the *Beiträge* was to differ from its predecessors, in particular, from the *Grundlagen*. Ten years earlier, in the mid-1880s, Cantor was still working almost entirely within the context of point sets, and thus any use of the term set had definite connotations. But in order to free himself from the particular character of point sets and to produce a completely general theory, he was compelled to make clear exactly what ought to be understood by the term set in the abstract.

Cantor had first tried to emphasize the independent importance of set as a mathematical concept in the *Grundlagen*. He did so, however, only in a note to Section 1 in an effort to explain what the term *Mannigfaltigkeitslehre* was supposed to represent. But in the process he also made clear what he meant by set. In his own words: "By an 'aggregate' or 'set' I mean generally any multitude which can be thought of as a whole, i.e., any collection of definite elements which can be united by a law into a whole."[3] He was primarily interested in sets as a whole because only in such terms could the transfinite numbers be defined. If one did not regard the set of integers N: 1,2,3, . . . as conceivable as an entirety, as a completed set, then there was no way to produce even the first transfinite numbers. That this was actually his motive in 1883 for insisting that sets be taken as a whole is clear from the additional remark in his note to Section 1. The sets he defined as whole unities were related to the Platonic *εἶδος* or *ἰδέα*, also termed *μικτόν* in Plato's dialogue *Philebos*. These were contrasted with the *ἄπειρον* which Cantor took to refer to an improper or merely potential but never completed infinity.[4] Cantor's emphasis upon the unitary or completed character of infinite sets was designed to ensure that they could indeed serve as a sound base for introducing the transfinite numbers. Thus it was important from the outset to stress this holistic character, and specifically, to define the idea of set as a *Ding für sich*.[5]

Powers, which Cantor had come to regard since the *Grundlagen* as cardinal numbers, were defined in the *Beiträge* as follows:

> *Definition:* We call the "power" or "cardinal number" of M that general concept which is derived from the set M by the help of our active powers of thought. The concept is abstracted from the character of the various elements m and from the order in which they occur in M.[6]

The only difference between this definition and the one given in 1887 was Cantor's appeal to active thought power in carrying out the process of abstraction from both the nature of elements and their order in any given well-defined set. Cantor used the same device he had introduced in 1887 to indicate this double process by denoting the cardinal number of a given set M with a pair of lines placed over the set symbol itself: thus the cardinal number, or power of a set M was represented by $\overline{\overline{M}}$. For the benefit of readers of the *Beiträge* however, he added the remark:

> Since a "unit" arises from every single element m, if one disregards its character, the cardinal number $\overline{\overline{M}}$ is also a definite set, comprised of nothing but units, which exists in our mind as an intellectual copy or projection of the given set $\overline{M}$.[7]

Here the connection between the complimentary roles of "active thought power" and "intellectual images" was significant. Cardinal numbers or powers claimed their reality in the *Beiträge* as intellectual images of sets in one's *mind*. When read in connection with the *Beiträge's* opening paragraph, this view of power helps to explain why Cantor chose to define the term set as he did. In 1887, sets were still taken to include "concrete things or abstract concepts," though power was defined as only a "general concept."[8] In dropping all reference to concrete objects, a certain homogeneity of presentation was achieved by defining sets as collected elements of intuition or thought. There was no longer any attempt to differentiate between the concrete reality of the numbers of sets on the one hand, and the intellectual or abstract reality of sets and their corresponding transfinite numbers on the other. At last Cantor accorded sets and transfinite numbers equal status as mutually existing in the intellect. The cardinal numbers (transfinite as well as finite) thereby existed intrinsically, inherently part of the very concept of a set.

This characterization of powers was tremendously important to Cantor in an ontological way. If all the elements of his set theory existed on the same level, with the same reality as thoughts and images of the mind, then there was no dependence upon real objects of any sort. Neither point sets, nor spatial factors, nor temporal sequences had any bearing upon the acceptability of his transfinite numbers. They followed necessarily and absolutely from the existence of sets themselves. The reality of sets as abstract objects in the mind then carried over directly to the transfinite numbers, and conferred upon them a similar sort of reality.

Equivalence between sets and their corresponding cardinal numbers was defined no differently in the *Beiträge* from previous presentations of Cantor's set theory. The concept of equivalence led in turn to the question of order in general, and in Section 2 he defined the relations of "greater than" and "less than" for cardinal numbers in terms of the comparability of their corresponding powers.[9] Given $\mathfrak{a} = \overline{\overline{M}}$ and $\mathfrak{b} = \overline{\overline{N}}$, then if there was no proper subset $M' \subset M$ such that $M' \sim N$, and if there was no proper subset $N' \subset N$ such that $N' \sim N$, then it was to be said that $\mathfrak{a} < \mathfrak{b}$ or $\mathfrak{b} > \mathfrak{a}$.

This was essentially the same definition of the order relation for cardinal numbers that Cantor had given in 1887. But in 1887 he had also claimed that if M and N were two nonequivalent sets, one always had to be equivalent to a proper subset of the other.[10] Moreover, he noted in consequence of his characterization of order among the powers of sets that whenever two sets M and N could be mapped (in a one-to-one fashion) to proper subsets of each other, so that $M \sim N' \subset N$ and $N \sim M' \subset M$, then M and N were necessarily equivalent. This same theorem appeared as Theorem B of the *Beiträge's* Section 2 and was one which Cantor had *never* been able to prove directly himself. It was later established independently by Felix Bernstein and E. Schroeder and has subsequently come to bear their names. Cantor noted in the *Beiträge* that the Bernstein-Schroeder Theorem, as well as three other theorems dealing with equivalency relations among sets, followed easily from a previously stated but *unproven* Theorem A: "If $\mathfrak{a}$ and $\mathfrak{b}$ are any two cardinal numbers, then either $\mathfrak{a} = \mathfrak{b}$ or $\mathfrak{a} < \mathfrak{b}$ or $\mathfrak{a} > \mathfrak{b}$."[11]

Cantor had already shown that, given any two cardinal numbers a and b, only one of the order relations could hold, but he was unable to prove that *exactly* one was always valid. The complication concerned the case of two cardinals represented by sets A and B where A was assumed equivalent to no part of B and B was assumed equivalent to no part of A. Cantor conjectured that this could only happen for finite sets where A and B were equivalent. But he was unable to show that the same could not occur for infinite sets. Consequently there was no way he could establish the necessary comparability of all cardinal numbers, finite and infinite.

Thus at the very outset an unwelcome flaw appeared. Only pages into the first article of the *Beiträge* a major question concerning Cantor's set theory and the transfinite numbers could not be resolved: were the transfinite cardinals strictly comparable? The matter was of critical importance because, if they were not, it would be impossible to arrange all cardinal numbers in an ordered sequence. and in turn, it would be impossible to say whether one of two given cardinals was necessarily larger than the other. This was a grave matter for Cantor's continuum hypothesis, for if the power of the continuum were not a comparable cardinal number, then he could never hope to show that it was in fact equivalent to the power of the second number class, which *was* a comparable cardinal by virtue of its definition in terms of a well-ordered set. Cantor always assumed that every set could be well-ordered, and consequently it followed that all cardinal numbers were necessarily comparable. But he never published any proof of these claims and chose not to raise the subject again in either part I of the *Beiträge* or in its successor of 1897.[12]

Like his definitions of order relations, rules for the addition and multiplication of any two cardinal numbers were first given in the *Mittheilungen zur Lehre vom Transfiniten*. In the *Beiträge* Cantor defined addition as he had earlier, in terms of two sets M and N representing two cardinal numbers $\mathfrak{a}$ and $\mathfrak{b}$ respectively.[13] The sum of $\mathfrak{a}$ and $\mathfrak{b}$ was then given in terms of the sets M and N (assumed to be disjoint) by forming what Cantor called their *Vereinigungsmenge* or their union, so that if $\mathfrak{a} = \overline{\overline{M}}$ and $\mathfrak{b} = \overline{\overline{N}}$, then $\mathfrak{a} + \mathfrak{b} = \overline{\overline{(M, N)}}$. Furthermore, he argued that the cardinal number of the sum of two cardinal numbers depended only upon the cardinals involved and in no way relied upon the sets in question, since given sets M and N such that $M' \sim M$, $N' \sim N$, then the cardinality of the union of the two sets was clearly independent of the sets in question and depended only upon the cardinality of the original sets themselves. Since the concept of power was abstracted from any properties of order among the elements of a given set, it followed immediately that the addition of cardinal numbers was commutative: $\mathfrak{a}+\mathfrak{b}=\mathfrak{b}+\mathfrak{a}$.

Multiplication of cardinal numbers was also presented in terms of the powers of corresponding sets. However, in the *Mittheilungen* Cantor had defined products originally in terms of a set formed from elements of the two sets M and N, where every element n in N was replaced by a set $M_n \sim M$, so that the collection of all such sets $\{M_n\}$ defined the product of M and N, written $(M \cdot N) = \{M_n\}$. With respect to the cardinal numbers of M and N, $\mathfrak{a}$ and $\mathfrak{b}$ respectively, Cantor wrote the product $\mathfrak{a} \cdot \mathfrak{b}$ as $\overline{\overline{(M \cdot N)}}$. But in the *Beiträge* he offered a somewhat different definition of the product of two cardinal numbers, and then showed that this new definition was equivalent to the one used previously.[14] In 1895 he preferred to define the multiplication of powers in terms of a *Verbindungsmenge* $(M \cdot N)$, consisting of all pairs of elements (m,n) generated from the elements of M and N taken one at a time. Thus $(M \cdot N) = \{(m,n)\}$ and the cardinal number of the product of two cardinals was correspondingly defined as $\mathfrak{a} \cdot \mathfrak{b} = \overline{\overline{(M \cdot N)}}$.

Cantor's reasons for advancing this alternative definition of multiplication were no doubt connected with his introduction of an entirely new but related concept in Section 4 of the *Beiträge*. There he presented his discussion of exponentiation for cardinal numbers and did so in terms of an idea that he had first used without explicit reference in 1891: the idea of the covering [*Belegung*] of a given set.[15] One might have expected Cantor to have regarded the exponentiation of cardinal numbers as merely a special case of multiplication, but unfortunately this could not be done with complete generality. Products were defined for cardinal numbers in terms of their corresponding sets; but the procedure could not be extended directly to include more than finitely many products. There was no difficulty in defining the product for $\mathfrak{b}$ factors of $\mathfrak{a}$, producing $\mathfrak{a}^{\mathfrak{b}}$ if $\mathfrak{b}$ were finite. But in 1891 Cantor had also shown that it was possible to represent the power of the continuum as an infinite exponentiation of the form $2^{\aleph_0}$, where $\aleph_0$ represented the cardinality of any denumerably infinite set. It was consequently of special importance to be able to define transfinite exponentiation, and this Cantor was able to do by generalizing his new definition of multiplication.

THE EXPONENTIATION OF POWERS

Cantor defined the covering of a set N with elements from M as follows:

> *Definition:* By a "*covering of the set N with elements of the set M*" or more simply, by a "*covering of N by M*," we mean a law by which every element n of N is associated with a definite element of M, where one and the same element of M can be applied repeatedly. The element of M associated with n is, so to speak, a single-valued [*eindeutige*] function of n, and may thus be designated by $f(n)$; let it be called the "*covering function* of n." Let the corresponding covering of N be called $f(N)$.[16]

The connection with Cantor's new definition of multiplication involving pairs of elements was thus made apparent. Moreover, Cantor offered two specific examples of coverings that might occur. If, for example, m_0 was a particular element from M, then the function $f(n) = m_0$ for all values of $n \in N$ was a covering involving only a single element. Drawing upon an example already used in the paper of 1891 to show that the set of single-valued functions was at least equivalent to the power of the continuum, Cantor suggested a covering by two elements of M, m_0 and m_1. If n_0 represented a particular element from N, then a covering function could be given such that $f(n_0) = m_0$ and $f(n) = m_1$ for all other values of $n \in N$ excepting the value n_0.

The set of all such covering functions served as the basis for defining exponentiation of cardinal numbers. Specifically, the set of all different coverings of a set N by the set M produced a set with elements $f(N)$, which Cantor denoted $(N|M)$ and called the *Belegungsmenge* or covering set.[17] Thus $(N|M) = \{f(N)\}$. Since the definition depended only upon the cardinal numbers $\mathfrak{a} = \overline{\overline{M}}$, $\mathfrak{b} = \overline{\overline{N}}$, then the cardinal number of $(N|M)$ served to define the exponentiation:

$$\mathfrak{a}^{\mathfrak{b}} = \overline{\overline{(N|M)}}.$$

Taking $\mathfrak{c} = \overline{\overline{P}}$, Cantor introduced the following laws:

$$\mathfrak{a}^{\mathfrak{b}}\mathfrak{a}^{\mathfrak{c}} = \mathfrak{a}^{\mathfrak{b}+\mathfrak{c}} \qquad ((M|N)\cdot(P|M)) \sim ((N,P)|M)$$

$$\mathfrak{a}^{\mathfrak{c}}\mathfrak{b}^{\mathfrak{c}} = (\mathfrak{a}\mathfrak{b})^{\mathfrak{c}} \qquad ((P|M)\cdot(P|N)) \sim (P|(M\cdot N))$$

$$(\mathfrak{a}^{\mathfrak{b}})^{\mathfrak{c}} = \mathfrak{a}^{\mathfrak{b}\mathfrak{c}} \qquad (P|(N|M)) \sim ((P\cdot N)|M).$$

The concept of covering was of great consequence because it enabled Cantor to define the exponentiation of cardinal numbers for transfinite values of the exponent. He was very much pleased with the conclusions he could suddenly obtain as a result and sent off word of his discovery on July 19, 1895, directly to Felix Klein, editor of *Mathematische Annalen*. By then the first part of the *Beiträge* was already in press, but Cantor was determined to include the new definition of exponentiation and to explain its ramifications in a freshly written fourth section. The language of his letter to Klein appeared nearly word for word without change in the *Beiträge,* which still bore the date March 1895, even though his newly added discoveries were not made until several months thereafter. As Cantor told Klein, "How rich these simple formulae are, when extended to powers, can be seen in the following example."[18] The example he offered was none other than an exact algebraic determination of the power of the continuum, expressed in terms of two other known cardinals of lesser degree. This was something he had never managed to do previously.

Cantor recognized that the power of the linear continuum, denoted $\mathfrak{o}$, could be represented as well by the set of all representations:

$$x = \frac{f(1)}{2} + \frac{f(2)}{2^2} + \ldots + \frac{f(\nu)}{2^\nu} + \ldots, \ (f(\nu) = 0 \text{ or } 1),$$

of the numbers of the continuum [0,1] as given in the binary system. Though the numbers $\frac{2\nu + 1}{2} < 1$ were represented twice, they were the only numbers in [0,1] that failed to have a unique representation, and they could be discounted in the question of cardinality, as Cantor showed, since they were only countably infinite in number. By denoting this set of elements with a double representation by $\{s_\nu\}$, the binary representations of all numbers on [0,1] were expressed by

$$2^{\aleph_0} = \overline{\overline{(X, \{s_\nu\})}}, \qquad (\text{where } X = [0,1]).$$

By taking $\{t_\nu\}$ a countable set in X, Cantor could simply write $X_1 = X - \{t_\nu\}$; then $X = (X_1, \{t_\nu\}) = (X_1, \{t_{2\nu-1}\}, \{t_{2\nu}\})$, and $(X, \{s_\nu\}) = (X_1, \{t_\nu\}, \{s_\nu\})$. Since $X_1 \sim X_1$, $\{t_\nu\} \sim \{t_{2\nu-1}\}$, $\{s_\nu\} \sim \{t_{2\nu}\}$, it followed that $X \sim (X, \{s_\nu\})$. Thus $2^{\aleph_0} = X = \mathfrak{o}$.

From the results of the arithmetic operations later defined for transfinite cardinal numbers in Section 6 of the *Beiträge,* Cantor could also conclude:

$$\mathfrak{o} \cdot \mathfrak{o} = 2^{\aleph_0} \cdot 2^{\aleph_0} = 2^{\aleph_0 + \aleph_0} = 2^{\aleph_0} = \mathfrak{o}; \quad \text{in general } \mathfrak{o}^\nu = \mathfrak{o}. \tag{8.1}$$

Furthermore:

$$\mathfrak{o}^{\aleph_0} = (2^{\aleph_0})^{\aleph_0} = 2^{\aleph_0 \cdot \aleph_0} = 2^{\aleph_0} = \mathfrak{o}. \tag{8.2}$$

These conclusions were most impressive. He expressed his delight in a pair of sentences at the end of the section:

> The formulae (8.1) and (8.2) have no other meaning than this: "The ν-dimensional, as well as the $\aleph_0$-dimensional continuum, have the same power as the one-dimensional continuum." Thus the *entire content* of my paper in volume *84* of Crelle's Journal, page 242 [Cantor's "*Beitrag*" of 1878] is derived purely algebraically *with these few strokes of the pen* from the *basic formulae of operating with powers.*[19]

Cantor was first impressed with the fact that the new formulae allowed purely algebraic derivation of results previously obtained only by geometric methods. Such applications of the concept of covering must have struck him as a tremendous advance, particularly because it seemed to have brought him one step closer to solving the problem of his continuum hypothesis. For the first time, algebraically, he had a firm grasp of what the power of the continuum must be. Even before he had introduced his new symbol for the smallest transfinite cardinal number $\aleph_0$, he was using it in print to show the light he hoped the exponentiation $2^{\aleph_0}$ would shed on his longstanding promise to establish the truth of his conjecture.

Second, the value of transfinite exponentiation was equally indicated by the ease with which it could be applied to substantiating earlier results. This also seemed to lend additional confirmation of the mathematical legitimacy of the transfinite numbers themselves, since they might be used to establish long-accepted theorems more directly and clearly. If earlier work, obtained without the use of transfinite numbers, could be directly and unambiguously transferred into the language of infinite sets, then the transfinite numbers seemed to justify themselves directly. But most of all he was glad for the purely algebraic conclusions which simultaneously revealed in an unmistakable way the value and importance of his transfinite numbers, allowing very complicated matters to be proven by merely "a few strokes of the pen."

THE FINITE CARDINAL NUMBERS

Finite numbers may be defined in one of two complimentary but opposite ways. Either they can be approached inductively, by successive addition of units, or, in contrast to infinite sets, they may be defined as those which are equivalent to no proper subset of themselves. Both definitions are equivalent and either can be derived if one assumes the other, but for Cantor's purposes in the *Beiträge* the inductive approach was clearly the more suitable. To have taken the latter path would have presupposed the definition of the infinite from the very beginning. Instead, he wanted to show how his entire transfinite set theory rested naturally upon the same principles as did the familiar theory of finite numbers and sets, and to stress implicitly the naturalness of producing cardinal numbers from sets by *abstraction*. If he could introduce the procedure convincingly for the definition of finite numbers, then the same procedure could be extended directly and easily to the definition of transfinite numbers by only considering infinite sets. In fact, Cantor went so far as to claim that, from the principles he had already outlined in the *Beiträge,* it was possible to give the most natural, the shortest, and the most rigorous foundations for the theory of finite numbers.[20]

Beginning with a single object e_0, Cantor explained how it could be subsumed under the concept of a set, $E_0 = (e_0)$. By then applying the process of abstraction already described to produce cardinal numbers, a new concept was produced, called "one," which was denoted: $1 = \overline{\overline{E_0}}$.[21] Cantor then joined another object e_1 with the set E_0, and called their union $E_1 = (E_0, e_1) = (e_0, e_1)$. This gave rise to the corresponding cardinal number "two": $\|2 = E_1$. By adding new elements, it was possible to generate an entire sequence of sets: $E_2 = (E_1, e_2)$; $E_3 = (E_2, e_3)$; . . . , producing in turn the sequence of finite cardinal numbers 3,4,5, ... In general:

$$\nu = \overline{\overline{E_{\nu-1}}};\ E_\nu = (E_{\nu-1}, e_\nu) = (e_0, e_1, \ldots, e_\nu).$$

From the definition of sums it then followed that $\overline{E_\nu} = \overline{E_{\nu-1}} + 1$. Thus every cardinal number except 1 was produced as the sum of its immediately preceding cardinal and 1.[22]

Cantor was well aware of the important difference which his own approach represented when compared with other theories and alternative foundations advanced for the number concept in the late nineteenth century. Mathematicians like Grassmann, Dedekind, Helmholtz, Kronecker, or Weierstrass would never, he argued, have discovered the transfinite numbers.[23] As Cantor claimed, this was because his method of abstraction was the key to everything he had accomplished. Only in terms of his analysis of the *forms* of sets were such comprehensive definitions of cardinal and ordinal numbers possible. Moreover, his definitions were so general that they applied equally well to both finite and infinite numbers. This in itself seemed convincing proof to Cantor of the correctness of his method. It was directly in terms of the process of abstraction from both the *nature* and *order* of elements that he began from a single unit to build the succession of finite cardinal numbers or, as he once reformulated his definition for Peano:

$$1 = \overline{\overline{(\mathfrak{a})}},\ 2 = \overline{\overline{(\mathfrak{a},\mathfrak{b})}},\ 3 = \overline{\overline{(\mathfrak{a},\mathfrak{b},\mathfrak{c},)}},\ \ldots\ .$$[24]

Each of these cardinal numbers was defined independently of any ordinal considerations. Thus the cardinal three was not to be regarded as an ordered sequence 1, 2, 3, but only as a triad, order being discounted entirely. This was why Cantor placed such heavy emphasis upon the cardinal numbers and the way in which they represented *powers* of finite sets. Unlike Dedekind, who had stressed the ordinal character of his "chains" in defining the finite numbers in his pamphlet *Was sind und was sollen die Zahlen,* Cantor needed to stress the cardinal concept of number for both technical and philosophical reasons.[25]

It was also true that Cantor needed to establish more than just the value and reasonableness of his *means* of defining number by abstraction. He needed to ensure as well a solid grounding of the properties of finite number. Taken as a whole, the totality of all such numbers N would subsequently serve as the indispensable foundation for the definition of Cantor's transfinite numbers. He was therefore careful to stress the distinct and unique generation of the finite cardinals. He did this in a series of theorems which began by showing that in the sequence of finite cardinals $1, 2, 3, \ldots, \nu, \ldots$, each was different from all the others.[26] Moreover, every number ν was greater than all the cardinals preceding it and less than every cardinal following it. Finally, he argued that between any two successive cardinals ν and $\nu+1$, there could never be another cardinal integer. Each of these theorems was proven in terms of two results dealing with the character of finite sets. The first claimed that if M were a set that was equivalent in power to none of its proper subsets, then the set (M,e) which was produced from M by adding a single element had the same property. Second, Cantor established that if N were a finite set with cardinal ν, and if N_1 were any proper subset of N, then the cardinal number of N_1 had to be equal to one of the numbers preceding ν. He was also able to conclude that among any set of finite cardinal numbers, there was always a least element, which in turn led to the conclusion that every finite set could be well-ordered.

Curiously enough, Cantor did not explicitly define the concept of finite number in the *Beiträge's* Section 5, though it was specifically devoted to the topic of finite numbers. Peano thought this most peculiar when he first read the article from proof sheets before it was published and wrote specifically to Cantor about the omission. In reply, Cantor said only that both the definition of finite set and the principle of induction, which he had also used without explicit justification to generate the finite numbers, could be found implicitly in Section 5 of the *Beiträge*.[27] Moreover, the fact that Cantor discussed finite numbers at all was a departure from his previous practice. In all his earlier presentations of the transfinite set theory he had been content to assume as given the nature of the natural numbers. The nearest Cantor ever came to presenting any exposition of the finite numbers before 1895 was in Section VIII of his *Mittheilungen zur Lehre vom Transfiniten*. In this paper the first eight paragraphs were devoted to a sketch of his theory of transfinite cardinal numbers, though he tried to give some completeness to his survey, as he put it, by adding a word about *finite* cardinal numbers.[28] The word was actually a lengthy footnote designed to argue the independent, ideal existence of each cardinal number from all others and free them entirely from any presuppositions involving ordinal, or dependent considerations. But he did write, as he would later in the *Beiträge*, that finite numbers were those produced from finite sets. In passing, he mentioned that his definition of a finite set could be found in another note several pages earlier. Turning to the page in question, one finds that he defined M to be a finite set if it could be generated from *one* original element by the successive addition of new elements such that it was always possible to return to the original element by successive removal of intervening elements.[29] In other words, finite sets following this procedure satisfied the conditions of mathematical induction. Assuming that M was a finite set equivalent to no part of itself, he went on to show its successor set, $M+1$, had the same property. This was exactly the same point he stressed in his reply to Peano.[30] He did not take the matter of mathematical induction as a principle which needed to be assumed. Instead, from his definition of the simply infinite succession 1, 2, 3, . . . , taken as cardinal numbers, each independent from all the others, he could derive the inductive principle and in essence define the finite numbers as those which satisfied the conditions of mathematical induction.

THE FIRST TRANSFINITE CARDINAL NUMBER, ALEPH NULL

Declaring simply that sets with finite cardinal numbers were to be called finite sets, Cantor defined all others to be transfinite sets and their cardinals transfinite cardinal numbers.[31] The first example of a transfinite cardinal number, then, was the power of the set of all finite cardinal numbers, represented as:

$$\aleph_0 = \overline{\overline{\{\nu\}}}.$$

When Cantor finally decided that the transfinite cardinals required a separate notation of their own, he felt that all the usual alphabets, the familiar Greek or Roman letters, were too widely used for other purposes both specific and general. His new numbers deserved something unique. He was always careful about the selection of notation, and it was to be expected that he would make the best choice possible in picking a new symbol for one of the most important concepts of his set theory. Not wishing to invent a new symbol himself, he chose the aleph, the first letter of the Hebrew alphabet. The choice was especially clever, as he was happy to admit, since the Hebrew aleph served simultaneously to represent the number one, and the transfinite numbers, as cardinal numbers, were themselves infinite unities.[32] In addition, it might be said that as the first letter of the Hebrew alphabet, the aleph could be taken to represent new beginnings, and he certainly believed that his theory of transfinite numbers represented a new beginning for mathematics. After all, his set theory had made it possible to deal successfully for the first time with the nature of mathematical infinity.

The evolution of Cantor's understanding of the transfinite cardinal numbers is itself instructive. Nowhere in the *Grundlagen,* for example, did he ever identify transfinite powers as cardinal numbers.[33] Instead, he always referred to the powers of infinite sets in terms that seemed calculated to avoid *any* suggestion that powers might also be numbers. Thus, in Section 2 of the *Grundlagen,* he explained that "the *power* of a set, as we see, is an *attribute* of same independent of ordering."[34]

By September 1883, when Cantor addressed the mathematicians at the GDNA meeting in Freiburg, he was much clearer about the significance of powers and equated them with cardinal numbers as follows: "But I call two sets M and N *equivalent* if the one can be associated with the other, element for element, by means of a one-to-one correspondence. To express this I use the shorter designation *valence* for power or cardinal number."[35] No symbol, however, was designated to represent even the least transfinite power corresponding to denumerably infinite sets, for which ω had already been introduced to denote the least transfinite *ordinal* number. The fact that special and distinct notation for order types was developed so quickly indicated how much more important they were, conceptually, than the transfinite cardinal numbers for the initial development of Cantorian set theory. After all, it was the ordinals that made precise definition of the transfinite cardinals possible. And until Cantor had introduced the order types of transfinite number classes, he could not define precisely any transfinite cardinal beyond the first power.

The evolution of notation for the transfinite numbers revealed their dependence upon transfinite order types. In June 1886 Cantor explained to Franz Goldscheider that if α denoted the order type of a well-ordered set, then he denoted the power of the same set by $\overset{*}{\alpha}$.[36] Since ω was the smallest transfinite number, then its power $\overset{*}{\omega}$ was the first transfinite power, the smallest transfinite cardinal number. Consequently,

$$\overset{*}{\omega} = \overset{*}{\omega + 1} = \overset{*}{\omega + 2} = \ldots = \overset{*}{\omega^2} = \ldots .$$

The set of all ordinal numbers equal in power to $\overset{*}{\omega}$ comprised the second number class (II), denoted Ω. In fact, $\overset{*}{\Omega}$ was the next transfinite power following $\overset{*}{\omega}$, as Cantor had shown in the *Grundlagen* but without benefit of the new symbols. In 1886 he could write that $\overset{*}{\omega} \neq \overset{*}{\Omega}$, but such notational conventions could not have remained satisfactory for very long. By November of the same year he was representing $\overset{*}{\omega}$ by $\mathfrak{o}_1$, and the power of the second transfinite cardinal number $\overset{*}{\Omega}$ by $\mathfrak{o}_2$. Clearly he was striving for a more homogeneous notation than $\overset{*}{\omega}$, $\overset{*}{\Omega}$, $\ldots$, which was too inflexible to describe the entire succession of transfinite powers with any generality.[37] Two years later he had decided to adopt the superimposed bar (which was the same notation he used to describe the order type of a given set M as $\overline{M}$) and denoted the power of order types ω as $\bar{\omega}$. Only in 1893 did he finally decide upon the alephs. Vivanti was engaged at the time in preparing a general account of set theory in Italian, and Cantor was anxious that the new notation be used, hoping that it might begin to find popular acceptance. It was then that he decided definitely upon the notation he would soon make standard in the *Beiträge;* in terms of his new alephs, Cantor wrote:

$$\overline{\omega} = \overline{\omega + 1} = \overline{\omega + 2} = \ \ldots = \overline{\alpha} = \aleph_1.$$

The last step in the evolution of the aleph notation was Cantor's decision in 1895 to denote the first transfinite cardinal number by $\aleph_0$.

Thus the eventual emergence of a distinctive notation for the transfinite cardinal numbers was surprisingly late in the history of Cantor's set theory. As long as he was limited to characterizing the various transfinite number classes in terms of the ordinal numbers, it was perhaps enough to designate their powers in terms of the initial order type of each class, by $\overset{*}{\omega}$ or $\overset{*}{\Omega}$. In fact, until there was some readily available means of proceeding easily up and down the scale of transfinite cardinals, the powers of the first two number classes were actually sufficient. This was particularly true so long as he limited his interest to point sets, since he had already shown that for such sets only two powers ever came into consideration—those of the first and second number classes.

But with Cantor's new idea of covering as he presented it in 1891, it was possible to define algebraically for the first time distinctions between powers, and thus it was imperative that a symbolism be adopted that would suit such manipulations. Not only had he been able to reconfirm his conclusion in the *Grundlagen* (based there upon the first and second principles of generation), that the increasing sequence of transfinite numbers was unending, but he could now show precisely that given any set L, it was always possible to construct a new set of greater power. In terms of cardinal numbers $\mathfrak{a}$, this amounted to statements of the form $2^{\mathfrak{a}} > \mathfrak{a}$. Above all, it seemed that the new idea of covering might help to provide the means necessary to solving the continuum hypothesis in straightforward algebraic terms. Cantor had noted this discovery as early in the *Beiträge* as he could, at the end of Section 4 devoted to exponentiation. Even before his readers knew what $\aleph_0$ was meant to represent, they knew he was ready to claim that the power of the continuum $\mathfrak{o}$ was given exactly by the equation $\mathfrak{o} = 2^{\aleph_0}$. In Section 6 of the *Beiträge* he was at last prepared to say something more specific about the first transfinite aleph, $\aleph_0$ itself.

Cantor began by showing that $\aleph_0$ *was* a transfinite number. That it was not equal to any finite number μ followed directly from the fact that if, to the set of all finite cardinal numbers $\{\nu\}$, one added a new element e_0, then the set $(\{\nu\}, e_0)$ was equivalent to the set $\{\nu\}$. Cantor offered as proof the simple correspondence $e_0 \longleftrightarrow 1$, $\nu_1 \longleftrightarrow 2$, $\nu_2 \longleftrightarrow 3$, ... Thus $\aleph_0 + 1 = \aleph_0$. But for finite cardinals, Cantor had already shown in Section 5 of the Beiträge that $\mu + 1$ was always different from μ. Thus $\aleph_0$ could not, by definition, be a finite number.[38]

Second, Cantor established that $\aleph_0$ was in fact larger than any finite number μ. This followed from the definitions of Section 3, and from the fact that considering any finite cardinal number $\mu = \overline{(1, 2, 3, \ldots, \mu)}$, no part of the set $(1, 2, 3, \ldots, \mu)$ was equivalent to the set $\{\nu\}$, though $(1, 2, 3, \ldots, \mu)$ was itself part of $\{\nu\}$. Hence $\aleph_0 > \mu$, whatever the value of the finite cardinal μ might be.

Finally, in order to characterize completely the nature of $\aleph_0$, Cantor established that it was also the least transfinite cardinal number. He reduced the problem to an equivalent proposition: if $\mathfrak{a}$ was any transfinite cardinal different from $\aleph_0$, then $\aleph_0 < \mathfrak{a}$. He proved this by first arguing that every transfinite set T must have a subset of power $\aleph_0$. This was labeled Theorem A, and its proof tacitly assumed that if one picked $\nu - 1$ elements from any infinite set T, it was always possible to pick another: t_ν. By such reasoning this sequence selected from T and the sequence of all finite cardinals were similar: $\{t_\nu\} \sim \{\nu\}$, and since $\{\nu\} = \aleph_0$, T itself contained a subset of power $\aleph_0$. But for this argument to have any general validity at all it was necessary to assume, as Cantor did, that every transfinite set could be well-ordered.[39]

Arguing along similar lines, Cantor pressed the validity of a Theorem B, that if one took S to be any transfinite set of cardinality $\aleph_0$, then any infinite subset S_1 of S must also be of power $\aleph_0$. Combined with his conclusion from Theorem A, he could then conclude that his new number, $\aleph_0$, was in fact the least of all transfinite cardinal numbers.

With $\aleph_0$ thus established as the pivotal element between the finite and transfinite domains, Cantor went on to introduce the sort of algebraic conclusions that only a special notation for the transfinite cardinal numbers could have facilitated:[40]

$$\aleph_0 + 1 = \aleph_0; \ \aleph_0 + \nu = \aleph_0; \ \aleph_0 + \aleph_0 = \aleph_0; \ \aleph_0 \cdot 2 = \aleph_0; \ \aleph_0 \cdot \nu = \aleph_0;$$
$$\aleph_0 \cdot \aleph_0 = \aleph_0.$$

As for the cardinality of the smallest transfinite sets, $\aleph_0$, Cantor noted that he would eventually show how it led directly and necessarily to an entire well-ordered sequence of transfinite powers that increased monotonically without end.[41] Each transfinite cardinal number was uniquely defined, and each had a definite successor. This in itself was not new; much the same result had already been given in the *Grundlagen* twelve years earlier. But the rigorous formulation had only been developed since then, and it was only upon his theory of order types that he believed the satisfactory foundations for all of transfinite set theory could be built.[42] Of central importance, as the *Mittheilungen zur Lehre vom Transfiniten* made clear, was the idea of well-ordered set. Though he had first stressed the significance of well-ordering in 1883, Cantor had not then recognized the importance of order types, as opposed to well-ordering, in presenting a satisfactorily rigorous treatment of the transfinite numbers. Nor did he turn immediately to explain such connections. Instead, he set aside the matter of his newly defined transfinite *alephs* with only a hint of those beyond $\aleph_0$, and brought Part I of the *Beiträge* to a close with five sections devoted to outlining his general theory of ordered sets. Later, the well-ordered sets would receive extensive treatment as a special theory to which all of Part II of the *Beiträge* was devoted. Until then, Cantor considered only the concept of order in general, and went on to present results that were previously unpublished, though of major significance in his arsenal of weapons designed to meet the challenge of mathematical continuity. As an indication of the progress he had made, Part I of the *Beiträge* ended with a completely general characterization of the type of order represented by the linear continuum.

ORDINAL TYPES OF SIMPLY-ORDERED SETS

In beginning to divorce his set theory from specific examples like point sets in the *Grundlagen,* Cantor was very much aware that the character of continuous domains had nothing to do with geometric considerations at all. Understanding the structural properties of the continuum depended, ultimately, upon purely ordinal features. From the very start, his research had implicitly recognized the inability of powers alone to make major distinctions clear between discrete and continuous sets. Even for only denumerably infinite sets, power was inadequate to describe the rich textures of various kinds of ordering. Thus, in addition to discovering the powers of various sets, it was clear that it was equally important to understand their purely ordinal character.

In the years following publication of the *Grundlagen* he had reached some impressive conclusions concerning the structure of sets, particularly the rationals and reals, by means of ordinal considerations alone. Some of these were published in Section VIII of his *Mittheilungen zur Lehre vom Transfiniten,* others were prepared for publication, specifically in his *Principien* but were withdrawn from press. The *Beiträge* therefore represented Cantor's first full-scale explanation of order types and of the related idea of simple order in any mathematical journal.[43]

A set was said to be simply-ordered if all its elements were ordered by some rule such that, given any two, one could always be said to precede the other.[44] Thus for any two elements m_1 and m_2 of a simply-ordered set, either $m_1 \prec m_2$ or $m_1 \succ m_2$. If $m_1 \prec m_2$, $m_2 \prec m_3$, then it was always true that $m_1 \prec m_3$.

As evidence that sets could easily be ordered in different ways, he turned to the set of all rational numbers p/q which represented a special kind of simple order when taken in their natural order by magnitude. But they could also be arranged in a sequence determined by the sum of each numerator and denominator. Thus, given two fractions p_1/q_1 and p_2/q_2, the lesser of the two sums p_1+q_1, p_2+q_2 determined which was to be taken as the lesser under this particular ordering. If $p_1+q_1=p_2+q_2$, then natural order among the two fractions would prevail to determine which was the smaller of the two. This type of order among the rational numbers Cantor denoted as the type R_0. The set was a denumerable one, and the order was represented as follows:[45] $R_0 = (r_1, r_2, \ldots, r_\nu, \ldots) = (\frac{1}{2}, \frac{1}{3}, \frac{1}{4}, \frac{2}{3}, \frac{1}{5}, \frac{1}{6}, \ldots)$, where $r_\nu \prec r_{\nu+1}$.

Having made clear the idea of simple order, Cantor went on to introduce the order type of any given set M, to be denoted $\overline{M}$. Like the definition of powers, it remained entirely abstract. Cantor insisted only that the elements of any given set be well-defined.

> *Definition:* To every ordered set M there corresponds a definite "order type" or, more simply, a definite "type," which we shall designate by $\overline{M}$; by this we mean the general concept which arises from $\overline{M}$ if we only abstract from the character of its elements m but retain the order in which elements occur in M.[46]

Consequently, the order types were actually ordered sets whose elements were taken, as Cantor said, to be "pure units" with the same order as their corresponding elements in M from which they were abstracted. The "similarity" of ordered sets was defined as before, though Cantor did simplify matters somewhat by introducing specific notation: $M \simeq N$. Two order types α and β were said to be equal if, for $\alpha = \overline{M}$ and $\beta = \overline{N}$, $M \simeq N$. Consequently, it was possible to write $\overline{M} = \overline{N}$ and $M \simeq N$, concluding that the one implied the other.

Having already abstracted the nature of elements of a given set to produce its order type $\overline{M}$, Cantor could now go a step further and note that by subsequently abstracting the order of elements one produced the cardinal number $\overline{\overline{M}}$ of the ordered set M of type $\overline{M}$. Sets of equal type were always of equal cardinality, hence $\overline{M} \simeq \overline{N}$ always implied that $\overline{\overline{M}} = \overline{\overline{N}}$, though clearly the converse was not generally true.

Cantor denoted ordinal numbers by lower case letters of the Greek alphabet. If α was a given order type, he had finally settled upon $\overline{\alpha}$ to represent its corresponding cardinal number. Finite ordinal types, by the time Cantor came to write the *Beiträge,* held no particular interest. As he had remarked repeatedly on previous occasions, for finite sets ordinal numbers were isomorphically identical with cardinal numbers. Thus the same laws of arithmetic held for both kinds of number, and though conceptually they were different, it was still possible to represent them by the same numbers, in fact, by the familiar sequence $1, 2, 3, \ldots, \nu, \ldots$.

It was only the transfinite case that allowed clear distinctions between the two kinds of number, and on infinite sets the differences were remarkable, as Cantor had demonstrated time and again. Taken in their entirety, all simply-ordered types corresponding to a given cardinal number $\mathfrak{a}$ constituted a particular class of types. As Cantor remarked: "Each of these type-classes is therefore determined by the transfinite cardinal number $\mathfrak{a}$, which is the same for all individual types belonging to the class. Thus we call it, for brevity's sake, the type-class $[\mathfrak{a}]$."[47] Of immediate interest was the relation of cardinal numbers to order types and to the powers of various number classes.

It was important to distinguish between the cardinal number $\mathfrak{a}$ which determined the class of types $[\mathfrak{a}]$ and that cardinal number $\mathfrak{a}'$ which was determined by the class of types $[\mathfrak{a}]$. The latter, $\mathfrak{a}'$, Cantor took to be the cardinal number (or power) of the class $[\mathfrak{a}]$, and it thus represented a well-defined set whose elements were all the type α of cardinality $\mathfrak{a}$. In general, $\mathfrak{a}$ and $\mathfrak{a}'$ were always distinct from one another, and in fact it was possible to show that $\mathfrak{a}'$ was always greater than $\mathfrak{a}$.[48]

In addition to the types of simply-ordered sets, Cantor went back to his earlier papers, specifically to the *Mittheilungen,* to introduce the inverse order types of a given set M. In reversing the order of elements, a new set was produced which Cantor denoted $*M$, with ordinal type $*\alpha$, assuming that $\alpha = \overline{M}$. For finite types it was always true that $*\alpha = \alpha$. This was true as well for certain transfinite sets, for example the set of rational numbers taken in their natural order of precedence on $(0,1)$, of order type η, for which $*\eta = \eta$.[49]

Given two simply-ordered sets, Cantor added that they could always be mapped onto one another in many ways. On the other hand, all finite types and all transfinite types of well-ordered sets (which determined the transfinite ordinal numbers) were such that they allowed only a single mapping onto themselves.[50] Cantor sought to make the differences clear with two examples. The ordinal type ω represented the type of any well-ordered set of the form: $(e_1, e_2, \ldots, e_\nu, \ldots,)$ where $e_\nu \prec e_{\nu+1}$, and where the indices ν represented all finite cardinal numbers.

Given any other well-ordered set of the same type ω: $(f_1, f_2, \ldots, f_\nu, \ldots)$, with the condition that $f_\nu \prec f_{\nu+1}$, it was possible to map the two sets onto one another in only one way: if similarity of order was to be preserved under the correspondence, e_ν and f_ν always had to be corresponding elements. Any other mapping would violate the assumed similarity between the two sets in terms of their respective orderings.

To show that simply-ordered sets, on the contrary, could be corresponded in any number of ways, Cantor introduced the set $\{e_{\nu'}\}$, where ν' represented all positive and negative integers, including zero, and where $e_{\nu'} \prec e_{\nu'+1}$.[51] This set had no largest or smallest element and was represented by the order type $*\omega + \omega$; it could be mapped onto other similarly ordered sets in uncountably many ways. He considered a set of the same type $\{f_{\nu'}\}$, where $f_{\nu'} \prec f_{\nu'+1}$. The two sets could then be corresponded in such a way that if ν_0' was any particular index ν', then to the element $e_{\nu'}$ of the first set, $f_{\nu_0'+\nu'}$ of the second could be matched. Since the choice of ν_0' was completely arbitrary, and could assume uncountably many values, the correspondence between $\{e_{\nu'}\}$, and $\{f_{\nu'}\}$, could be made in an infinity of different ways.

Once again Cantor had shown how his new theory could sharpen the differences between properties of finite and infinite sets. But he had also indicated a most significant link between finite sets and the ordinal properties of a special group of simply-ordered sets. All finite sets could be mapped in only one way onto other finite sets and retain similarity of ordering. The same was true for infinite well-ordered sets, and this fact afforded Cantor another opportunity for stressing the reasonableness of calling the types of well-ordered infinite sets "transfinite ordinal numbers." Similarly, just as every finite ordinal number was for all intents identical with its corresponding cardinal, he could urge the same comparison on infinite sets to conclude the reasonableness of terming the powers of infinite sets "transfinite cardinal numbers." The important difference, of course, was that there was no unique ordinal number bound to every transfinite cardinal; in fact, each transfinite cardinal determined entire classes of ordinal numbers of the form $[\mathfrak{a}]$, the most important in the *Beiträge* being $[\aleph_0]$.

To make possible the combination of various order types, arithmetic operations were also introduced.[52] These followed the familiar definitions given earlier in both the *Grundlagen* and the *Mittheilungen*, and included the ubiquitous caveats concerning the noncommutative character of transfinite operations wherever order types were concerned.

Finally, Cantor summarized the connection between his definitions of cardinal numbers and order types of simply-ordered sets. Proceeding directly from their respective definitions, it was clear that the cardinal number of the sum of two order types was equal to the sum of the cardinal numbers of the two order types themselves. The same was true of the product of order types, expressed directly in terms of cardinal numbers: $\overline{(\alpha+\beta)} = \overline{\alpha}+\overline{\beta}$, and $\overline{(\alpha\cdot\beta)} = \overline{\alpha}\cdot\overline{\beta}$. Thus all the arithmetic results obtained for ordinal types were equally valid if every order type were replaced by its corresponding cardinal number. This was an immediate consequence of the fact that cardinal numbers were defined from the same sets as their corresponding order types but without respect for the order of elements.[53] At last, with the basic arithmetic properties at hand for the types of simply-ordered sets, Cantor was ready to turn his attention to two major results.

THE ORDER TYPE η

The set of all rational numbers R was particularly fascinating. Though it was countable, and therefore of cardinality $\aleph_0$, it produced distinctly different order types depending upon how the precedence of elements was defined. Cantor had earlier shown that the rationals could be ordered according to the magnitude of the sum of numerator and denominator to produce an ordered set he denoted R_0 with order type ω. But taken in their natural order on $(0,1)$, this simply-ordered set of rationals exhibited a very different and very special sort of order designated η.

In characterizing the properties of order peculiar to R, Cantor noted that there was neither a lowest nor a highest element and that between any two elements of R there were infinitely many others belonging to the set.[54] These properties, he claimed, were the necessary and sufficient properties determining the order type η, exemplified by the rationals for one, and this Cantor intended to prove as a separate theorem:

Theorem: If one has a simply-ordered set M which fulfills the three conditions:

1. $\overline{\overline{M}} = \aleph_0$;
2. M has no least and no greatest element;
3. M is everywhere-dense;

then the order type of M is equal to η: $\overline{M} = \eta$.[55]

Cantor's proof of the unique character of the type of order represented by η required the Axiom of Choice (though it was not mentioned explicitly), and it rested on the assumption that every set could be well-ordered. In fact, his very first step assumed that since $\overline{\overline{M}} = \aleph_0$, it was possible to order M as a denumerable sequence similar in type to ω. He denoted such an ordering of M as

$$M_0 = (m_1, m_2, \ldots, m_\nu, \ldots).$$

Now it was essential to show that somehow M could be corresponded with R in such a way that the order of elements in M was coincident with order in R; symbolically, $M \simeq R$. He did this by considering two elements from R and M taken to correspond, r_1 and m_1. Then r_2 in R had a definite relation to r_1 in R, while there were infinitely many elements m_ν of M with the same relation to m_1 in M as r_1 to r_2. In order to pick one of these m_ν, Cantor turned to the ordering among terms of M in M_0 and chose the element m_ν of lowest index, denoted it m_{ι_2} and let it correspond with r_2.

This procedure repeated itself with respect to r_3 in R. Again, there were infinitely many elements m_ν of M with the same relation to m_1 and m_{ι_2} in M as r_3 to the elements r_1 and r_2 of R. But M_0 could be used again to determine a least element, denoted m_{ι_3}, which was then correlated with r_3 and had the same order in M with respect to m_1 and m_{ι_2} as r_3 assumed with respect to r_1 and r_2.

Analogously, to ν elements of R Cantor could correlate definite elements of M: $m_1, m_{\iota_2}, m_{\iota_3}, \ldots, m_{\iota_\nu}$, where both sequences were similar to one another. To the element $r_{\nu+1}$ he could correlate that element with smallest index in M_0 of those elements with the same relation of order to the elements m_1, m_{ι_2}, $m_{\iota_3}, \ldots, m_{\iota_\nu}$ in M as had $r_1, r_2, \ldots, r_\nu$, to $r_{\nu+1}$ in R. This would determine $m_{\iota_{\nu+1}}$.

In this manner Cantor claimed to have correlated definite elements m_ι of M

will *all* elements r_ν of R in such a way that every element m_ι in M had the same relation of order among all elements of M as did the entirety of corresponding elements r_ν in R. What he had yet to show was that the elements m_{ι_ν} included *all* elements m_ν of M, or, equivalently, that the sequence of subscripts $1, \iota_2, \iota_3, \ldots, \iota_\nu, \ldots$ was nothing more than a permutation of the sequence $1, 2, 3, \ldots, \nu, \ldots$. He was able to do this through an appeal to complete induction.[56]

As a consequence of this correspondence, the entire set of elements M was mapped onto the whole set R in such a way that elements in each set retained their order of precedence under the correspondence itself. Thus M and R were similar sets, and he could claim to have demonstrated conclusively that the three characteristics listed for the simply-ordered set M did indeed describe completely the order type η of which the set of rational numbers in their natural order on $(0,1)$ was but one example.

The theorem itself, once established, led to a number of related results concerning the order type η. These were not entirely new, for some had appeared previously, but the level of generality which Cantor achieved in the *Beiträge* was striking. Earlier, when he had outlined the conditions requisite for a set M to be of type η, he had not bothered to elaborate the sort of proof just given. Instead, in the *Principien einer Theorie der Ordnungstypen,* he had thought it sufficient to describe the properties of the order of the rational numbers on $(0,1)$.[57] Thus he was able to characterize with certainty the nature of the general type η, though he did not go on to show that any everywhere-dense, denumerably infinite set having no greatest or least element was *necessarily* of type η. Instead, he simply went on to remark that arithmetic combinations might also produce sets of the same type η: $\eta+\eta=\eta$; $\eta\cdot\eta=\eta$; $(1+\eta)\eta=\eta$; $(\eta+1)\eta=\eta$; $(1+\eta+1)\eta=\eta$; $\eta\cdot\nu=\eta$; $\eta^\nu=\eta$. However, other combinations did not conform in type.[58] Assuming $\nu > 1$, then $1+\eta$, $\eta+1$, $\nu\cdot\eta$, and $1+\eta+1$ were all different in type from one another and from η itself. Though Cantor had shown that $\eta+1+\eta=\eta$, he added that $\eta+\nu+\eta\neq\eta$.

In elaborating the necessary and sufficient properties for any set to be of order type η, Cantor had managed to characterize completely and in general the type of order exhibited by a large number of simply-ordered sets of cardinality $\aleph_0$. Simply-ordered sets of higher cardinality, and especially sets equivalent in power to that of the continuum, were a special challenge. Among them was the type θ of the continuum. Though shortly after the appearance of the *Grundlagen* in 1883 he had managed to state the most general conditions determining simple order of the type η, he had not done so for the type θ.[59] Over the years he had investigated various ways of characterizing sets, and later denoted even their order types as being, for example, adherent, coherent, or perfect. But he had not published any general characterization of the type of order exemplified by the linear continuum. The last section of the *Beiträge,* Part I, provided just such an analysis. It was clearly the pinnacle of Cantor's theory of simply-ordered sets, and above all, it was the firm and specific clarification of the ordinal nature of continuous domains.

THE ORDER TYPE Θ OF THE LINEAR CONTINUUM

Before Cantor could proceed with any discussion of the type of order found on continuous sets, he had first to introduce an important feature of such sets that was missing from simply-ordered aggregates of denumerable power. Specifically, he had to express the idea of limit point for point sets but solely in the language of order. Earlier he had accomplished such a translation by introducing the concept of principal elements of simply-ordered transfinite sets in the unpublished *Principien*.[60] In the *Beiträge* he sharpened the concept considerably by introducing the idea of fundamental sequences to the discussion.

For simply-ordered sets M, it was clear that any subset of M was also an ordered set. In the attempt to study in greater detail the order type $\overline{M}$, he found that those subsets of M with types ω and $*\omega$ were particularly helpful. Such sets he labeled specifically as "fundamental sequences" [*Fundamentalreihen*] of the first order in M.[61] Ascending fundamental sequences were those of type ω, while $*\omega$ denoted the type of descending fundamental sequences. For ascending fundamental sequences $\{a_\nu\}$, $a_\nu \prec a_{\nu+1}$; for descending sequences $\{b_\nu\}$, $b_\nu \succ b_{\nu+1}$. Two ascending sequences were said to be coherent [*Zusammengehörig*], written $\{a_\nu\} \| \{a_\nu'\}$, if for every element a_ν there were elements a_λ' such that $a_\nu \prec a_\lambda'$; and also for every element a_ν' there were elements a_μ such that $a_\nu' \prec a_\mu'$. Equivalent formulations were given as well for the coherence of descending fundamental sequences. Finally two fundamental sequences, one ascending $\{a_\nu\}$, the other descending $\{b_\nu\}$, were *coherent,* written $\{a_\nu\} \| \{b_\nu\}$, if for all indices ν and μ, $a_\nu \prec b_\mu$, and if there was at most one element m_0 of M (but possibly none) such that for all indices ν, $a_\nu \prec m_0 \prec b_\nu$.

Coherence was a transitive relation; if $\{a_\nu\} \| \{b_\nu\}$ and $\{b_\nu\} \| \{c_\nu\}$, then $\{a_\nu\} \| \{c_\nu\}$. Moreover, whenever two fundamental sequences were either both ascending or descending, if one was contained as part of the other, they were necessarily coherent to each other:

> *Definition:* If there exists in M an element m_0 such that, with respect to the monotonically increasing fundamental sequence $\{a_\nu\}$,
>
> 1. $a_\nu \prec m_0$ for all ν;
> 2. For every element m of M such that $m \prec m_0$, there exists a number ν_0 such that $a_\nu \prec m$ for $\nu \geq \nu_0$;
>
> then we shall call m_0 a "limit element of $\{a_\nu\}$ in M," or a "principal element of M."[62]

Cantor offered a similar definition for the limit elements of descending fundamental sequences $\{b_\nu\}$ and noted that a fundamental sequence could have only one such element m_0 in M. (In general, M would have an unlimited number of such limit elements m_0.) Following the definitions, he went on in a series of theorems to show that all coherent fundamental sequences had the same limit elements, assuming such limits existed. Moreover, if M and M' were taken as two similarly ordered sets, so that $\overline{M} \simeq \overline{M}'$, then he was also able to conclude that every fundamental sequence in M had a corresponding fundamental sequence in M', and that the limit elements of one coincided with the limit elements of the other. At last he was equipped to draw a number of familiar and basic conclusions concerning the role limit elements might play in any simply-ordered set.

If all the elements of a given set M were principal elements, then the set was said to be dense-in-itself. Similarly, if every fundamental sequence in M had a limit element that was also contained in M, then M was said to be closed. Simply-ordered sets that were both dense-in-themselves and closed were called perfect sets. By virtue of the fact that if M were either dense-in-itself, closed, or perfect, then any set M' similar to M also shared such features, Cantor was able to conclude that the same could be said of their order types. Thus to such sets corresponded types of order which were said to be types dense-in-themselves, closed types, or perfect types. He was able to offer some familiar examples of each. The type η was dense-in-itself, and though it was also everywhere-dense, it was not a perfect type. Neither the type ω nor its inverse type $*\omega$ possessed limit elements. Both $\omega+\nu$ and $\nu+*\omega$ had a principal element, and both were in fact closed types. The type $\omega\cdot 3$ had two limit elements but was not closed, while $\omega\cdot 3+\nu$ had three principal elements and was a closed type. The most important example to which Cantor had been leading all along, however, concerned types that were both dense-in-themselves and perfect.

Initially, Cantor limited his analysis to the set of real numbers on the closed linear interval $X=[0,1]$. Drawing upon the elements of his theory of real numbers, he recalled that every fundamental sequence $\{x_\nu\}$ in X had a limit element x_0 in X, and conversely, every element of X was the limit element of some coherent fundamental sequence in X. Thus it was immediately clear that X was a perfect set and that θ was a perfect type.[63] He had shown as early as the *Grundlagen* that this was by no means a complete characterization of the type θ. X, as the set of all real numbers on $[0, 1]$, contained as a subset the set of all rational numbers R of type η, and thus it was true that between any two elements x_0 and x_1 of X, there were an infinite number of additional elements of X, which meant that X was also dense-in-itself.

Collecting all these properties together, Cantor claimed that quite apart from the specific examples of the linear continuum X, such properties were both sufficient and necessary to characterize the ordinal type θ of any linearly continuous domain. He formulated this as follows:

Theorem: If an ordered set M is of such a character that:

1. it is perfect;
2. it contains a set S of the cardinality $\overline{\overline{S}} = \aleph_0$, which is so related to M that between any two elements m_0 and m_1 of M elements of S occur;

then $M = \theta$.[64]

Cantor's proof proceeded along lines very similar to those he had laid out in characterizing sets of type η. The problem, as before, was to show that the set M could be corresponded with the linear continuum X in such a way that both sets were similar in terms of their respective orderings.

He began by assuming that S had no largest or smallest element (if such did exist, they could be discarded without altering the property of S described in condition 2) and thus was of type η, since it was both everywhere-dense and denumerably infinite, that is, $\overline{\overline{S}} = \aleph_0$. Consequently S and R (the set of all rational numbers of type η) were similar. Taking any mapping of R onto S, Cantor then set out to show how such a correspondence produced a definite mapping of X onto M.

To all elements of X which belonged to R, Cantor corresponded those elements of M which were also elements of S. Supposing x_0 to be an element of X which did not belong to R, x_0 could be considered as a limit element of some fundamental sequence $\{x_\nu\}$ of X, and this sequence then had a corresponding coherent fundamental sequence $\{r_{\kappa_\nu}\}$ in R. To $\{r_{\kappa_\nu}\}$ corresponded as image a fundamental sequence $\{s_{\lambda_\nu}\}$ in S and M which, because M was taken to be a perfect set, was limited by some element m_0 of M that did not belong to the set S itself.

Cantor then let this element m_0 of M be the image of x_0 in X. Conversely, to every m_0 in M which did not occur in S, there corresponded a definite element x_0 of X which was not contained in R and of which m_0 was the image. Thus he had managed to provide a one-to-one correspondence between M and X. It was still necessary to show that the two sets under the given correspondence were similar.

For those elements of X which belonged to R, and those of M belonging to S, this was immediately true. Therefore, the only case of interest involved those in which one considered an element r of R and an element x_0 of X, where x_0 did not belong to R.

Cantor took s and m_0 in M to correspond with r and x_0 in X. If $r < x_0$, then there was a fundamental sequence of elements $\{r_{\kappa_\nu}\}$ which had x_0 as limit element and such that, from a certain index ν_0 on, $r < r_{\kappa_\nu}$ for $\nu \geq \nu_0$. Thus the image of $\{r_{\kappa_\nu}\}$ in M was an ascending fundamental sequence $\{s_{\lambda_\nu}\}$, which was limited by some m_0 of M, and therefore, $s_{\lambda_\nu} \dashv m_0$ for every value of ν, and yet $s \dashv s_{\lambda_\nu}$ for all values $\nu \geq \nu_0$.

It thus followed from the basic character of the order types of simply-ordered sets that $s \prec m_0$. If $r > x_0$, then it was similarly true that $s \succ m_0$. Finally, if two elements x_0 and x_0' of X were assumed *not* to belong to R, Cantor took their corresponding elements in M to be m_0 and m_0'. As before, he could argue that if $x_0 < x_0'$, then $m_0 \prec m_0'$. At last he could assert the truth of his original claim: $\overline{M} = \theta$.

Part I of the *Beiträge* had come to an end on a strong and significant note.

CONCLUSION

Cantor's earliest characterization of the continuum was the distinction he introduced in 1874: the real numbers were nondenumerable. This example was concrete, produced in terms of the set of real numbers on linear intervals. But at that same time there was nothing to suggest the structural properties of continua that made them special. A decade later, when the *Grundlagen* made its appearance, Cantor distinguished numerous special properties of point sets and in fact defined continua in the most general terms then possible: as sets that were both perfect and connected. But the result was confined to point sets, and the definition involved the metric concept of distance between points in order to determine the connectedness of the sets in question. As he came to draft the *Principien* (which was never published in his lifetime), he began to suggest the advances made possible by a general consideration of order types, concepts generalized from point sets but free from those properties that were unnecessary for an abstract study of continuity.[65] He had yet to reach any detailed conclusions concerning the specific type of order represented by the linear continuum as contrasted with other kinds of serial order. A decade later, the *Beiträge* brought such ideas to an impressive conclusion by offering as detailed an account of the continuum as Cantor was capable of giving.

It was two years before Cantor produced the sequel to the *Beiträge's* Part I. Sometime in 1895 he discovered the first of the paradoxes of set theory, specifically the paradox of the largest ordinal and cardinal numbers.[66] Presumably, he had come upon them in the course of trying to establish his comparability theorem for transfinite cardinal numbers, and in the attempt to deal with the related questions of whether every transfinite power was necessarily an aleph, and whether every set could in fact be well-ordered. Consideration of the complex connections between discovery of the antinomies and the problem of Cantor's well-ordering theorem, including the diverse implications of each for all of Cantorian set theory, is found in Chapter 11. But it is significant that Cantor first began to worry about such matters in 1895. This may help to account for his delay in forwarding the second half of the *Beiträge* to Felix Klein for printing. It may also explain why Part I made few references to well-ordered sets and went no further than to introduce the first transfinite aleph, $\aleph_0$.

Discussion of simply-ordered types, however, introduced Cantor's readers to a basic concept of set theory without raising the controversial problems of well-ordered sets in general, and the paradoxes to which such ideas led, as indicated above. Even if he could not yet specify the power of the continuum as being $\aleph_1$, he could at least offer a complete description of its type θ.

It was already clear in 1895 that the *Beiträge,* Part I, was an outstanding contribution to the mathematical literature of its time. Whatever its shortcomings, whatever problems it was unable to resolve, it was still an impressive achievement. It gave in a careful and polished form the many theoretical advances which Cantorian set theory had made over the preceding twenty-five years. Above all, it succeeded in dealing with the theory of order in great generality. In Part II, Cantor presented the very important particularization of the general theory, and offered a survey, though in some respects an imperfect one, of his special theory of well-ordered sets.

CHAPTER 9

The *Beiträge*, Part II: The Study of Well-Ordered Sets

Although Cantor had nearly completed Part II of his *Beiträge zur Begründung der transfiniten Mengenlehre* within six months of the publication of Part I, the sequel did not appear until 1897.[1] He was determined to verify his continuum hypothesis, establish the equality $2^{\aleph_0} = \aleph_1$ and include this result in the second half of the *Beiträge*.[2] Thus there is little doubt that one of Cantor's major mathematical preoccupations following the publication of Part I in 1895 was the continuum hypothesis and that his failure to produce a solution contributed to his delay of more than a year and a half in bringing the second part into print. By March 1897, despite the lack of any satisfactory proof that the power of the continuum was equal to $\aleph_1$, Cantor decided to wait no longer, and he published the second half of the *Beiträge* in the *Mathematische Annalen*.[3]

Part II presented the bulk of Cantor's important theory of the transfinite ordinal and cardinal numbers, which would carry his readers beyond $\aleph_0$ to $\aleph_1$, the first of the nondenumerable, transfinite *alephs*. In Part II he introduced his special theory of well-ordered sets, whose order types were the finite and transfinite ordinal numbers. $\aleph_1$, the power of the second number class, was defined and a good deal of transfinite arithmetic was explored. In due course Cantor also introduced the process of transfinite induction to make possible the definition of transfinitely many multiplications and transfinite exponentiation. But despite the seeming thoroughness of it all, the special results of transfinite arithmetic did not compensate for the major flaw in Cantor's entire presentation. The continuum hypothesis remained unresolved, as did the questions of whether every transfinite power was an aleph, whether the transfinite cardinal numbers were all comparable, and whether every set could actually be well-ordered. Throughout the entire presentation, in fact, there was a nagging sense that something was not quite right. Though in Part I of the *Beiträge* Cantor had enthusiastically noted that from $\aleph_0$ there proceeded an unlimited sequence of cardinal numbers:

$$\aleph_0, \aleph_1, \aleph_2, \ldots, \aleph_\nu, \ldots,$$

the sequence was never mentioned again.[4] He had also promised in Part I to establish the existence of even greater cardinal numbers, $\aleph_\omega$, $\aleph_{\omega+1}$, ..., and to show that they formed a well-ordered succession without any end whatsoever. But by 1897 he offered neither his proof for the existence of $\aleph_\omega$, nor did he even consider any transfinite order types. He was also careful to deal only with certain well-ordered sets and never with the entirety of any well-ordered set in general. Such restrictions were clear indications that serious difficulties were yet to be satisfactorily resolved, ones which were fundamental and rooted in the very nature of Cantor's entire theory of transfinite sets.

Nevertheless, Part II forged ahead with confidence that any imperfections could eventually be removed. Cantor was certain that the major outline of his theory was certain and true, and therefore he went on to explain the very important fundamentals of his theory of well-ordered sets and their application in determining the transfinite cardinal number $\aleph_1$. As a final contribution, he offered an impressive analysis of the elaborate number theory which complemented his entire formulation of the transfinite numbers.

WELL-ORDERED SETS

As early as the *Grundlagen* Cantor had realized that well-ordered sets were of unique importance to the theory of transfinite numbers. The ordinal types of such sets constituted, in fact, the finite and transfinite ordinal numbers, without which it was impossible to make any rigorous advances in transfinite set theory. Consequently, Cantor devoted all of Part II of the *Beiträge* to the detailed theory of transfinite ordinal numbers and the higher cardinals or transfinite alephs which corresponded to infinite sets. The first step was to define the concept of well-ordered sets, introduced as special kinds of simply-ordered sets:

> *Definition:* We call a simply-ordered set F *well-ordered* if its elements f increase (monotonically) from a least element f_1 in a definite succession so that the following two conditions are fulfilled:
> 1. There is a *least* element f_1 in F;
> 2. If F' is any subset of F and if F consists of one or more elements greater than all elements of F', then there is an element f' of F which follows *next* after the entire set F', so that no elements in F exist which fall between F' and f'.[5]

Cantor stated immediately, in a note, that this definition was exactly the same as that given in the *Grundlagen,* if not word for word, certainty in substance.

Among ordinal numbers, the sequence most prominent in Cantor's set theory in Part II of the *Beiträge* was the well-ordered succession:

$$1,2,3, \ldots, \omega, \omega+1, \ldots, 2\omega, \ldots, \nu\omega, \ldots, \omega^2, \ldots, \omega^\omega, \ldots, \omega^{\omega^\omega}, \ldots.$$

In the *Grundlagen* this sequence of finite and transfinite ordinal numbers had been generated by appeal to both a first and second principle of generation, and divisions were shown to exist naturally in the sequence by appeal to a limiting process which Cantor used to define the second and higher number classes. It was in terms of the second number class (II) that Cantor eventually defined the second transfinite power or cardinal number $\aleph_1$. Similarly, the second number class (II) assumed special significance in the early sections of the *Beiträge*, Part II.

In order to illustrate the concept of well-ordered sets directly, Cantor offered three examples:

$$(a_1, a_2, a_3, \ldots, a_\nu, \ldots)$$
$$(a_1, a_2, \ldots, a_\nu, \ldots; b_1, b_2, \ldots, b_\mu, \ldots),$$
$$(a_1, a_2, \ldots, a_\nu, \ldots; b_1, b_2, \ldots, b_\mu, \ldots; c_1, c_2, c_3),$$

where $a_\nu \prec a_{\nu+1} \prec b_\mu \prec b_{\mu+1} \prec c_1 \prec c_2 \prec c_3$.[6]

In the first and second examples neither set had a greatest element, while in the third case c_3 was largest. In the first case, every element had a successor, and there was no element without a predecessor. In the second, b_1 had no predecessor, and in the third case, neither b_1 nor c_1 had immediate predecessors. Later Cantor would appeal to these properties of certain elements of any well-ordered set in order to introduce concepts comparable to the first and second principles of generation used to produce well-ordered sets in the *Grundlagen*. Cantor extended the notation $\prec$ to sets, indicating that all elements of one preceded every element of the other. Thus $M \prec N$ or $N \succ M$ meant simply that in terms of their respective orderings every element of M preceded, or was lower in rank to, every element of N.

Cantor then went on to establish that every subset F_1 of a well-ordered set F had a lowest element. Similarly, if a simply-ordered set F was so constituted that it, as well as every one of its subsets, had a lowest element, then F was itself a well-ordered set. Consequently, every subset F_1 of a well-ordered set F was itself well-ordered. Moreover, he was able to show that every set that was similar to a well-ordered set was also a well-ordered set, and finally that if every element of a well-ordered set was replaced by a well-ordered set, then the resulting collection was likewise well-ordered. But these were consequences which followed directly from the concept of similarity of simply-ordered sets when applied to well-ordered sets. Before Cantor could advance to a more satisfying definition of ordinal numbers than had been possible in the *Grundlagen*, he needed the concept of segments [*Abschnitte*] of well-ordered sets.

Definition: If f is an element of the well-ordered set F *different* from the first element f_1 [*Anfangselement*], then we want to call the set A of all elements of F which $\dashv f$ a "segment of F," and in fact the segment of F *determined by the element* f. On the other hand, let the set R of all remaining elements of F (including the element f) be called a "remainder of F," and in fact the remainder of F *determined by the element* f. By Theorem C (Section 12), it follows that the sets A and R are well-ordered, and we can write (according to Sections 8 and 12):

$$F = (A, R),$$
$$R = (f, R'),$$
$$A \dashv R.$$

R' is the part of R following after the first element f, and is reduced to 0 if R has no other element than f.[7]

In order to make the definition clearer, he offered the example of a well-ordered set F introduced earlier:

$$f = (a_1, a_2, \ldots, a_\nu, \ldots; b_1, b_2, \ldots, b_\mu, \ldots; c_1, c_2, c_3),$$

where a_3 determined the segment (a_1, a_2) and the remainder $(a_3, a_4, \ldots; b_1, b_2, \ldots; c_1, c_2, c_3)$; b_1 determined the segment $(a_1, a_2, \ldots, a_\nu, \ldots)$ and the remainder $(b_1, \ldots, b_\mu, \ldots; c_1, c_2, c_3)$; c_2 determined the segment $(a_1, a_2, \ldots, a_\nu, \ldots; b_1, b_2, \ldots; c_1)$ and the remainder (c_2, c_3).

Inequalities were introduced between segments as follows. Given two segments A and A' determined by two elements f and f', respectively, of F, A' was said to be a segment of A if $f' \dashv f$. In such cases, A' was said to be the lesser segment, written $A' < A$. Clearly, for every segment A of F, $A < F$.

Cantor went on to explain that if two segments A and B of similar sets F and G were similar, then the two elements f and g determining the similar segments A and B, respectively, were mapped to one another under the necessarily unique correspondence of two similar, well-ordered sets.[8] He then gave:

Theorem B: A well-ordered set F is similar to none of its segments A.[9]

From this Theorem B and the fact that two different segments of any well-ordered set could not be similar to each other, it followed that given two similar well-ordered sets F and G, they could be corresponded in only a single, unique way.[10] Of particular importance for what followed were Cantor's Theorems K, L, and M:

Theorem K: If for every segment A of a well-ordered set F there is a segment B, similar to A, of another well-ordered set G, and conversely, to every segment B of G there is a similar segment A of F, then $F \simeq G$.[11]

Theorem L: If for every segment A of a well-ordered set F there is a segment B, similar to A, of a well-ordered set G, and if, on the other hand, there is at least one segment of G to which there is no similar segment of F, then there is a definite segment B_1 of G such that $B_1 \simeq F$.[12]

Theorem M: If the well-ordered set G has at least one segment to which there is no similar segment in the well-ordered set F, then every segment A of F must have a similar segment B in G.[13]

The major result of Section 12 of the *Beiträge* was then introduced as Theorem N. In terms of segments, Cantor was able to state clearly the relations that were possible under similar correspondences between any two well-ordered sets F and G. The results of this theorem were translated directly into important conclusions concerning the order of any two ordinal numbers in general, once they had been defined in the next section of the *Beiträge*. Cantor's theorem stating the similarity of any two well-ordered sets read as follows:

Theorem N: If F and G are any two well-ordered sets, then either:
1. F and G are similar; or
2. There exists a definite segment B_1 of G which is similar to F; or
3. There is a definite segment A_1 of F which is similar to G;

each one of these three cases excludes the possibility of the other two.[14]

The proof considered one by one the three ways in which F and G might be related. Theorem M excluded the possibility of there being both a segment of F with no similar segment in G, and a segment of G with no similar segment in F. Only three other possibilities remained:

1. To every segment A of F there was a similar segment B of G, and conversely; then by Theorem K it followed that $F \simeq G$.

2. To every segment A of F there was a corresponding similar segment B in G, but the converse was not true. Then there had to be at least one segment of G to which there was no corresponding segment in F. Thus by Theorem L there was a segment B_1 of B such that $B_1 \simeq F$.

3. The final possibility involved the existence of at least one segment in F which had no similar corresponding segment in G, though to every segment B in G there was a correspondingly similar segment A in F. But then there was a definite segment A_1 of F such that $A_1 \simeq G$.

To show that each of these possibilities was mutually exclusive, Cantor argued that if $F \simeq G$ and $F \simeq B_1$ simultaneously, then $G \simeq B_1$, which would have been at odds with Theorem B; for the same reason $F \simeq G$ and $G \simeq A_1$ were not simultaneously possible. Finally, it was not possible to have both $F \simeq B_1$ and $G \simeq A_1$, since $F \simeq B_1$ implied the existence of a segment B_1' of B_1 such that $A_1 \simeq B_1'$. But then $G \simeq B_1'$, which would again have been contrary to the conclusion of Theorem B.

Theorem N would serve in a major capacity as soon as the ordinal numbers of well-ordered sets had been defined. Moreover, the segments of well-ordered sets would again reappear in helping define the ordinal numbers of the second class, the types of denumerably infinite well-ordered sets. All the conclusions involving segments and the similarity of well-ordered sets were essential to the introduction of unique transfinite ordinal numbers which, themselves, formed a well-ordered set.

THE ORDINAL NUMBERS OF WELL-ORDERED SETS

The ordinal numbers of well-ordered sets were introduced just as they had been for simply-ordered sets: by abstracting all individual properties of the elements of a given set M, while allowing them to retain their original order, thereby defining a sequence of pure units whose order defined the order type $\overline{M}$ of M. Only sets similar to each other could have the same order type; the type of a well-ordered set F was called its ordinal number. Given any two sets F and G, such that $\overline{F}=\alpha$ and $\overline{G}=\beta$, then Theorem N concerning the relations possible between segments of similar sets ensured that only three mutually exclusive possibilities could occur.[15] Either:

1. $F \simeq G$, in which case $\alpha=\beta$.
2. G contained a definite segment B_1 such that $F \simeq B_1$; then $\alpha<\beta$.
3. There was a definite segment A_1 of F such that $G \simeq A_1$; then $\alpha<\beta$.

Moreover, because of the comparability of segments, it followed immediately that if α and β were any two ordinal numbers, then exactly one of three possibilities was necessarily true: either $\alpha<\beta$, $\alpha=\beta$, or $\alpha<\beta$. The nature of segments also ensured that such relations were transitive: among any three ordinal numbers, if $\alpha<\beta$ and $\beta<\gamma$, then $\alpha<\gamma$, and it followed that all of the ordinal numbers, taken in their order of magnitude, constituted a simply-ordered set.

Arithmetic operations, Cantor noted, when applied to the ordinal numbers, were the same as those already defined for the order types of simply-ordered sets. Thus, for any two well-ordered sets F and G such that $\alpha=\overline{F}$ and $\beta=\overline{G}$, $\alpha+\beta=(\overline{F,G})$. Since the union set (F,G) was clearly a well-ordered set whenever both F and G were well-ordered, Cantor concluded that the sum of any two ordinal numbers was also an ordinal number. Since F was a segment of (F,G), Cantor could also conclude that $\alpha<\alpha+\beta$.[16] Since G, however, could not be taken as the remainder of G, it was entirely possible that G might be similar to the union set (F,G) itself. Otherwise, G was necessarily similar to some segment of (F,G) and thus it was always true that $\beta \leq \alpha+\beta$.

Multiplication was also introduced exactly as before, by substituting a set F_g of type α for every element g of a set G of type β, thereby defining a new set H which determined the product $\alpha\cdot\beta$. Cantor noted that the order type of H was completely determined by the separate types α and β and that since H was clearly a well-ordered set whenever F and G were, then $\alpha\cdot\beta=\overline{H}$. Thus he could conclude that the product of two ordinal numbers was always an ordinal number.

Cantor had to define subtraction rather carefully, just as he had in the *Grundlagen,* since for transfinite ordinal numbers arithmetic operations were not in general commutative.[17] Assuming for any two ordinals α and β that $\alpha<\beta$, Cantor showed that there was always a number, to be called $\beta-\alpha$, satisfying the equation $\alpha+x=\beta$. That such differences were uniquely and definitely determined followed from the properties of a segment B in the well-ordered set G such that $\overline{G}=\beta, \overline{B}=\alpha$. The corresponding remainder in G defined by the segment B was denoted S; then $G=(B,S)$ and $\beta=\alpha+\overline{S}$; consequently $\beta-\alpha=S$. Since the segment B of G was uniquely determined, then so too was the remainder S and its corresponding order type, in this case the difference $\beta-\alpha$. In terms of the arithmetic rules governing subtraction, Cantor could then write that:

$$(\gamma+\beta)-(\gamma+\alpha)=\beta-\alpha;$$
$$\gamma(\beta-\alpha)=\gamma\,\beta-\gamma\,\alpha.$$

Both formulae were true for any γ, assuming that $\beta>\alpha$.

The fact that any infinity of ordinal numbers produced sums which depended upon the order of the summands themselves was also a consequence of the noncommutative nature of transfinite ordinal arithmetic. Such sums, however, were uniquely determined.[18] To show this, Cantor considered any denumerable sequence of ordinal numbers $\beta_1,\beta_2, \ldots,\beta_\nu, \ldots$, stipulating that $\beta_\nu=G_\nu$, and thus $G=(G_1, G_2, \ldots, G_\nu, \ldots)$ was also a well-ordered set representing the sum of the denumerably many ordinals β_ν. Consequently $\beta_1+\beta_2+\ldots+\beta_\nu+\ldots=\overline{G}=\beta$. The distributive law for ordinals could be applied as follows: $\gamma\cdot(\beta_1+\beta_2+\ldots+\beta_\nu+\ldots) = \gamma\cdot\beta_1+\gamma\cdot\beta_2+\ldots+\gamma\cdot\beta_\nu+\ldots$. Cantor then put $\alpha_\nu=\beta_1+\beta_2+\ldots+\beta_\nu$, where $\alpha_\nu=\overline{(G_1, G_2, \ldots, G_\nu)}$. Since $\alpha_{\nu+1}>\alpha_\nu$, he could appeal to the definition of subtraction to express the ordinals β_ν in terms of α_ν: $\beta_1=\alpha_1$; $\beta_{\nu+1}=\alpha_{\nu+1}-\alpha_\nu$. Thus the sequence $\alpha_1,\alpha_2, \ldots,\alpha_\nu, \ldots$ represented any infinite sequence where the only requirement was that $\alpha_{\nu+1}>\alpha_\nu$. Such a sequence Cantor called a "fundamental sequence"[19] following the idea as discussed in detail in Section 10 of the *Beiträge,* Part I. Cantor first made clear the relations subsisting between the sum β and the ordinal numbers α_ν:

1. The ordinal number β was greater than any of the ordinals α_ν, for all values of the index. This followed immediately from the fact that the set $(G_1, G_2, \ldots, G_\nu)$ with ordinal α_ν was a *segment* of the entire set G of ordinal β. Likewise,

2. Given any ordinal β' less than β, then for all sufficiently large values of the index ν, $\alpha_\nu>\beta'$. This followed from the fact that since $\beta'<\beta$, there was a segment of G with ordinal β'. But the element which

determined this segment had to be contained in some segment of G, and Cantor took this to be G_{ν_0}. Then B' was clearly a segment of $(G_1, G_2, \ldots, G_{\nu_0})$ and therefore $\beta' < \alpha_{\nu_0}$. For all indices $\nu \geq \nu_0$, it was consequently true that $\alpha_\nu > \beta'$.

By combining these two characterizations of the relation between β and the elements of the fundamental sequence $\{\alpha_\nu\}$, Cantor was prepared to introduce the definition of limit elements among sequences of ordinal numbers.[20] The fact that β was the ordinal number following next after all the ordinals α_ν made it quite natural to define β as the "limit" of the sequence of ordinals α_ν. Thus he could denote such limits as follows:

$$(9.1)\quad \operatorname*{Lim}_{\nu} \alpha_\nu = \alpha_1 + (\alpha_2 - \alpha_1) + (\alpha_3 - \alpha_2) + \ldots + (\alpha_{\nu+1} - \alpha_\nu) + \ldots .$$

In summary, he formulated the preceding conclusions in a single theorem:

> *Theorem I:* To every fundamental sequence $\{\alpha_\nu\}$ of ordinal numbers there corresponds an ordinal number $\operatorname*{Lim}_{\nu} \alpha_\nu$, which in magnitude follows immediately after all the α_ν; they are represented by the formula (9.1).[21]

For any given ordinal γ it was then a direct conclusion from earlier results that: $\operatorname*{Lim}_{\nu} (\gamma + \alpha_\nu) = \gamma + \operatorname*{Lim}_{\nu} \alpha_\nu$; $\operatorname*{Lim}_{\nu} \gamma \cdot \alpha_\nu = \gamma \cdot \operatorname*{Lim}_{\nu} \alpha_\nu$.

Cantor was at last prepared to give a thorough treatment of the ordinal numbers of the second number class (II). First he recalled a statement from Part I of the *Beiträge,* where he had mentioned that all simply-ordered sets of a given finite cardinality ν were identical in order type.[22] He now believed he could prove the assertion, based on the fact that every finite set, if simply-ordered, was also well-ordered. Thus the types of simply-ordered, finite sets were the finite ordinal numbers. Moreover, two distinct ordinals α and β could not belong to the same finite cardinal number ν. Supposing $\alpha < \beta$, and $\overline{G} = \beta$, then there was a segment B of G such that $\overline{B} = \alpha$. If both were of equal cardinality, then B would have the same cardinal number ν as one of its parts, B. The impossibility of such a situation had already been established in the *Beiträge,* Part I, and hence he could immediately conclude that the properties of the finite ordinal numbers must necessarily coincide with those of the finite cardinals. The case was very different, however, for transfinite ordinals, as he had emphasized on numerous previous occasions.

For transfinite sets, Cantor described the relation between the cardinal and ordinal numbers for well-ordered sets as follows:

> It is entirely different with transfinite *ordinal numbers;* to one and the same transfinite cardinal number a there is an infinite number of ordinal numbers, which comprise a homogeneous, coherent [*einheitliches zusammenhängendes*] system, which we call the "number class $Z(\mathfrak{a})$." It is a part of the type class $[\mathfrak{a}]$.[23]

Cantor had already used the first number class (I) of finite ordinal numbers ν to determine the first transfinite cardinal number $\aleph_0$. Similarly, in order to introduce the second transfinite cardinal number $\aleph_1$, he had to establish securely the set of transfinite ordinals comprising the second number class (II). Thus, the major focus of attention for all that followed in Part II of the *Beiträge* concerned the second number class $Z(\aleph_0)$. It was in terms of the development of the higher ordinal numbers that Cantor was to go well beyond the earliest foundations of the number theory of the transfinite numbers as sketched in the *Grundlagen*. The last sections of the *Beiträge* were remarkable for their generality and for their presentation of concepts that would not find immediate appreciation among mathematicians but would eventually prove to be of the utmost significance.

THE NUMBERS OF THE SECOND NUMBER CLASS $Z(\aleph_0)$

In the *Grundlagen* Cantor had gone to great lengths to justify his introduction of the first transfinite number. In terms that were not entirely satisfying, he advanced the new concept ω as the first number following all the finite numbers $1, 2, 3, \ldots$. Succeeding ordinals were then produced by means of his first and second principles of generation, but he was not convinced of the logical soundness of this approach. Within a year of the *Grundlagen's* appearance he had decided that the concept of order type provided the surest means of introducing the transfinite numbers.[24] Except for Sections I and VIII of his *Mittheilungen zur Lehre vom Transfiniten,* however, he had never published any account of this new theory. Moreover, the brief exposition of the *Mittheilungen* was not calculated to meet the demands of a thorough mathematical presentation. Not only was the *Mittheilungen* published in a journal devoted to philosophy, it also failed to offer the details of how the concept of order types of well-ordered sets could be used to produce a rigorous theory of transfinite *cardinal* numbers.[25] Therefore, for a strictly mathematical readership, Part II of the *Beiträge* was Cantor's first publication to provide in full detail his theory of transfinite numbers based upon the concept of order types. Unlike the *Grundlagen,* where various principles of generation had been used to produce the higher number classes of transfinite ordinals, the *Beiträge* defined the second number class as follows:

> *Definition:* The second number class $Z(\aleph_0)$ is the totality [*Gesamtheit*] $\{\alpha\}$ of all order types α of well-ordered sets of cardinality $\aleph_0$.[26]

Since Cantor relied upon the entire class $Z(\aleph_0)$ to define the transfinite power $\aleph_1$, it was imperative that the well-ordered character of $Z(\aleph_0)$ be established. Otherwise, he could never proceed beyond definition of $\aleph_0$ which had also relied upon the well-ordered character of the natural sequence of positive integers for its own definition. With this in mind, he produced a sequence of theorems leading to the conclusion that $Z(\aleph_0)$ was in fact well-ordered.

> *Theorem A:* The second number class has a smallest number $\omega = \operatorname{Lim}_{\nu} \nu$, $\nu = 1, 2, 3, \ldots$.[27]

By ω Cantor understood the type of the well-ordered set

$$F_0 = (f_1, f_2, \ldots, f_\nu, \ldots),$$

where ν assumed all finite ordinal values, and $f_\nu \dashv f_{\nu+1}$. Consequently $\omega = \overline{F}_0$, and it followed from Part I of the *Beiträge* that $\overline{\overline{\omega}} = \aleph_0$. Thus ω was clearly an ordinal number of the second number class (II). That ω was also the least such number followed from consideration of any ordinal number γ less than ω. Such an ordinal γ had to correspond to some segment of F_0. But the only segments of F_0 were of the form $A = (f_1, f_2, \ldots, f_\nu)$, with finite index and therefore *finite* ordinal number ν. Thus $\gamma = \nu$. As a result, there could be no transfinite numbers less then ω, and therefore ω had to be the least of all transfinite ordinal numbers. By virtue of the way in which $\operatorname{Lim}_{\nu} \alpha_\nu$ was defined in Section 14 of the *Beiträge* (Part II), it followed that $\omega = \operatorname{Lim}_{\nu} \nu$.

Making similar appeals to the nature of segments, Cantor went on to show that if α was any number of the second number class, then $\alpha + 1$ was the number next following, with no ordinal numbers falling between the two.[28] It was also clear (assuming the Axiom of Choice, as Cantor did) that given any fundamental sequence of numbers of the first or second number class, then the number defined as $\operatorname{Lim}_{\nu} \alpha_\nu$ following next in order of magnitude was also a number of the second number class.[29]

Exactly as before, for simply-ordered sets, Cantor then introduced the conditions under which two fundamental sequences $\{\alpha_\nu\}$ and $\{\alpha'_\nu\}$ of numbers of the first or second number class could be said to be coherent,[30] written $\{\alpha_\nu\} \| \{\alpha'_\nu\}$. For every ν he required that there be finite ordinal numbers λ_0 and μ_0 such that $\nu'_\lambda > \alpha_\nu$ for $\lambda \geq \lambda_0$, and $\alpha_\mu > \alpha'_\nu$ whenever $\mu \geq \mu_0$. Following this characterization of coherent fundamental sequences, he could translate all the results of his study of segments of well-ordered sets directly into the language of the limits of fundamental sequences:

> *Theorem D:* The two limit numbers $\operatorname{Lim}_{\nu} \alpha_\nu$ and $\operatorname{Lim}_{\nu} \alpha'_\nu$ corresponding to two fundamental sequences $\{\alpha_\nu\}$, $\{\alpha'_\nu\}$, are equal if and only if $\{\alpha_\nu\} \| \{\alpha'_\nu\}$.[31]

This theorem was subsequently used to prove several results needed to establish his claim that the set of all transfinite ordinal numbers was in fact well-ordered. First he noted that if α were any number of the second number class, and ν_0 any finite ordinal number, then it was always true that: $\nu_0 + \alpha = \alpha$; $\alpha - \nu_0 = \alpha$; $\nu_0 \omega = \omega$; $(\alpha + \nu_0)\omega = \alpha\omega$.[32]

Having already shown that the set $Z(\aleph_0)$ was simply-ordered, he then took another step toward proving that it was also well-ordered:

> *Theorem H:* If α is any number of the second number class, then the totality $\{\alpha'\}$ of all numbers α' of the first and second number classes which are less than α comprises, in order of magnitude, a well-ordered set of type α.[33]

Cantor's proof involved segments of the aggregate F which he took to be well-ordered from the start. By assuming that $F = \alpha$, he denoted f_1 as the least element of F. Taking α' as any ordinal number less than α, then there was a segment A' of F such that $A' = \alpha'$. Moreover, every segment A' of F determined by its type $A' = \alpha'$ an ordinal number of the first or second number class that was always less than α. Since $\overline{F} = \aleph_0$, then the cardinality of any segment of F, $\overline{\overline{A'}}$, was either finite or at most $\aleph_0$.

The segment A' of F was determined by some $f' \dashv f_1$ in F. Likewise, every element $f' \dashv f_1$ in F determined a segment A'. Letting f' and f'' be two elements of F, both greater then f_1, and assuming that $f' \dashv f''$, then it followed that, for their corresponding segments, $A' < A''$, and for their corresponding order types, $\alpha' < \alpha''$ respectively. By putting $F = (f_1, F')$, and by making the element f' of F correspond with the element α' of $\{\alpha'\}$, then a mapping of the two aggregates was produced. Therefore, $\overline{\{\alpha'\}} = \overline{F'}$. By virtue of the nature of subtraction among transfinite ordinal numbers, $\overline{F'} = \alpha - 1 = \alpha$. Consequently, $\overline{\{\alpha'\}} = \alpha$.

Cantor therefore concluded that the collection of ordinal types preceding any transfinite ordinal number was itself well-ordered. As a result, the sequence of all such order types could be represented by an ordinal number, while a definite power or cardinal number could also be ascribed to it, uniquely. But there were several additional consequences of Theorem H that Cantor could establish by adding the simple observation that since $\overline{\overline{\alpha}} = \aleph_0$, then it was also true that $\overline{\overline{\{\alpha'\}}} = \aleph_0$. Thus the power of the set $\{\alpha'\}$ of numbers α' of the first and second number classes which were less than a given number α of the second number class was of cardinality $\aleph_0$. Moreover, there was an important conclusion to be drawn about the method by which each number of the second number class could be said to have been generated:

> *Theorem K:* Every number α of the second number class is of such a nature that it either arises from a next smaller predecessor α_{-1}: $\alpha = \alpha_{-1} + 1$; or a fundamental sequence $\{\alpha_\nu\}$ of numbers of the first or second number class can be given, such that $\alpha = \operatorname{Lim}_{\nu} \alpha_\nu$.[34]

The first case followed directly if one assumed that $\alpha = \overline{F}$, and that F contained an element g that was greater in terms of order than all other elements of F. Then g determined a segment in F, and $F = (A, g)$. Consequently, $\alpha = \overline{A} + 1 = \alpha_{-1} + 1$, and the ordinal number corresponding to the segment A produced the number next smaller than α denoted by Cantor in the theorem as α_{-1}.

The case in which F was assumed to contain no largest element required a more complicated proof. If one considered the set of all ordinals of the first and second number classes less than α, Cantor had already shown in Theorem H that $\{\alpha'\}$ was similar to the set F and hence there was no largest element to be found in $\{\alpha'\}$. He had also shown that $\{\alpha'\}$ was denumerable; thus it could be ordered in the form of a simply infinite sequence. Whatever this order might be, by beginning with α_1', though α_2', α_3'', . . . might be less than α_1', there would eventually come terms of $\{\alpha_\nu'\}$ that were greater than α_1'. Otherwise, α_1' would be the largest of all elements of the set $\{\alpha'\}$, and this would contradict the theorem's original assumptions. In this manner, beginning from α_1' and proceeding to the next larger, and then successively larger elements of $\{\alpha'\}$, Cantor produced a fundamental sequence $\alpha_1', \alpha_{\rho_2}', \alpha_{\rho_3}', \ldots, \alpha_{\rho_\nu}', \ldots$, where the following inequalities always held: $1 < \rho_2 < \rho_3 < \ldots < \rho_\nu < \rho_{\nu+1} < \ldots$; $\alpha_1' < \alpha_{\rho_2}' < \alpha_{\rho_3}' > \ldots < \alpha_{\rho_\nu}' < \alpha_{\rho_{\nu+1}}' < \ldots$; $\alpha_\mu' < \alpha_{\rho_\nu}'$ if $\mu < \rho_\nu$.

Since $\nu \leq \rho_\nu$, it was always true that $\alpha_\nu' \leq \alpha_{\rho_\nu}'$. Consequently it was clear that for every number α_ν' (and thus for every $\alpha' < \alpha$) there were always numbers α_{ρ_ν}' greater than α_ν' for sufficiently large values of the index ν. However, since α was taken as the number next following in magnitude all preceding numbers α', it was also the number next largest following all the numbers α_{ρ_ν}'. Cantor then put $\alpha_1' = \alpha_1$, $\alpha_{\rho_{\nu+1}}' = \alpha_{\nu+1}$, and concluded directly that $\alpha = \operatorname{Lim}_\nu \alpha_\nu$.

He was therefore able to show that the ordinal numbers of the second number class arose in two distinct ways. There were those which proceeded directly from a predecessor by addition of a unit; these were called numbers of the first kind, for which $\alpha = \alpha_{-1} + 1$. Numbers of the second kind were preceded by fundamental sequences so that $\alpha = \operatorname{Lim}_\nu \alpha_\nu$. Cantor distinguished between these two modes of generation for transfinite ordinal numbers by naming them, respectively, "the first and second principles of generation of the numbers of the second number class."[35]

He had first called upon such principles of generation in the *Grundlagen*, but to somewhat different effect. In the absence of any developed theory of order types, he had depended upon the first and second principles of generation to define the transfinite ordinal numbers. Shortly thereafter he had already seen the advantages of relying upon order types to secure the same results, and by the time Part II of the *Beiträge* was written, he had worked out his entire theory of segments and well-ordered sets which provided a rigorous means of characterizing the first transfinite ordinal number ω, as well as its successors.

The introduction of segments underscored a significant innovation, one which clearly distinguished the *Beiträge* from Cantor's earlier presentations of the transfinite numbers. In the attempt to be as mathematically rigorous as possible, the *Beiträge* eschewed models from which, by analogy, the transfinite numbers might be generated. Earlier, such advance by analogy had been a valuable source of inspiration, but Cantor realized that too much reliance upon the examples of point sets or natural numbers might needlessly obscure the purely logical basis of his arguments. Though such models of specific well-ordered sets like the natural numbers largely saved him from errors in the *Grundlagen* concerning the properties of transfinite numbers, they did not suggest the generality and necessity of the concepts being advanced. However, the *Beiträge* was meant to establish the theory of transfinite sets along lines that were completely independent of any particular examples, though the ultimate application of the theory to specific cases was an indication of both its correctness and utility. Thus, instead of defining transfinite ordinals like ω as an expression for the totality of integers N taken in their natural order, thereby assuming a role analogous to that of the "next largest" integer after all the numbers $1, 2, 3, \ldots, \nu, \ldots$, Cantor took a somewhat different approach in the *Beiträge*. By regarding ω as only representing the type of order of any denumerably infinite well-ordered set, he avoided having to bind the definition of ν to any one set of elements.

As a result, Cantor's definition of ω in the *Beiträge* (in terms of the segments of well-ordered sets and fundamental sequences) made the modes of generation exhibited by transfinite ordinals a consequence of the type of order exhibited by well-ordered sets generally. Rather than assert that the transfinite ordinal numbers were defined in only two ways, either by addition of units or through a "limit" process designated to create yet another element following any well-ordered but unending sequence of ordinals, Cantor was able to show in Theorem K that given any transfinite ordinal number α, the type of a particular well-ordered set, then it was either the limit of a fundamental sequence $\{\alpha_\nu\}$, or it was the successor of a definite and unique predecessor α_{-1}. In this way the *Beiträge* avoided the vague use of a second principle of generation, called upon to define a new number, or as Cantor had expressed himself in the *Grundlagen*, whenever "one comes to no greatest number, one imagines a new one."[36]

Having defined in detail the nature and character of the order type which comprised the second number class $Z(\aleph_0)$, Cantor was at last prepared to return to the subject he had avoided in Part I of the *Beiträge*. From the well-ordered sequence of ascending transfinite ordinal numbers, he was ready to produce its sister sequence of the transfinite alephs, the cardinal numbers of the second and higher number classes.

ALEPH-ONE

Before Cantor could actually determine the power of the second number class, he had first to show that the set $Z(\aleph_0)$ taken in its entirety was in fact a well-ordered set. His proof of Theorem H, Section 15, had been a step in this direction. There he had argued that given any number α of $Z(\aleph_0)$, the set of all numbers $\{\alpha'\}$ of the first and second number classes *less* than α formed a well-ordered set. But this was a slightly different result and could not be carried over immediately to the question of the totality of elements $Z(\aleph_0)$, since there was no largest element α of the class. Thus Theorem H was not directly applicable in determining the matter of the power of $Z(\aleph_0)$ itself. Cantor remedied this with the following theorem:

> *Theorem A:* The totality $\{\alpha\}$ of all numbers α of the second number class comprises, in its order of magnitude, a well-ordered set.[37]

In proving the theorem, he began with the results from the earlier Theorem H. Letting A_α represent the segment of ordinals of the second number class which were all smaller than a given number α, taken in their order of magnitude, they formed a well-ordered set. Since Theorem H had included both numbers of the first and second number classes, he had to discount the part played by numbers of the first class before applying Theorem H to Theorem A in concluding that the set A_α was well ordered and of type $\alpha-\omega$. This could be done easily by representing the set of numbers $\{\alpha'\}$ from Theorem H in terms of Theorem A, so that: $\{\alpha'\}=(\{\nu\}, A_\alpha)$. Therefore $\overline{\{\alpha'\}}=\overline{\{\nu\}}+\overline{A_\alpha}$. Since $\overline{\{\alpha'\}}=\alpha$, $\overline{\{\nu\}}=\omega$, it followed that $\overline{A_\alpha}=\alpha-\omega$.

Cantor then considered any subset J of $\{\alpha\}$ such that there were numbers in $\{\alpha\}$ larger than all the numbers of J. Letting α_0 represent one such number of $\{\alpha\}$ greater than all elements of J, it was clear that J was also a part of $A_{\alpha_{0+1}}$. Since it followed from Section 12 of the *Beiträge* that $A_{\alpha_{0+1}}$ must also be well ordered, then there had to be some number α' of $A_{\alpha_{0+1}}$, and therefore also of $\{\alpha\}$, which followed directly after all the numbers of J. But this fulfilled the second condition for well-ordering given in Section 12, since any part of $\{\alpha\}$ was immediately followed by a unique and definite next larger element. Since the first condition for well-ordering was also fulfilled, because the set $\{\alpha\}$ contained a least element ω, the proof was complete.

Once it had been established that the set of all transfinite ordinal numbers of the second number class was well-ordered, then all the results of Section 12 of the *Beiträge* dealing with well-ordered sets in general could be directly applied. Therefore Cantor concluded that every set of different numbers of the first and second number classes had a least element, and second, that every such set when arranged in ascending order of magnitude formed a well-ordered set. At last Cantor was ready to establish one of the most basic theorems concerning transfinite cardinals in all of Cantorian set theory:

> *Theorem D:* The power of the entirety $\{\alpha\}$ of all numbers α of the second number class is not equal to $\aleph_0$.[38]

Following his earlier approaches to questions of nondenumerability, Cantor began by assuming that $\{\alpha\}$ was equal in power to $\aleph_0$. Writing $\{\alpha\}$ in the form of a denumerably infinite sequence $\gamma_1, \gamma_2, \ldots, \gamma_\nu, \ldots$, he showed that $\{\gamma_\nu\}$ would represent all elements of $\{\alpha\}$ in an order different from the order of magnitude in $\{\alpha\}$, but like $\{\alpha\}$, $\{\gamma_\nu\}$ would contain no largest element. Beginning with the first element of $\{\gamma_\nu\}$, γ_1, he considered the term γ_{ρ_2} of least index but next larger than γ_1; similarly the element γ_{ρ_3} of least index but next larger after γ_{ρ_2}, and so on. Thus he produced an infinite sequence $\gamma_1, \gamma_{\rho_2}, \ldots, \gamma_{\rho_\nu}, \ldots$, always increasing, of ordinals in such a way that

$$1 < \rho_2 < \rho_3 < \ldots < \rho_\nu < \rho_{\nu+1} < \ldots;$$
$$\gamma_1 < \gamma_{\rho_2} < \gamma_{\rho_3} < \ldots < \gamma_{\rho_\nu} < \gamma_{\rho_{\nu+1}} \ldots;$$
$$\gamma_\nu < \gamma_{\rho_\nu}.$$

From results previously established (Theorem C, section 15), it then followed that there had to be some definite, uniquely determined number δ of the second number class, specifically $\delta = \operatorname{Lim}_\nu \gamma_{\rho_\nu}$, such that δ was greater than every term γ_{ρ_ν}. Thus $\delta > \gamma_\nu$ for all indices ν. But since the set $\{\gamma_\nu\}$ was assumed to contain *all* numbers of the second number class, then δ was also included. For some definite index ν_0, $\delta = \gamma_{\nu_0}$. But this contradicted the inequality, already confirmed, that $\delta > \gamma_{\nu_0}$. Consequently the original assumption that the totality of numbers of the second number class was denumerable was impossible, and $\overline{\overline{\{\alpha\}}} \neq \aleph_0$.

The argument was familiar, but instead of applying it to the set of real numbers, as he had first done in 1873, Cantor now used the method in a more general context. Though he had shown that the set of transfinite ordinals of the second number class was nondenumerable, and therefore of a higher power, he had yet to prove that the cardinal number associated with the set $Z(\aleph_0)$ was in fact the *next* one larger than $\aleph_0$. Before showing that in fact this was the case, he interposed an intermediate step, in which he reasoned that if $\{\beta\}$ were any transfinite set of different numbers of the second number class, then it was either of cardinality $\aleph_0$, or of cardinality $\overline{\overline{\{\alpha\}}}$ of the second number class taken in its entirety. This was the essential step, from which the desired proof that $\overline{\overline{\{\alpha\}}}$ followed directly after $\aleph_0$ would necessarily follow:

> *Theorem E:* A given set $\{\beta\}$ of different numbers β of the second number class, if it is infinite, has either the cardinal number $\aleph_0$ or the cardinal number $\overline{\overline{\{\alpha\}}}$ of the second number class.[39]

Since the set $\{\beta\}$ could be well-ordered as part of the well-ordered set $\{\alpha\}$, then it also had to be similar to some segment A_{α_0} representing the set of all numbers of the second number class less than α_0, or $\{\beta\}$ had to be similar to the entire set $\{\alpha\}$. Cantor had already shown in Theorem A that $\overline{\overline{A_{\alpha_0}}} = \alpha_0 - \omega$, and it followed that either $\overline{\{\beta\}} = \alpha_0 - \omega$ or $\overline{\{\beta\}} = \overline{\{\alpha\}}$. Thus the cardinality of $\{\beta\}$ was either $\overline{\overline{\{\beta\}}} = \alpha_0 - \omega$ or $\overline{\overline{\{\beta\}}} = \overline{\overline{\{\alpha\}}}$. But in the preceding section of the *Beiträge*,[40] he had made it clear that $\alpha_0 - \omega$ was either a finite cardinal number or equal to $\aleph_0$. Since $\{\beta\}$ was assumed to be a transfinite set, it was only possible to conclude that the cardinal number $\overline{\overline{\{\beta\}}}$ was either $\aleph_0$ or $\overline{\overline{\{\alpha\}}}$.

The theorem itself, however, was not enough to ensure that the power of the second number class was indeed the next larger following $\aleph_0$. Cantor had only managed to establish that the cardinality of $Z(\aleph_0)$ was greater than $\aleph_0$, and that any infinite set of numbers of the second number class was either denumerable, or had the cardinal number $\overline{\overline{\{\alpha\}}}$. The final step was a short one, but he still had to show that there were no other cardinals possible between $\aleph_0$ and $\overline{\overline{\{\alpha\}}}$. The proof of this was nearly an immediate consequence of Theorem E:

Theorem F: The power of the second number class $\{\alpha\}$ is the second smallest transfinite cardinal number aleph-one.[41]

Supposing there were some cardinal number $\mathfrak{a}$ greater than $\aleph_0$ but less than $\overline{\overline{\{\alpha\}}}$, Cantor argued that there must then be a transfinite subset $\{\beta\}$ of $\{\alpha\}$ such that $\overline{\overline{\{\beta\}}} = \mathfrak{a}$. But Theorem E insisted that $\{\beta\}$, as a subset of $\{\alpha\}$, must have either the cardinal number $\aleph_0$ or the cardinal number $\overline{\overline{\{\alpha\}}}$. Thus the cardinal number $\overline{\overline{\{\alpha\}}}$ was necessarily the next largest transfinite cardinal number following $\aleph_0$, and it was thereby properly denoted $\aleph_1$. Cantor ended his remarks concerning transfinite cardinal numbers in the *Beiträge* with a remark reflecting what he had believed all along: "We therefore have, in the second number class $Z(\aleph_0)$, the natural *representative* for the second-smallest transfinite cardinal number *aleph-one*."[42]

It is remarkable that the transfinite cardinal number $\aleph_1$ only appeared once more in the *Beiträge*.[43] Little was said concerning cardinal numbers at all after the important theorems of Section 16 had finally succeeded in defining the extent and properties of $\aleph_1$ itself. The entire painstaking process of defining the order types of the second number class had been ostensibly undertaken to secure a rigorous definition and foundation for higher order transfinite cardinal numbers. And yet, once this had been done for the first nondenumerably infinite power $\aleph_1$, Cantor pressed no further but returned instead to concentrate his last energies in the *Beiträge* upon certain number-theoretic properties of the second number class.[44]

EXPONENTIATION IN $Z(\aleph_0)$ AND TRANSFINITE INDUCTION

The last four sections of the *Beiträge* were devoted entirely to discussion of the arithmetic behavior of the numbers of the second number class. Here their form as types of order was almost forgotten, and the ease with which they could be manipulated as ordinal numbers was stressed. At first Cantor paid special interest to the numbers of $Z(\aleph_0)$ which could be expressed as whole algebraic functions of finite degrees of ω. In fact, such numbers could always be uniquely expressed in the form:

$$\phi = \omega^{\mu}\nu_0 + \omega^{\mu-1}\nu_1 + \ldots + \nu_{\mu},$$ [45]

assuming μ, ν_0 were finite and non-zero, though $\nu_1, \nu_2, \ldots, \nu_{\mu}$ might equal zero. Later he was able to generalize such representations for ordinals of the second number class without restricting the degree μ to finite values. But until he had introduced the idea of transfinite induction and defined the product of infinitely many terms, he could do no more than restrict himself to ordinals represented by polynomials of finite degree.

In order to define the addition of any two such transfinite ordinals, the fact that arithmetic in $Z(\aleph_0)$ was generally noncommutative required that definitions be introduced case by case, just as Cantor had done earlier in establishing similar procedures in the *Grundlagen*.[46] For example, given $\psi = \omega^\lambda \rho_0 + \omega^{\lambda-1}\rho_1 + \ldots + \rho_\lambda$, where λ, ρ_0 were assumed to be finite and non-zero, then

1. If $\mu < \lambda$, then $\phi + \psi = \psi$.
2. If $\mu = \lambda$, then the sum $\phi + \psi = \omega^\lambda(\nu_0 + \rho_0) + \omega^{\lambda-1}\rho_1 + \ldots + \rho_\lambda$.
3. But finally, if $\mu > \lambda$, then
$\phi + \psi = \omega^\mu \nu_0 + \omega^{\mu-1}\nu_1 + \ldots + \omega^{\lambda+1}\nu_{\mu-\lambda-1} + \omega^\lambda(\nu_{\mu-\lambda} + \rho_0) + \omega^{\lambda-1}\rho_1 + \ldots + \rho_\lambda$.

Similar formulations for multiplication were slightly more complicated but reflected the process of repeated addition. Cantor noted in particular that $\phi\omega = \omega^{\mu+1}$, hence $\phi\omega^\lambda = \omega^{\mu+\lambda}$, and by following the distributive law for transfinite multiplication:

$$\phi \cdot \psi = \phi\omega^\lambda \rho_0 + \phi\omega^{\lambda-1}\rho_1 + \ldots + \phi\omega\rho_{\lambda-1} + \phi\rho_\lambda.$$

He could therefore conclude that given ϕ and ψ as above, then:

1. If $\rho_\lambda = 0$, then
$\phi \cdot \psi = \omega^{\mu+\lambda}\rho_0 + \omega^{\mu+\lambda-1}\rho_1 + \ldots + \omega^{\mu+1}\rho_{\lambda-1} = \omega^\mu\psi$;
2. If $\rho_\lambda \neq 0$, then
$\phi\psi = \omega^{\mu+\lambda}\rho_0 + \omega^{\mu+\lambda-1}\rho_1 + \ldots + \omega^{\mu+1}\rho_{\lambda-1} + \omega^\mu \nu_0 \rho_\lambda + \omega^{\mu-1}\nu_1 + \ldots + \nu_\mu$.

This much allowed Cantor to produce what he termed a "remarkable factorization [*Zerlegung*] of the numbers ϕ."[47] Taking

$$\phi = \omega^\mu \kappa_0 + \omega^{\mu_1}\kappa_1 + \ldots + \omega^{\mu_\tau}\kappa_\tau, \text{ where } \mu > \mu_1 > \ldots > \mu_\tau \geq 0 \text{ and}$$

$\kappa_0, \kappa_1, \ldots, \kappa_\tau$ were assumed to be non-zero finite numbers, then:

$$\phi = (\omega^{\mu_1}\kappa_1 + \omega^{\mu_2}\kappa_2 + \ldots + \omega^{\mu_\tau}\kappa_\tau)(\omega^{\mu-\mu_1}\kappa_0 + 1).$$

Repeated application of this procedure produced:

$$\phi = \omega^{\mu_\tau}\kappa_\tau\,(\omega^{\mu_{\tau-1}-\mu_\tau}\kappa_{\tau-1} + 1)\,(\omega^{\mu_{\tau-2}-\mu_{\tau-1}}\kappa_{\tau-2} + 1) \ldots (\omega^{\mu-\mu_1}\kappa_0 + 1).$$

Since $\omega^\lambda\kappa + 1 = (\omega^\lambda + 1)\kappa$, assuming κ a non-zero, finite number, Cantor could likewise produce the unique factorization:

$$\phi = \omega^{\mu_\tau}\kappa_\tau\,(\omega^{\mu_{\tau-1}-\mu_\tau} + 1)\kappa_{\tau-1}\,(\omega^{\mu_{\tau-2}-\mu_{\tau-1}} + 1)\kappa_{\tau-2} \ldots (\omega^{\mu-\mu_1} + 1)\kappa_0.$$

Since the individual factors $\omega^\lambda + 1$ were all prime factors, the product form Cantor had given ϕ was also a unique prime factorization. Nevertheless, his results concerning the arithmetic features of transfinite ordinal numbers in both the *Grundlagen* and in Section 17 of the *Beiträge* were restricted to polynomials of finite degree.[48] Though he could introduce the entire sequence of ordinal numbers of the form $\omega^{\mu}\nu_0 + \omega^{\mu-1}\nu_1 + \ldots + \nu_{\mu}$, where ν was a finite index, there was no way he could determine the limit number, or transfinite ordinal following all such algebraic numbers of finite degree. The next ordinal, in fact, could not be expressed by any finite exponentiation ω^{ν}. No transfinite ordinals of the form ω^{ω}, for example, could be included in his transfinite number theory until he had established a satisfactory means of introducing the product of transfinitely many ordinal numbers. To do so, he invented the process of transfinite induction, which was similar, but by no means the same as the familiar mathematical or complete induction on well-ordered sets of type ω like N.

At first glance it might seem impossible to extend one's reach beyond finitely inductive situations in a rigorous way. Complete induction was based on the idea of counting from 0,1,2,3, . . . , establishing the criteria of a definition or theorem for some lowest number and then indicating how one could proceed from the assumed truth of the case of $n-1$ to show that it must hold as well for the nth case too. Cantor, however, showed how this procedure might be extended to provide a similarly inductive means of "counting" infinite sets in a successive way.

Given any definition or proposition $F(n)$ which was assumed true for some least value of n, transfinite induction proceeded to show that if $P(n)$ was assumed true for all values $n<\alpha$, then it was also true for the ordinal α, whatever the cardinality of α might be. By introducing a suitable limit process in terms of fundamental sequences and their limit numbers, it was consequently possible to establish the multiplication of transfinitely many factors and transfinite exponentiation as well.

Letting ξ be a variable of either the first or second number class, Cantor considered two constant ordinals δ, γ of the same domain, assuming that $\delta > 0, \gamma > 1$. By relying upon transfinite induction he could then prove the following theorem:

Theorem A: There is one, entirely definite, single-valued function $f(\xi)$ of the variable ξ which satisfies the following conditions:

1. $f(0) = \delta$;
2. If ξ' and ξ'' are any two values of ξ, and if $\xi' < \xi''$, then $f(\xi') < (\xi'')$;
3. For every value of ξ, $f(\xi+1)=f(\xi)\gamma$;
4. If $\{\xi_\nu\}$ is any fundamental sequence, then $\{f(\xi_\nu)\}$ is also a fundamental sequence, and if $\xi = \operatorname{Lim}_{\nu} \xi_\nu$, then $f(\xi) = \operatorname{Lim}_{\nu} f(\xi_\nu)$.[49]

First he established the fact that the function $f(\xi)$ was completely determined for all $\xi<\omega$. Assuming the theorem true for all values of $\xi<\alpha$, where α was any number of $Z(\aleph_0)$, he then went on to assert its validity for all $\xi \leq \alpha$. If α were of the first kind, then $f(\alpha) = f(\alpha_{-1})\gamma > f(\alpha_{-1})$, which clearly satisfied the conditions of the theorem for all $\xi \leq \alpha$. If α were of the second kind, where $\operatorname{Lim}_{\nu} \alpha_\nu = \alpha$, then it followed from his analysis of fundamental sequences that $\{f(\alpha_\nu)\}$ was also a fundamental sequence, so that $f(\alpha) = \operatorname{Lim}_{\nu} f(\alpha_\nu)$. Considering another fundamental sequence $\{\alpha'_\mu\}$ where $\operatorname{Lim}_{\nu} \alpha'_\nu = \alpha$, then the two sequences $\{f(\alpha_\nu)\}$ and $\{f(\alpha')\}$ were necessarily coherent, and therefore $f(\alpha) = \operatorname{Lim}_{\nu} f(\alpha'_\nu)$. Consequently, $f(\alpha)$ was uniquely determined if α was a number of the second kind. Given any number $\alpha' < \alpha$, it was clear that $f(\alpha') < f(\alpha)$. Thus the conditions of the theorem were also met by all values $\xi \leq \alpha$, and for all such values there could be only one function $f(\xi)$ which satisfied the terms of Cantor's Theorem A.

By specifying $\delta = 1$, and denoting $f(\xi) = \gamma^\xi$, he was able to translate the general existence proof of Theorem A into a specific one used to define the exponentiation of transfinite ordinal numbers. Hence, Theorem B:

> *Theorem B:* If $\gamma > 1$ is any constant of the first or second number class, then there is an entirely determined function γ^ξ of ξ such that:
> 1. $\gamma^0 = 1$;
> 2. If $\xi' < \xi''$, then $\gamma^{\xi'} < \gamma^{\xi''}$;
> 3. For every value of ξ, $\gamma^{\xi+1} = \gamma^\xi \gamma$;
> 4. If $\{\xi_\nu\}$ is a fundamental sequence, then $\{\gamma^{\xi_\nu}\}$ is also a fundamental sequence, and, if $\xi = \operatorname{Lim}_{\nu} \xi_\nu$, then $\gamma^\xi = \operatorname{Lim}_{\nu} \gamma^{\xi_\nu}$.[50]

Cantor went on to show that if $f(\xi)$ were the function whose existence was established in Theorem A, then $f(\xi) = \delta\gamma^\xi$, and the exponentiation in question was necessarily unique by virtue of the uniqueness of $f(\xi)$. Formulae were immediately forthcoming: $\gamma^{\alpha+\beta} = \gamma^\alpha\gamma^\beta$; $\gamma^{\alpha\beta} = (\gamma^\alpha)^\beta$. Moreover, for all values of ξ, assuming $\gamma > 1$, he established that $\gamma^\xi \geq \xi$. Of particular interest to the theory of transfinite numbers was the special case of Cantor's ϵ-numbers, those numbers ξ of the second number class for which $\omega^\xi = \xi$. But before he went on to investigate the very special properties of such numbers, he introduced the "normal form" of numbers of the second number class.[51] He had already discussed the representation of numbers $\phi = \omega^\mu\omega_0 + \omega^{\mu-1}\nu_1 + \ldots + \nu_\mu$ of *finite* degree. Now he could extend the same concept to include transfinite degrees. This was made possible by his introduction of transfinite exponentiation, and as a result it was no longer necessary to restrict consideration to ordinal numbers of finite degree ω^ν, but at last even transfinitely valued exponents of the second number class were admissible.

Given a number α of $Z(\aleph_0)$, it then followed that $\omega^\xi > \alpha$ for a sufficiently large value of ξ. The least of such exponents ξ Cantor denoted β, and added that it could never be a number of the second kind.[52] Denoting β_{-1} by α_0, then $\beta = \alpha_0 + 1$, and he could assert the existence of a completely determined number α_0 of either the first or second number class such that: $\omega^{\alpha_0} \leq \alpha$, with $\omega^{\alpha_0}\omega > \alpha$. Therefore, it was clear that the relation $\omega^{\alpha_0}\nu \leq \alpha$ could not hold for all finite values of ν, for then $\operatorname{Lim}_{\nu} \omega^{\alpha_0}\nu = \omega^{\alpha_0}\beta \leq \alpha$. Consequently, it was possible to say that a least value ν had to exist for which $\omega^{\alpha_0}\nu > \alpha$. This value of ν he denoted $\kappa_0 + 1$, and there was correspondingly a definite, determined finite number κ_0 such that $\omega^{\alpha_0}\kappa_0 \leq \alpha$ and $\omega^{\alpha_0}(\kappa_0 + 1) > \alpha$. Setting $\alpha' = \alpha - \omega^{\alpha_0}\kappa_0$, then $\alpha = \omega^{\alpha_0}\kappa_0 + \alpha'$.

Cantor summarized this result in a separate Theorem A,[53] which stated that under the conditions $0 \leq \alpha' < \omega^{\alpha_0}$ and $0 < \kappa^0 < \omega$, every number α of $Z(\aleph_0)$ was uniquely representable in the form $\alpha = \omega^{\alpha_0}\kappa_0 + \alpha'$. If α' were a number of the second number class, the procedure could be reapplied, so that $\alpha' = \omega^{\alpha_2}\kappa_2 + \alpha''$, and similarly for α'', and thereafter. But the sequence of $\alpha', \alpha'', \ldots$ was necessarily finite, since it was monotonically decreasing, and were it infinite then no least term could exist, which contradicted the fact that every collection of numbers of the first and second number classes *always* contained a least number. Thus, for some value τ, $\alpha^{(\tau+1)} = 0$. Collecting all these steps together, Cantor presented the normal form for any number α of $Z(\aleph_0)$, where α_0 was called the degree of α, and α_τ its exponent:

> *Theorem B:* Every number α of the second number class can be represented in only one way in the form
>
> $$\alpha = \omega^{\alpha_0}\kappa_0 + \omega^{\alpha_1}\kappa_1 + \ldots + \omega^{\alpha_\tau}\kappa_\tau,$$
>
> where $\alpha_0, \alpha_1, \ldots, \alpha_\tau$ are numbers of the first or second number class which satisfy the conditions $\alpha_0 > \alpha_1 > \alpha_2 > \ldots > \alpha_\tau > 0$, while κ_0, $\kappa_1, \ldots, \kappa_\tau$, $\tau + 1$ are all non-zero numbers of the first number class.[54]

As he had already done for the previously characterized polynomial forms of finite degree, Cantor was able to produce generalized formulae for sums and products of transfinite ordinals represented in their normal form. He also gave a similar theorem establishing the unique prime factorization of *any* number of the second number class in terms of the product:

$$\alpha = \omega^{\gamma_0}\kappa_\tau(\omega^{\gamma_1} + 1)\kappa_{\tau-1}(\omega^{\gamma_2} + 1)\kappa_{\tau-2} \ldots (\omega^{\gamma_\tau} + 1)\kappa_0, \text{ where}$$
$$\gamma_0 = \alpha_\tau,\ \gamma_1 = \alpha_{\tau-1} - \alpha_\tau;\ \gamma_2 = \alpha_{\tau-2} - \alpha_{\tau-1};\ \ldots;\ \gamma_\tau = \alpha_0 - \alpha_1,$$

and where the values $\kappa_0, \kappa_1, \ldots, \kappa_\tau$ were all taken to be non-zero finite numbers.[55] Finally, Cantor was prepared to introduce the last and one of the most interesting of the new results brought forward in Part II of the *Beiträge*. His epsilon numbers of the second number class were a fitting conclusion to his entire presentation of the foundations of his theory of transfinite ordinal numbers.

THE ϵ-NUMBERS OF THE SECOND NUMBER CLASS

In terms of Cantor's definition of exponentiation, made possible by his use of transfinite induction, Cantor could advance beyond his algebraic ordinals of finite degree and introduce the completely general normal form of the *Beiträge's* Section 19. Therefore, following all numbers of the form $\varphi = \omega^{\mu}\nu_0 + \omega^{\mu-1}\nu_1 + \ldots + \nu_{\mu}$ for $\mu < \omega$, he could now list the sequence ω^{ω}, $\omega^{\omega} + 1$, $\omega^{\omega} + 2, \ldots$. Moreover, by inspecting the sequence $\omega^{\omega}, \omega^{\omega^2}, \omega^{\omega^3}, \ldots$, it was clear that $\operatorname{Lim}_{\nu} \omega^{\omega^{\nu}} = \omega^{\omega^{\omega}}$. Considering the sequence of all ordinal numbers which could not be produced by any finite combinations of numbers of lower order, he noted that in terms of the well-ordered sequence 1, ω, ω^{ω}, $\omega^{\omega^{\omega}}, \ldots$, a limit could again be assigned which he denoted ϵ_0. Thus, according to the definition he had introduced for transfinite exponentiation,

$$\omega^{\epsilon_0} = \operatorname{Lim}_{\nu} (\omega, \omega^{\omega}, \omega^{\omega^{\omega}}, \ldots) = \epsilon_0.$$

By considering only those numbers which could not be defined by any finite combination of elements less than themselves, like $\omega, \omega^{\omega}, \ldots$, Cantor was led to discover and to investigate the remarkable properties of what he called the ϵ-numbers of the second number class.[56]

The first reference he ever made to ϵ-numbers was in a letter he sent to Franz Goldscheider on October 11, 1886, in which he called solutions to the equation $\omega^x = x$ "Giganten" of $Z(\aleph_0)$.[57] The smallest of these was defined in terms of $\gamma_1 = \operatorname{Lim}_{\nu} (\omega, \omega_1, \omega_2, \ldots)$, where $\omega_1 = \omega^{\omega}$, $\omega_2 = \omega^{\omega_1}$, $\omega_3 = \omega^{\omega_2}$, $\ldots, \omega_{\nu} = \omega^{\omega_{\nu-1}}, \ldots$, or, equivalently, $\gamma_1 = \operatorname{Lim}_{\nu} (\omega, \omega^{\omega}, \omega^{\omega^{\omega}}, \ldots)$. Then came the next *Giganten*, and in fact, a well-ordered succession of type Ω: $\gamma_1, \gamma_2, \ldots, \gamma_{\omega}, \gamma_{\omega+1}, \ldots, \gamma_{\alpha}, \ldots$. Cantor then went on to establish various specific properties of his *Giganten* (Cantor's letter to Goldscheider of October 11, 1886, may be found in Appendix A).

Thus as early as 1886 Cantor had already discovered many of the most essential characteristics of the numbers he later termed the ϵ-numbers of the second number class. In Section 19 of the *Beiträge* he had shown that any number α could be uniquely represented in its normal form $\alpha = \omega^{\alpha_0}\kappa_0 + \omega^{\alpha_1}\kappa_1 + \ldots + \omega^{\alpha_{\tau}}\kappa_{\tau}$, where $\alpha_0 > \alpha_1 > \ldots$, and $0 < \kappa_{\nu} < \omega$. In Section 20, he went on to discuss the ϵ-numbers of the second number class.

In terms of its normal form, it was clear that the degree α_0 of any number $\alpha \in Z(\aleph_0)$ could never be so large that $\alpha_0 > \alpha$. The question of real significance, as Cantor noted, concerned the possibility of cases in which $\alpha_0 = \alpha$. Then the normal form of α would reduce directly to the first term alone, more precisely, α would be a root of the equation $\omega^{\xi} = \xi$. Otherwise, $\alpha = \omega^{\alpha}\kappa_0 + \omega^{\alpha_1}\kappa_1 + \ldots > \omega^{\alpha}$. Conversely, it was just as clear that every root α of $\omega^{\xi} = \xi$ would have the normal form ω^{α}. The cases of interest then were those in which the degree of α was equal to itself.

Cantor set out to determine the roots of $\omega^\xi = \xi$ in their totality.[58] First he established that such numbers as the ϵ-numbers did in fact exist. Taking any number γ of the first or second number class which did *not* satisfy the equation $\omega^\xi = \xi$, Cantor showed that the fundamental sequence $\{\gamma_\nu\}$: $\gamma_1 = \omega^\gamma$, $\gamma_2 = \omega^{\gamma_1}, \ldots, \gamma_\nu = \omega^{\gamma_{\nu-1}}, \ldots$, produced, in the limit, $\operatorname{Lim}_\nu \gamma_\nu = E(\gamma)$, an ϵ-number. The least of all such ϵ-numbers was $\epsilon_0 = E(1) = \operatorname{Lim}_\nu \omega_\nu$, where $\omega_1 = \omega$, $\omega_2 = \omega^{\omega_1}, \ldots, \omega_\nu = \omega^{\omega_{\nu-1}}$. As he had done in 1886, he also established in the *Beiträge* that for two *consecutive* ϵ-numbers ϵ', ϵ'', for any γ such that $\epsilon' < \gamma < \epsilon''$, then $E(\gamma) = \epsilon''$. Moreover, for any ϵ-number ϵ', its immediate successor was $E(\epsilon' + 1)$. The infinite sequence of all such numbers $e < e' < \epsilon'' < \ldots < \epsilon^\nu < \epsilon^{\nu+1} < \ldots$ formed a well-ordered set of type Ω, consequently of power $\aleph_1$, such that $\operatorname{Lim}_\nu \epsilon^\nu$ was also an ϵ-number. In fact, it was the very next ϵ-number following all the numbers of the succession ϵ^ν. The final result of the entire *Beiträge*, Part II, concerned the completeness of Cantor's characterization of the ϵ-numbers. His last proof confirmed the fact that for any number α of the second number class, $\alpha^\xi = \xi$ had no other roots than the ϵ-numbers greater than α.[59]

Why were the ϵ-numbers of such importance that they should have been chosen to bring the entire *Beiträge* to a conclusion? Their significance can be directly appreciated by considering the connections between arithmetic operations among the transfinite numbers, specifically addition, multiplication, and exponentiation. For each of these operations there were certain numbers, *Hauptzahlen,* which could be used to characterize the nature of each operation itself.[60]

For addition, *Hauptzahlen* could be characterized as those numbers $\gamma > 0$ for which $\gamma = \alpha + \gamma$ was always true for all α such that $0 \leq \alpha < \gamma$. Denoting such additive *Hauptzahlen* as π, then clearly $\pi > 0$ and for all $\alpha < \pi$, $\pi = \alpha + \pi$.[61] Since $1 = 0 + 1$ satisfied these conditions, 1 was an additive *Hauptzahl,* and in fact the only finite *Hauptzahl*. All others were necessarily limit numbers, and the first of these was ω. Clearly for all $\nu < \omega$, $\omega = \nu + \omega$. Each additive *Hauptzahl* was irreducible; it was impossible to find numbers $\alpha < \pi$, $\beta < \pi$ such that $\alpha + \beta = \pi$. Thus, each additive *Hauptzahl* was an additive *prime* number. Once again the centrality of addition to transfinite number theory clearly emerged. Addition was operationally characterized (whereas multiplication was not) by the fact that *Hauptzahlen,* irreducible numbers, and prime numbers were all equivalent. Every transfinite number, as Cantor had shown, could be represented in exactly one way as the sum of additive *Hauptzahlen*. This of course meant nothing more than the normal form for any given number α.

By contrast, transfinite ordinal multiplication was operationally much less simple and unified than addition. *Hauptzahlen* could be defined as numbers $\gamma > 1$ such that $\gamma = \alpha \cdot \gamma$ for all values of α such that $1 \leq \alpha < \gamma$. For example, $2 = 1 \cdot 2$ showed that 2 was a multiplicative *Hauptzahl* and was in fact the only

Hauptzahl that was not a limit number. But for multiplication, the *Hauptzahlen* represented only a part of the irreducible numbers, and of these only a small number were actually prime numbers, like the finite prime numbers and ω. Among transfinite multiplicative *Hauptzahlen,* called δ-numbers, each was also an additive *Hauptzahl,* and the least of all such multiplicative *Hauptzahlen* was ω.[62] Every limit number with exactly two factors, the irreducible limit numbers, were also δ-numbers. Thus the irreducible limit numbers were already identified. But unlike the analogous case for additive prime numbers, there were other irreducible numbers besides the finite multiplicatively prime numbers which had immediate predecessors. These were numbers of the first kind, specifically $\pi+1$. Thus numbers greater than ω which were not also limit numbers were multiplicatively irreducible only if they were the immediate successor of an additive *Hauptzahl* of the form $\pi+1$. Multiplicatively irreducible numbers were therefore of three kinds:[63] finite prime numbers; the multiplicative *Hauptzahlen;* numbers of the form $\pi+1$. The only number to satisfy all of these requirements was the finite number 2. As a result it was possible to produce unique factorizations of any number β as a product of finitely many irreducible numbers, which followed directly from the additive decomposition Cantor had given earlier:

$$\alpha = \omega^{\gamma_0}\kappa_\tau(\omega^{\gamma_1}+1)\kappa_{\tau-1}(\omega^{\gamma_2}+1)\kappa_{\tau-2}\ldots(\omega^{\gamma_\tau}+1)\kappa_0.$$

Nevertheless, since the additive factorization consisted of additively irreducible factors which were also prime factors, while the multiplicatively irreducible factors were not necessarily multiplicative prime numbers, the additive and multiplicative factorizations were not necessarily comparable.

Finally, all these considerations help make clear the special and significant position of the ϵ-numbers in the number theory of transfinite numbers. They were none other than the *Hauptzahlen* of exponentiation: any number $\epsilon > 2$ such that $\epsilon = \alpha^\epsilon$ for all numbers α such that $1 < \alpha < \epsilon$. The first such number was ω. Every ϵ-number greater than ω, as Cantor showed, was the limit of a sequence of multiplicative *Hauptzahlen,* or δ-numbers. If ≥ 2 and $\beta = \alpha^\beta$, then β could only be a limit number, otherwise $\alpha^\beta > \beta$. As Cantor demonstrated,[64] the necessary and sufficient condition determining the ϵ-numbers was the equation $\epsilon = 2^\epsilon$. This set them apart from additive *Hauptzahlen,* for which $2+\beta=\beta$ only characterized those *Hauptzahlen* $\beta \leq \omega$, but not just additive *Hauptzahlen*. $2\beta=\beta$ didn't even ensure that β was a δ-number, only that β was a limit number. But $2^\beta=\beta$ characterized fully the ϵ-numbers, and this had been Cantor's concluding statement of the *Beiträge*.[65]

One of the most important features of Cantor's transfinite numbers was succinctly characterized by the ϵ-numbers. Clearly central to his introduction of transfinite cardinals were those "initial elements" which could not be reached or produced by any arithmetic or exponential combination of elements preceding them. Initial elements, like ω and Ω, were transfinite numbers which were preceded by no numbers of equal power. Moreover, to every transfinite power there was only one such number. And every "first number" was necessarily an ϵ-number. Accordingly, there was a very central connection, a direct correlation between the succession of transfinite alephs and the ϵ-numbers Cantor had introduced at the very close of the *Beiträge,* Part II.

CONCLUSION

Cantor's presentation of the principles of transfinite set theory in the *Beiträge* was elegant but ultimately disappointing. One might have thought that at long last, having given the extensive and rigorous foundations for the transfinite ordinal numbers of the second number class, Cantor would then have gone on to discuss the higher cardinal numbers in some detail. In particular one might have expected him to fulfill his promise made in Part I to establish not only the entire succession of transfinite cardinal numbers $\aleph_0, \aleph_1, \ldots, \aleph_\nu, \ldots$, but to prove as well the existence of $\aleph_\omega$, and to show that in fact there was no end to the ever-increasing sequence of transfinite alephs. But instead, as we have already noted, once the transfinite ordinal numbers of $Z(\aleph_0)$ had been constructed and duly investigated, the proofs concerning the power of $Z(\aleph_0)$ were presented but were then left to impress the reader in silence as Cantor went on to devote the remaining sections of the *Beiträge* to an analysis of transfinite ordinals. The entire manner of Cantor's handling of the transfinite cardinals in the *Beiträge* was fundamentally unsatisfying because it seemed so anticlimactic.

In the *Grundlagen* the transfinite ordinal numbers had served as little more than handmaidens to an adequate determination of the power next larger after $\aleph_0$. Cantor's lack of interest in the numbers of the second number class themselves was reflected in the final sentence of the sixth and last paper of his *Punktmannigfaltigkeitslehre:* "Thus with the help of the theorems proven in Section 13 [of the *Grundlagen*] it follows that the linear continuum has the power of the second number class (II)."[66]

Throughout the period of research and publication on linear point sets, including the *Grundlagen,* the major goal was always clear: the existence and validity of the transfinite numbers was stoutly defended while Cantor repeatedly promised that his new set theory had been required to deal successfully with the nature of continuity and the infinite. It would more than justify itself in a soon-to-be-published proof of the continuum conjecture. New techniques, new distinctions, number-theoretic comparisons, all were designed, ultimately, to provide new weapons with which to attack the continuum hypothesis. But as his research progressed, Cantor found that the properties of the transfinite ordinal numbers were intriguing in themselves. Their variety and properties were much richer than those of the transfinite alephs, and increasingly, he became interested in the general properties of order itself, in the order types of simply-ordered sets, and in those of greater importance to the study of transfinite cardinal numbers, the types of well-ordered sets.

By the time he wrote the *Beiträge,* the solution of the continuum hypothesis seemed as elusive as ever, despite the tantalizing hope that coverings, which led to the formulation $2^{\aleph_0} = \aleph_1$, might provide the key for which Cantor had searched so long. But by 1897 the discovery of the paradoxes of set theory, his inability to establish directly the comparability of all cardinal numbers, and the lack of any proof that every set could be well-ordered seemed to leave him with no alternative than the one he chose: rather than to produce complete, absolutely certain solutions to the outstanding problems his set theory had raised, he was forced to accept something less. Instead, he sought to present the elements and internal workings of his theory of transfinite sets as rigorously as he could. In its abstractness, in its independence from point sets and physical examples, the *Beiträge* represented Cantor's last effort to present mathematicians with the basic elements of his transfinite set theory. He hoped that at last the theory would speak for itself, and that its utility and interest would be acknowledged accordingly.

Following publication of the *Beiträge* and its translation almost immediately into Italian and French, Cantor's ideas became widely known and circulated among mathematicians the world over. What is more important, the value of his transfinite set theory was quickly recognized, and soon Cantor's ideas were stimulating heated polemics between widely divided camps of mathematical opinion. Though Cantor never seemed able to avoid controversy and division over the nature of his work, he was, after 1895, increasingly defended by younger and more energetic mathematicians. No longer was he left to face the opposition alone. Though Kronecker's dissent was kept alive by critics of similar persuasion, like Poincaré, Cantor could begin to count an impressive and growing array of mathematicians ready to support transfinite set theory. For him, the crusade was nearly over, and though the theoretical difficulties were by no means satisfactorily resolved, recognition that Cantor had contributed something of lasting significance to the world of mathematics had been achieved.

CHAPTER 10

The Foundations and Philosophy of Cantorian Set Theory

Hypotheses non fingo,
—Isaac Newton

Neque enim leges intellectui aut rebus damus ad arbitrium nostrum, sed tanquam scribae fideles ab ipsius naturae voca latas et prolatas excipimus et describimus

—Francis Bacon

Veniet tempus, quo ista quae nunc latent, in lucem dies extrahat et longioris aevi diligentia.

—I Corinthians

Cantor's *Beiträge zur Begründung der transfiniten Mengenlehre* was prefaced with three aphorisms, each one meant to reveal something about his new work.[1] Consequently, the three epithets offer helpful clues to his own interpretation of transfinite set theory. Closer examination suggests that they reflected his belief that the *Beiträge* was much more than a straightforward exposition of set theory and the transfinite numbers. To understand fully what Cantor had in mind, it is necessary to go beyond definitions, theorems, and proofs. In fact, the secret to appreciating how Cantor regarded the very nature of mathematics lies hidden in the three aphorisms with which he opened the *Beiträge,* Part I.

However, before proposing any interpretation of the meaning Cantor attached to the three Latin mottos, it is necessary first to introduce the more philosophical but perhaps less obvious side of the *Beiträge*. In certain fundamental ways it was quite different in character, despite obvious similarities in content, from all his earlier work. In addition to the mathematical results of Cantor's various theorems and definitions published in the *Beiträge,* it is also important to appreciate the special points of philosophical and methodological interest. What soon came to be seen as crucial for the future course of mathematics was controversy over the nature and behavior of sets as Cantor had defined them, and as he had used them, in developing his theory.

FREGE'S CRITIQUE OF CANTORIAN SET THEORY

One of Cantor's earliest critics was Gottlob Frege, who even before publication of the *Beiträge* had questioned Cantor's whole approach to establishing the transfinite numbers. Though Frege was in large part sympathetic to Cantor's achievement, he was tremendously concerned over what he felt were its weak and consequently unacceptable foundations. Frege was among the first to accept the idea of Cantor's actually infinite numbers as mathematically permissible, but he also raised substantial questions concerning the soundness of Cantor's methods by which his transfinite set theory had been produced.

Frege's approach to mathematics was substantially different from Cantor's, and though Cantor could appreciate and even praise Frege's aims, he could never bring himself to adopt the austerity of Frege's program.[2] More than anything else, Frege believed that the principles upon which arithmetic must be founded were essentially logical in character. Thus his analysis of arithmetic stressed the need to base all theorems and deductions upon purely logical definitions. In his view, no sharp line could be drawn between logic and arithmetic. Thus Frege directed his attention to definitions in evaluating Cantor's transfinite set theory.

Both Cantor and Frege could agree upon one essential point: pure mathematics had nothing at all to do with feelings, sensations, or perceptions. Frege had emphasized this as early as 1885 during a meeting of the Gesellschaft für Medizin und Naturwissenschaften in Jena, where he stressed that it was possible to count anything that was an object of thought.[3] Mathematically nothing more was required of the number concept than a sharpness of demarcation between the objects counted accompanied by a certain logical completeness. This was the key to Frege's belief that mathematics and logic were strictly connected. The basic elements of arithmetic had to encompass everything thinkable; such general reasoning fell within the bounds of theoretical logic. Neither arithmetic nor the concept of number should be tied to any narrow domain of particulars—point sets, for example. Mathematics, in this interpretation, was taken to express completely general truths, and in the words of David Hilbert, it ought to apply with equal validity to "tables, chairs, and beer mugs."[4]

In 1885, the same year in which Frege had commented upon the concept of number before his local Society for Medicine and Science in Jena, Cantor reviewed Frege's *Die Grundlagen der Arithmetik.*[5] Basically he found that he was in complete agreement with the latter's approach to arithmetic. Though he was critical of Frege's definition of number as the "extension of a concept,"[6] he nevertheless gave favorable assent to Frege's approach. Frege was absolutely correct, he maintained, to have omitted all psychological considerations from his attempt to establish the logical rigor of arithmetic. This was the only means by which logical purity could be won, thereby ensuring the applicability of mathematics to any perceived object of thought. Cantor seemed to agree fully with Frege's attempt to free arithmetic from the context of any particulars whatsoever, concentrating instead upon its purely logical content.[7]

And yet, in his next publications which dealt with such matters, Cantor was not entirely clear about such basic issues, as Frege was to emphasize in his review of Cantor's *Mittheilungen zur Lehre vom Transfiniten*.[8] In laying out the basic elements for his systematic development of a theory of order types, Cantor first tried to explain how his transfinite numbers were the products of abstraction from sets:

> By the *power* or *cardinal number* of a set M (which consists of distinct, conceptually separate elements m, m', . . . and is to this extent determined and limited), I understand the general concept or generic character (universal) which one obtains by abstracting from the character of the elements of the set, as well as from all connections which the elements may have (be it between themselves or other objects), but in particular from the order in which they occur, and by reflecting only upon that which is common to all sets which are equivalent to M.[9]

Frege was critical from the beginning. Sometime after 1890 he drafted a review which was never published and which diverged substantially from the remarks he did print in the *Zeitschrift für Philosophie und philosophische Kritik* in 1892.[10] The unpublished draft was considerably more caustic and was specifically concerned with the whole question of abstraction and the misuse to which Frege believed Cantor had put the entire process. It was not entirely clear who among mathematicians Frege meant to include when he likened them and their awe of philosophical expressions (like abstractions) to primitive tribesmen: "When natives in darkest Africa see a telescope for the first time or a pocket watch, they are inclined to assume that these objects have very wonderful magical properties. It is the same for many mathematicians with philosophical expressions."[11] Certainly Cantor must have figured prominently in Frege's mind as a mathematician skilled in producing transfinite magic. Reliance upon abstraction was mere hocus-pocus—abstraction was only the means by which the transfinite numbers were conjured. Frege was no kinder in his description of Cantor's methods:

> If, for example, one finds a property of a thing upsetting, one abstracts it away. If one wants to order a stop, however, to this destruction, so that properties which one wants to see retained are not obliterated, then one

> reflects upon these properties. Finally, if one painfully misses properties of the thing, one adds them back by definition. Possessing such magical powers, one is not very far from omnipotence.[12]

Perhaps the best example of Frege's point is what might be called his "pencil of arbitrary cardinality."[13] In trying to demonstrate how indefinite and unclear Cantor's use of abstraction actually was, Frege imagined an ordinary pencil. He then pictured a group of people trying to "abstract its properties" in order to answer the question, "What general concept do you find?" Frege conjectured a number of responses:

Someone, being unmathematical, might reply: "Pure being."

Another, presumably more attuned to Frege's skepticism of what abstraction from all specific properties of a given object might produce, answered: "Pure nothing."

A third (one Frege suspected of being a student of Cantor's) found: "The cardinal number 'one'."

Even more alarming, another of Cantor's assumed protégés (prompted by what Frege termed some "inner voice") regarded graphite and wood as "constituent elements" of the pencil and thus discerned the cardinal number "two."

Frege summarized his objection to Cantor's entire procedure quite bluntly: "Now why shouldn't one person come out with this, another with that?"[14] Frege was convinced that Cantor's entire orientation was untenable for a rigorous theory of sets and number, even though he found no fault in the *results* of Cantor's transfinite arithmetic. He believed that no theory could claim acceptability as part of mathematics until its most fundamental elements had been set out directly, simply, self-evidently, and then the entire theory built firmly, step by step, upon logical and consistent foundations. But Cantor's theory, because it began by trying to abstract the number concept from objects collectivized as sets, had gone wrong at the very outset. Definiteness and rigor, hallmarks for Frege of mathematical thinking, could not be guaranteed through the vague, indeterminate application of a psychological principle like abstraction.

But there was more to Frege's critique than his objection to the process of abstraction. Even if one were to admit the use of abstraction in determining numbers, whether cardinal or ordinal, abstraction was still of no utility. Frege imagined, for example, a large sand pile. Here again the sharpness of his criticism, the sarcasm of his tone was exuberant:

> Well, let's take a sand pile then! Ah, someone is about to touch one grain after another. "You don't want to count them do you? That is completely forbidden! You must obtain the number through a single act of abstraction."[15]

At this point Frege's manuscript broke off. Perhaps he felt he was being overly levitous with Cantor's view of abstraction. After all, Cantor never claimed *he,* or anyone, could abstract from the grains of sand in the universe to produce in some magical way the cardinality of the set of all sand grains. But whatever his reasons may have been for scrapping the first draft of his review of Cantor's *Mittheilungen* of 1890, Frege's argument was perfectly clear. Abstraction was not a justifiable procedure in rigorous mathematics. Logically, it lacked definiteness. Depending upon individual viewpoints, one could plumb presumably any cardinality from any object imaginable. It was all too vague and too imprecise to suit Frege's standards of logical exactness.

In this, Frege was a pathfinder for a whole generation of mathematicians who read Cantor's *Beiträge*. They realized its value but were dismayed by its faults, which in large measure were foundational in character. Frege, like Peano, Whitehead, Russell, and Zermelo, believed that strictly logical considerations had to be made precise before one could accept mathematics, finite or transfinite, as rigorously established. In order to do so, the intuitive levels on which Cantor had worked his entire life were abandoned in favor of strictly logical analysis. This was territory in which Cantor would have felt foreign and uncomfortable. It was a different sort of mathematics, with a different personality shaped by a later generation which Cantor would encourage but of which he was not a part. Though he could praise the attempt to axiomatize arithmetic, it was not his style. He believed in his intuition, and the successful path his inner voice had illuminated since his youth, when he had first followed its call and had decided to become a serious mathematician.

Frege's published review of Cantor's *Mittheilungen,* except for its opening paragraph, abandoned the earlier draft entirely.[16] Although he remained critical of Cantor's method of abstraction, Frege placed greater emphasis upon the proper form of definitions and raised a fundamental problem that later proved to be a major source of trouble once the antinomies of set theory had been discovered. Frege paid special attention to Cantor's use of the term set. He felt that it was entirely inadequate to the demands of a logically sound theory. In fact, Frege objected to Cantor's failure to provide satisfactory definitions for any of his most basic concepts.

Devoting only his first paragraph to positive notice of Cantor's successful routing of traditional opposition to actual infinities, Frege launched four pages of critique by saying: "Cantor is less fortunate where he defines."[17] Charging Cantor with intellectual laziness, complaining that he should have *thought* about the meaning of Frege's *Grundlagen* and not just *reviewed* it, Frege insisted that Cantor's view of powers and cardinal numbers coincided exactly with his own use of the "extension of a concept." But this was in no way a concession to Cantor's use of powers as logically rigorous. On the contrary, Frege believed Cantor's definitions revealed an outmoded point of view:

> He demands impossible abstractions and is unclear about what ought to be understood by a "set," although a glimmer of the correct [definition] shines through when he says "a set is indeed completely restricted in that

everything belonging to it is determined in itself and is fully distinguished from all those not belonging to it."[18]

Frege believed that any definition was merely the product of specific characteristics and thus was nothing but the definition of a concept. This view was directly expressed in his *Die Grundlagen der Arithmetik,* where Frege wrote simply that "the assertion of number includes a statement of a concept."[19] But as he had already noted in his draft review, he did not regard Cantor's concept of number as sufficient to define anything. It was Frege's belief that the ultimate end of complete abstraction was "pure nothing," and this was echoed in his reference to Cantor's "unfortunate units" which were supposedly different, though not differentiated from each other by anything, once Cantor's thorough abstractions were complete.

Frege's criticisms of Cantor's definition of *finite* sets summarized the nature of his major objections. He referred to Cantor's view as equivalent to his own definition of finite numbering [*Anzahl*], but was quick to add that Cantor had erred in his approach.[20] Cantor defined a set M as finite if it "arises from *one* original element by successive addition of new elements in such a way that the original element can be retrieved by successively removing in reverse order the elements of M."[21] Frege wanted to know how one ought to understand "can be," and he insisted that there could be insurmountable hindrances to counting backwards to the original unit: the shortness of one's life, for example. But Frege was not trying to advocate a Kroneckerian sort of finitism in opposition to Cantor's transfinite numbers. He was actually concerned with stressing the inadequacy of Cantor's methods to present the most important ideas of set theory with satisfactory clearness and logical soundness. Above all, Frege regarded Cantor's successive removal of elements as a psychological process and therefore not to be used in rigorous mathematics. Frege preferred, even insisted upon strictly logical formulation of such procedures at every step.

Frege's final salvo was aimed at Cantor's order types. In part, his view was reminiscent of his earlier allusion to the sand pile and to the impossibility of abstracting instantaneously the number of grains it was assumed to contain. Frege stressed that Cantor's instructions to produce order types by abstracting everything but order from a given set of elements could not be taken as a legitimate definition. Either what was to be defined had to be assumed as known, or the process of abstraction could not be uniquely determined. Frege's fundamental objection continued to underscore the fact that abstraction was a psychological, not a formally logical process. This was the key to understanding the gulf which separated Frege and Cantor. Cantor had always relied upon the strength and sharpness of his intuition. He did not regard the process of abstraction as a muddled, illegitimate procedure because he applied it to cases that were familiar; as Frege suggested, the results were already known. Abstraction for Cantor was less a method for defining than it was a means of making his theory completely general. In the end, though he was unhappy with Cantor's presentation of set theory and the concept of number in particular, Frege was also convinced that Cantor's ideas could be formulated in a completely satisfactory way.[22] Frege had even indicated how it might be done, and one might almost take his concluding paragraph as a claim that his own foundations for arithmetic, using purely logical principles, could invest Cantor's own theory with the rigorous logical framework it required. In particular, Frege suggested that the concept of order types might well turn out to be of great importance for the future of mathematics.

Though in fundamental outlook, method, and interest there were clear differences between their respective positions, Frege emphasized an issue upon which he and Cantor were in perfect agreement: they both believed in the mathematical legitimacy of the infinite. Frege also sympathized with Cantor's resentment of the "academic positivistic skepticism" which Cantor believed was infiltrating even arithmetic to disastrous and pernicious effect. Frege was ready, as well, to predict that a crucial point was about to be reached.[23] The infinite could not be denied existence for long, excluded from arithmetic despite its powerful consequences for analysis. Frege realized that Cantor's work, once the logical rigor of his principles became clear, would speak unmistakably in favor of transfinite arithmetic. And once the infinite had attained its proper stature, Frege predicted that the conflict between the irreconcilable camps would begin. The infinite would make plain the gulf between those who proclaimed its existence and logical soundness and those who denied it. This was the issue, said Frege, over which mathematics would be wrecked: "Here is the reef on which it will founder. For the infinite will eventually refuse to be excluded from arithmetic, and yet it is irreconcilable with that [finitist] epistemological direction. Here, it seems, is the battlefield where a great decision will be made."[24] He could not have known in 1892 how soon the struggle would commence, nor how destructive the battle would be.

SETS, FINITE NUMBERS, AND MATHEMATICAL INDUCTION

Though there is no evidence to suggest that Frege's criticisms of the *Mittheilungen* had any influence upon Cantor as he prepared the *Beiträge* for publication, there were nevertheless changes in the latter with which Frege would have agreed. For example, the first paragraph of the *Beiträge* defined in explicit terms what the word set was to mean. Frege had complained specifically that Cantor had been unclear in the *Mittheilungen* as to how the term should be construed, and before the publication of the *Beiträge* it was true that Cantor had never bothered to give the concept any special prominence. In the *Beiträge,* however, it was imperative to make clear what sets actually *were,* since they constituted the initial concept basic to all of Cantor's transfinite set theory.

There may have been an even more important reason why Cantor should have chosen to stress the definition of set at the very beginning. He had discovered the inconsistencies of his transfinite numbers sometime in 1895. Except for a letter to Hilbert on the subject in 1896, he does not seem to have discussed the paradoxes at length with anyone until 1899, when he took up the issue in a series of letters with Dedekind. Cantor's solution to the matter was to restrict the definition of set in such a way that the paradoxes could be avoided. He was careful to say that a set must not only be well defined, but that it must also be conceivable as a "collection into a whole."[25] If Cantor first discovered the antinomies early in 1895, then this definition of set in Part I of the *Beiträge* might well have been calculated to exclude cases where paradoxes might arise.

Other matters were not so easily reconciled. Frege complained in his review of 1892 that Cantor had improperly developed the related ideas of finite set and number. In fact, three years later, one notable feature in the *Beiträge,* Part I, was its devotion of an entire section to the exclusive treatment of finite cardinal numbers, a subject to which he had not given much separate attention before. As with sets, so too with finite numbers, he had always taken their existence and properties for granted, used them as silent constituents as he worked out the details of more advanced ideas. All along they were actually first principles which he assumed were also sound. But if Cantor's presentation in the *Beiträge* was to be complete, he could not avoid giving more detailed consideration to the finite numbers upon which, after all, the transfinite numbers relied for their own definition.

Cantor defined (see above, Chapter 8) the finite cardinal numbers by first subsuming any single object e_0 under the concept of a set, producing $E_0 = (e_0)$, which he then denoted as "one." More generally: $\nu = \overline{\overline{E_{\nu-1}}}$, $E_\nu=(E_{\nu-1},e_\nu)=(e_1, e_2, e_3, \ldots, e_\nu)$.[26] He then proceeded to establish certain basic properties of the sequence of finite numbers and went on to define his first and least transfinite cardinal number $\aleph_0$. Even so, Frege would have been no happier with Cantor's lengthier treatment of finite number in the *Beiträge* than he had been with its comparative neglect in the *Mittheilungen*. In particular, Frege had been very suspicious of Cantor's claim that a finite set M was any set generated by successive addition of elements from an *original* element in such a way that, by proceeding backwards by successive removal of elements from a finite set, it was possible to return to the original element.[27] Frege discredited the entire argument as being psychological rather than logical in character. Instead, he preferred his own treatment of finite number which made the nature of succession clear among the integers, while displaying as well its intimate connection with mathematical induction.[28] Though Cantor made no effort to expose the essential features of either complete induction or the process of succession in any separate or special way in the *Beiträge,* he did omit any reference to the possibility of working backwards to original units of finite sets and stressed instead the fact that finite sets were those equivalent to none of their parts.

Like Frege, there was another prominent mathematician who wondered why Cantor had not been more definite in the *Beiträge* about the basic principles upon which the finite cardinal numbers depended. Late in the summer of 1895 Cantor received a letter from the Italian mathematician Giuseppe Peano, who asked about certain fundamental shortcomings in Cantor's introduction of the transfinite numbers and raised specific questions about Cantor's treatment of the finite numbers. In reply, Cantor admitted that he had not given explicitly any definition of finite cardinal numbers, nor had he justified the principle of induction he had used to generate all the natural numbers.[29] But he did contend that these were both direct and straightforward consequences of results drawn largely from Sections 1 and 5 of the *Beiträge*. By simply beginning with any single element subsumed under the concept of set, augmented repeatedly by unitary elements, the definitions by which cardinal numbers had been introduced in Section 1 as the powers of sets produced the natural numbers $1, 2, 3, \ldots$. Above all, Cantor was anxious to counter a point Peano had suggested in wondering whether the induction principle (upon which Cantor seemed to rely) wasn't really an unproven "postulate." Instead, Cantor insisted that the principle of induction followed "with absolute necessity directly from our definition of the sequence of finite numbers."[30] Cantor took this to mean that each cardinal ν was defined from $\nu-1$ in an entirely permissible way. Since his definition of all such finite numbers ν linked each, by clear and certain rules, to all those preceding, the return from ν backwards through all intermediary elements of the chain to 1 was just as possible as the process by which ν had been generated. Thus, one could go down from $\nu, \nu-1, \nu-2, \ldots, 1$ just as easily as one could go up from $1, 2, \ldots, \nu$. But as soon as one came to the first transfinite cardinal number $\aleph_0$, such regression was no longer possible. While this may throw some additional light upon Frege's reasons for objecting in 1892 to such ideas, it also makes it clear that Cantor had by no means given them up but was still willing to regard them as useful in characterizing finite sets and mathematically satisfactory in elucidating their nature.

The principle of mathematical induction could be taken to follow directly from the type of order which the natural numbers N exhibited as Cantor had defined them. Peano had chosen another approach in his *Formulario* and stated the induction principle as an axiom in his own presentation of the natural numbers.[31] It was then perfectly legitimate to define finite numbers as those which could be produced with the help of the principle of induction. The important distinction for Cantor was the fact he emphasized in a letter to Peano (December 14, 1895): finite numbers obeyed the principle, but infinite numbers did not. Consequently, in defining those numbers beginning from 1 and ascending by successive addition of units from ν to $\nu+1$, Cantor was not taking the principle of induction as a postulate as Peano suggested. But it was particularly remarkable, in light of Frege's earlier emphasis upon the desirability of making explicit both the definition of finite sets and the use to which the induction principle was put, that Cantor did not do so. Perhaps he was more interested in making clear the nature of order between finite cardinal numbers in N and thus left aside other features which were unnecessary for the major interests of the *Beiträge*. He only needed the order of N to provide a rigorous foundation for his definition of the first transfinite ordinal number, the order type ω. And he only needed to ensure that the well-ordered set N was also completely determined to define $\aleph_0 = \overline{\overline{\{\nu\}}}$, and this was the subject to which he turned directly after finishing with the characteristics of N outlined in Section 5.

Cantor was equally unmoved by Frege's critique of the role abstraction played in his definitions generally. The entire method was carried over virtually unchanged from the *Mittheilungen* to the *Beiträge*. If there was any point of agreement at all, it was linked to Frege's opinion that anything could be counted that was an object of thought. In a similar way, Cantor defined sets in 1895 as composed of any definite, separate objects of *thought* which could be comprehended as a whole. His abstractions were meant to ensure that the elements of his set theory were all objects of equal status in the mind, where every power or type of order existed as "an intellectual copy or projection of the given set M."[32] To appreciate fully Cantor's insensitiveness to Frege's dislike of abstraction in mathematics, it is important to emphasize his own rather special understanding of the transfinite numbers and the nature of their existence. Though he never explained in detail his own personal views of mathematics in any of his publications, he fortunately confided them to his friend, the French mathematician Charles Hermite. Cantor elaborated his thoughts on such matters in a letter of November 1895.

CANTOR'S CORRESPONDENCE WITH HERMITE CONCERNING THE NATURE AND MEANING OF THE TRANSFINITE NUMBERS

Cantor's letter to Hermite is a curious document.[33] Mathematically, it is of interest chiefly because he devoted nearly two pages to epistemological questions concerning the nature of numbers and their mode of existence. Hermite had expressed his own opinion on such matters in an earlier letter:

> The (whole) numbers seem to me to be constituted as a world of realities which exist outside of us with the same character of absolute necessity as the realities of Nature, of which knowledge is given to us by our senses.[34]

Cantor went even further by insisting that the reality and absolute legitimacy of the natural numbers were much greater than any based on their existence in the real world and perceived through one's senses. He explained to Hermite that this was so for one very simple reason. Both separately and collectively as an infinite totality, the natural numbers "exist in the highest degree of reality as eternal [*ewige*] ideas in the *Intellectus Divinus*."[35] Cantor added that this knowledge came to him as no new revelation. In fact, he had expressed similar views at the very outset of his career, in 1869, in one of three theses at the end of his *Habilitationsschrift*. The third of these (which he publicly defended) read as follows:

> Numeros integros *simili modo atque corpore coelestia totum quoddam* legibus et relationibus compositum efficere.[36]

Years later, Cantor said he realized that Saint Augustine had expressed much the same idea in his *De civitate Dei;* he even went so far as to footnote an entire excerpt in the *Mittheilungen,* specifically part of Augustin's Chapter 19 of Book XII: "Contra eos, qui dicunt ea, quae infinita sunt, nec Dei posse scientia comprehendi."[37] In attempting to counter a variety of views opposed to the existence of the infinite, Cantor concluded that the transfinite numbers were just as possible and existent as the finite numbers: "Therefore the transfinites are just as much at the disposal of the intentions of the Creator and His absolutely infinite [*unermesslichen*] Will as are the finite numbers."[38]

Thus the efficacy of Cantor's theory was ultimately referred to the Divine Intellect where the *Transfinitum,* all the transfinite numbers, existed as eternal ideas. This was a strong form of Platonism but one to which Cantor returned repeatedly for support. As noted earlier, the religious implications were as significant to him as the mathematical ones, and in the final analysis, Cantor could be sure of the propriety of his abstractions because they found their ideal representation in the mind of God. Hermite placed the reality of number in the concrete and absolute reality he found in the perceptual world.

Cantor needed to go much further and found his most important source of inspiration and reassurance in God's infinite capabilities.[39] As the all-knowing and all-powerful Divine Intellect, God played a key part in determining the ontological status of the *Transfinitum* as Cantor regarded it. Consistency alone was the determining factor in any question of mathematical existence, since God could realize any "possibility," and by possibility Cantor meant that the ideas capable of realization *in concreto* be only consistent.[40] In this respect he seems very much a latter-day Leibnizian, believing that it would have contradicted God's omnipotence had he been unable to realize any possible, that is, consistent idea. But Cantor did not go so far as to insist that such possibilities actually had a physical existence somewhere in the phenomenal world. He would only say that if ideas were consistent, then they were possibilities, and as possibilities they had to exist in the mind of God as eternally true ideas; this was sufficient to confer upon them the right to mathematical existence. Consequently, none of the criticism leveled against his theory of the transfinite numbers could ever shake his belief in their necessary, even absolute reality.

Though Cantor would only argue that the *possibility,* in terms of the consistency of an idea was necessary before one could be certain that God would comprehend and thereby ensure its reality, he saw nothing inherently implausible about the existence of the actually infinite *in concreto*. He made this clear in responding to an article in which Axel Harnack had argued that the existence of the infinite *in concreto* was impossible.[41] Assuming the infinite to have a real existence led to contradictions. Cantor wrote to Harnack and wanted to know his reasons, since he could find no proof or justification for the opinion expressed in Harnack's paper. Cantor's letter is of interest because it makes explicit his position with respect to the actual infinite.[42] He referred directly to the nature of number as it was embodied in the concept of powers abstracted from the objects of any set *taken as a whole*. He emphasized these last four words as constituting the significant element Harnack had overlooked.

In Cantor's view the transfinite numbers arose from infinite sets just as the finite numbers arose from finite sets. He even recalled that this similarity was consonant with the approach he had originally taken in trying to formulate his infinite numbers by abstracting specific qualities from particular infinite sets to produce the concept of transfinite cardinal numbers and order types. It was in this sense, as abstractions from particulars, that Cantor referred to the powers of sets as being the only *representatio generalis* which could arise by abstracting all specific characteristics, including order, from any given set.[43] This explains why he never doubted that his theory of the transfinite cardinal numbers, as powers of infinite sets, was necessarily and absolutely the only true theory possible.

Cantor realized that mathematicians, nonetheless, differed widely with respect to their interpretations of something so basic as the system of finite numbers. He had alluded to this problem when he admitted in his letter to Harnack that matters of mathematics *in concreto* might always be subject to different opinion and points of view.[44] The only way in which the reasonableness of various approaches could be decided was by appealing to other than the merely concrete factors involved. The correctness of his own view of number, he believed, was established through the additional consideration of the ideal, theoretical connections between the objects of thought, whether they were finite numbers or his entire transfinite set theory. Similarly, Cantor was careful to make clear the differences which distinguished numbers in and of themselves as part of the Absolute Intelligence from numbers as they were generally regarded, obscurely and without full comprehension, as part of one's own limited capacity for comprehension.

In the phenomenological world one might be used to thinking of numbers as linked in succession, from a given n proceeding to $n + 1$, but Cantor thought the numbers used in mathematics had an entirely separate existence, one wholly distinct and independent from any considerations of place in a chain of numbers beginning from 0 to 1 and never ending. For the sake of convenience he allowed that one might think of numbers in such terms, but he insisted that they also existed as simple concepts, as unities. As such, one could proceed from sets, not with arbitrariness but with absolute necessity to the results he had discovered over the course of his lifetime.[45] Such beliefs only helped strengthen his resolve that despite all criticism, regardless of any opinion to the contrary, no rival theories were even *possible*.

Thus he explained how mathematicians who concentrated upon different aspects of number as it was represented in the real world might hold views at variance with his own. One prominent example involved divergence of opinion over whether the ordinal numbers were logically prior to the cardinal numbers or not.[46] In order to carry out his definition of the transfinite numbers, Cantor needed the cardinal property which regarded each number as being a separate unity unto itself. The ordinal conception of number was not enough, Cantor stressed, in opposing his detractors. Inductive procedures were sufficient for definition of finite numbers, but they would never have permitted the production of even Cantor's first transfinite number ω. Such numbers were necessarily beyond reach by virtue of their very definition, and hence completely inadmissible if the inductively ordinal form of number was taken as logically necessary before any cardinal distinctions could be allowed. Thus Cantor needed to be sure of at least the independence of the cardinal conception from any ordinal view of number. The ultimate authority to which Cantor turned in support of this view was again the Absolute Intelligence of God as the *Intellectus Divinus*. Cantor emphasized exactly this point in correspondence with Veronese:

> One must distinguish numbers as they are in and of themselves, in and of the Absolute Intelligence, and those same numbers as they appear in our limited discursive mental capacity and are defined (in different ways) by us for systematic or pedagogic purposes . . . The (cardinal numbers) are all independent from one another (taken absolutely), all are equally good and equally necessary metaphysically.[47]

Just as God had earlier confirmed the necessary reality of the transfinite numbers, Cantor made a similar appeal to ensure the correctness of his own belief in the independent origin of the cardinal numbers. Once more Cantor had invoked God to guarantee the absolute truth of the principles upon which set theory was based.

Nothing reveals as profoundly the importance Cantor attributed to the inextricable bond he felt between his mathematical ideas and his religious beliefs than a letter he wrote to the neo-Thomist priest I. Jeiler in 1888.[48] He was anxious that his view of the infinite be carefully studied by the Roman Catholic Church in order that it might be taken constructively and not be interpreted as being at variance with accepted dogma. Specifically, he was worried that unless the Church took notice of the implications of his research concerning the nature of infinity, it might fall into dangerous theological errors:

> In any case it is necessary to submit the question of the truth of the Transfinitum to a serious examination, for were it the case that I am right in asserting the truth or possibility of the Transfinitum, then (without doubt) there would be a sure danger of religious error in holding the opposite opinion, for: *error circa creaturas redundat in falsam de Deo scientiam (Summa contra gentiles II,* 3*)*.[49]

Cantor thus emerged as a modern Galileo. Each of them felt it was his duty to make the Church accept the reality of a world which God had created, and to face the fact that man had been given the capacity to understand the universal order as it had been created. The Church could only fall into error if it stubbornly ignored the necessary logic of the reasoning mind. Like Galileo, Cantor was certain that there could be no doubt of the correctness of his theory once it had been carefully studied and understood. Moreover, Cantor also believed in the absolute truth of his set theory because it had been revealed to him, as he once told Mittag-Leffler, from God directly.[50] Thus he may have seen himself not only as God's messenger, accurately recording and transmitting the newly revealed theory of the transfinite numbers but as God's ambassador as well. In fact, Cantor did regard his work as a sacred mission. He believed that God had actively *chosen* to keep him in Halle, rather than allow him to have a post in Göttingen or Berlin. By denying his hopes for a better university position, God had ensured that Cantor would better serve Him and the Roman Catholic Church.[51] He had been given faith in the necessary truth of his mathematical discoveries by virtue of their having come from, and their being part of, God's infinite wisdom. In return, Cantor hoped to aid the Church in correctly understanding the problem of infinity. He drew upon the *Summa contra gentiles* of St. Thomas to support his contention that misunderstandings about the nature of the real world could easily lead to false knowledge of God.[52] God, however, had transmitted the true understanding of the infinite to Cantor, and reciprocally, Cantor believed that his study of the *Transfinitum* could foster improved and proper understanding of God.

In view of the very fundamental link between such ideas and Cantor's own perception of mathematics, it is surprising that his deep religious convictions have received so little attention in discussions of his development of set theory.[53] The early literature dealing with Cantor's life and work does contain sporadic mention of his religiosity, but nothing concerning its significance. Cantor not only found encouragement and support from his faith in God, but he also believed that he was destined to put that knowledge into service for the greater understanding of God and nature. That Cantor took such matters seriously, with the utmost sincerity, is further suggested by yet another aspect of the *Beiträge,* one which has not been previously discussed. But first, Cantor's rejection of a rival theory of the infinite, one criticized sharply in the *Beiträge,* must be considered.

CANTOR'S REJECTION OF INFINITESIMALS AND VERONESE'S *FONDAMENTI DI GEOMETRIA*

Cantor's certainty that his transfinite numbers were reflected in "real ideas" drawn absolutely and necessarily from the nature of sets greatly aided him in combating rival theories. Viewed as such, he was convinced that his characterization of the infinite was the *only* characterization possible, and he was consequently prepared to argue that no rival theory could ever be true unless its results were identical with his own.[54] This attitude was repeatedly reflected in his eagerness to represent transfinite set theory as absolutely certain, open neither to variant opinions nor to opposing interpretations. In this connection the work of the Italian mathematician Giuseppe Veronese was particularly unwelcome, and Cantor devoted some of his most vituperative correspondence, as well as a portion of the *Beiträge,* to attacking what he described at one point as the "infinitary Cholera-bacillus of mathematics,"[55] which had spread from Germany, through the work of Thomae, Paul du Bois-Reymond, and Stolz, to infect Italian mathematics. Mostly at issue, but not exclusively, was the question of infinitesimals.

Veronese had just published a German translation (in 1894) of his *Fondamenti di Geometria,* and Cantor felt it was timely to warn everyone of its manifold errors.[56] Very early in his career Cantor had denied any role to infinitesimals in determining the nature of continuity, and by 1886 he had devised a proof (see Chapter 6) that the existence of such entities was impossible.[57] Thus any attempt to urge their legitimacy could be interpreted as a direct challenge to one of the most basic principles of Cantor's set theory, since it was in terms of the character of his transfinite numbers that he argued the impossibility of infinitesimals. Moreover, as Vivanti put the matter to Cantor in a letter not long after Veronese's book had appeared in Italian, any acceptance of infinitesimals necessarily meant that Cantor's own theory of number was incomplete.[58] Thus to accept the work of Thomae, du Bois-Reymond, Stolz, and Veronese was to deny the perfection of Cantor's own creation. Understandably, he launched a thorough campaign to discredit Veronese's work as vigorously as he could. The very nature of Cantor's own view of his set theory determined the sort of attack he would make.

Cantor believed that his introduction of order types, together with cardinal numbers or powers included *everything* capable of being numbered that was thinkable. To Cantor this meant that no further generalizations of the concept of number were conceivable.[59] Moreover, there was nothing in the least way arbitrary about his definitions of number; cardinals and order types were perfectly natural extensions of the number concept. This had always been one of the mainstays of Cantor's position. Equally free from arbitrariness was his condition for the equality of two order types, which was given in terms of their similarity. This condition, too, followed with absolute necessity from the concept of order type and hence permitted no alteration. Cantor claimed that Veronese's failure to understand this absolute character of the transfinite numbers was the major source of error in his misguided attempt to establish a different sort of infinite number in the *Fondamenti di Geometria*.

Veronese, in defining the equality of "numbers of ordered groups," (which Cantor took to be nothing more than a poorly plagiarized form of his own definition for order types of simply-ordered sets), had made a "logical mistake."[60] "Numbers are equal," Veronese had written, "if their units correspond uniquely and in the same order, and *if the one is not a part of the other or equivalent to a part of the other*."[61] Cantor dismissed the efficacy of Veronese's definition on the grounds that it was viciously circular and therefore meaningless. To employ a concept of inequality ("not a part") in a definition of equality *presupposed* that one already knew what was meant by equality. This *petitio principii* rendered Veronese's entire approach suspect and therefore mathematically unsound.[62]

Even if the circularity could have been removed, Cantor was ready to add that the definition, in any case, was arbitrary. He complained to Peano that Veronese believed the definition of equality, both for numbers and for order types, was entirely at the mercy of one's choice, which was a heretical suggestion from Cantor's point of view. Instead, Cantor emphasized his usual refrain: "The definition of the equality of order types follows *with necessity* from the definition itself of *order types*."[63] For Cantor the necessary and only definition possible concerned the *similarity* of types in question. As he said in the *Beiträge,* Veronese had failed to grasp the obvious fact that transfinite arithmetic followed not from arbitrary principles but naturally from certain necessary principles derived from sets themselves. Because Veronese's definition of equality was circular and arbitrary, it was impossible to establish any meaningful comparison of numbers. "Consequently," wrote Cantor, "one should not be surprised at the lawlessness with which he operates with his pseudo-transfinite numbers, and with which he ascribes properties to them which they cannot possibly possess for the simple reason that, in the form imagined by him, they have absolutely no existence except on paper."[64]

For Cantor, the proper foundations for arithmetic were indisputable because they were necessary, drawn directly from his own view of the nature of sets. But there were additional reasons for rejecting Veronese's theory of number, ones which stressed the impossibility of the results Veronese had obtained. Examining Veronese's transfinite numbers and his infinitesimals, Cantor easily spotted what seemed to him obviously erroneous conclusions. Not surprisingly, his criticisms were based on the incompatibility of Veronese's work with his own.

Concerning Veronese's infinitely large numbers, Cantor once commented that as soon as he had seen $2 \cdot \infty_1 = \infty_1 \cdot 2$, he knew that the entire theory was necessarily false.[65] Assuming the absolute character of his own theory of transfinite numbers, Cantor had concluded that any theory of the infinite would have to be comparable with his, and one major requirement was the noncommutativity of arithmetic operations for infinite ordinal numbers. Since Veronese's numbers clearly violated such necessary laws, they were inadmissible. To Cantor's way of thinking, it was as simple as that.

Infinitesimals, at the opposite extreme, constituted an even greater horror in Cantor's list of mathematical "ghosts and chimeras."[66] He accused Veronese of letting false principles lead him to the equally impossible assertion of infinitely small numbers. Cantor's proof of the contradictory nature of infinitesimals (Chapter 6) involved the Axiom of Archimedes. Basically, Cantor refused to regard it as an axiom at all. Instead, he argued that it followed directly from the concept of linear number.[67] Numbers were linear if a finite or infinite number of them could be added together producing yet another linear magnitude. But since he had also assumed what Hilbert, in his *Grundlagen der Geometrie* later called the Axiom of Continuity, all Cantor's assertions followed directly.[68] Since the Axiom of Archimedes could be deduced from the Axiom of Continuity (and vice versa), in Cantor's view there was no difficulty in asserting that the Archimedean "Axiom" could be proven. In terms of Cantor's assumptions, it was as provable as the nonexistence of infinitesimals. Moreover, had Cantor agreed that the Archimedean property of the real numbers was merely axiomatic, then there was no reason to prevent the development of number systems by merely denying the axiom, so long as consistency was still preserved. But to have allowed this would have left Cantor open to the challenge that, if infinitesimals could be produced without contradiction, then his own view of the continuum was lacking and the completeness of his own theory of number would have been contravened. On the other hand, were the Axiom of Archimedes *not* an axiom at all, but a theorem which could be proven from other accepted principles, then he could rest assured that one could not merely *deny* the proposition and produce a consistent theory of infinitesimals. Cantor was so persuasive, in fact, with his disavowal of infinitesimals, that he was able to convince Peano, who wrote an article on the subject in his *Rivista di Matematica*.[69] Bertrand Russell went further than Peano and argued in his *Principles of Mathematics* that mathematicians, completely understanding the nature of real numbers, could safely conclude that the nonexistence of infinitesimals was firmly established.[70] He was wise to add, however, that if it were ever possible to speak of infinitesimal numbers, it would have to be in some radically new sense.

Finally, there was an additional argument, one which Cantor found equally persuasive in rejecting the attempts Thomae, du Bois-Reymond, Stolz, and Veronese had made to develop logically sound theories of the infinitesimal. In writing to Veronese on the subject, he accused the infinitesimalists of talking nonsense, since in the realm of the possible (which he explained as encompassing all of nature in the broadest sense) there *were* no infinitely small entities. Cantor stressed that he linked the veracity of his transfinite numbers with *real ideas* produced directly from *sets,* and he once challenged Veronese to show any *real ideas* corresponding to the supposed infinitesimals.[71] Until Veronese could do so, Cantor promised to continue his opposition, and would furthermore regard any deviation from the Axiom of Archimedes, which he took as proven, to be an error of the greatest seriousness.

Thus any theory of number at variance with Cantor's basic results was inevitably wrong. If anyone produced an ordinal arithmetic of infinite numbers that was commutative, there was no possibility of its being correct. Should infinitesimals be proposed, they were demonstrably impossible. Unlike infinite numbers, whose nature and properties could be taken from the properties of *sets,* there were *no* sets from which infinitesimals could be directly abstracted as could ω or $\aleph_0$. Such attempts were entirely lacking in any content, and were utterly meaningless. These were all necessary and absolute conclusions about which there could be no debate. Cantor felt that Veronese and his sympathizers should recognize the compelling logic behind his research. Veronese, Cantor counseled, ought to admit his mistakes and begin again with the only true theory of the infinite that was possible.[72] And that, Cantor believed, was his own.

HYPOTHESES NON FINGO

The three Latin quotations set at the very beginning of the *Beiträge* characterized in varying ways the significance of transfinite set theory in Cantor's own mind. Each said something very special about Cantor as a person and as a mathematician. Taken together, they offer a key to understanding an otherwise hidden bond between his private life and his public mathematics.

"Hypotheses non fingo," was Newton's famous line from the General Scholium to the second edition of his *Principia*.[73] The words were even reflected directly in the language of one passage in particular from Cantor's *Beiträge,* and thus Newton's dictum occupied a unique position among the three aphorisms. The allusion occurred in Section 7 of Part I, but no one reading Jourdain's translation would ever have recognized his use of "imagined" for Cantor's *fingierten*. The original passage from the *Beiträge* read as follows:

> Nachdem Herr V[eronese] auf solche Weise das unentbehrliche Fundament für die Vergleichung von Zahlen sozusagen *freiwillig preisgegeben* hat, darf man sich über die Regellosigkeit nicht wundern, in welcher er des weiteren mit seinen pseudo-transfiniten Zahlen operiert und den letzteren Eigenschaften zuschreibt, die sie aus dem einfachen Grunde nicht besitzen können, weil sie, in der von ihm *fingierten* Form [italics mine], selbst keinerlei Existenz, es sei denn auf dem Papiere, haben.[74]

The connection with Veronese was explicit, precisely in terms of the controversy over his rival theory of the transfinite. Depsite Cantor's advice that Veronese abandon a view of number that was demonstrably impossible, Veronese was by no means convinced that Cantor's arguments were so compelling. His most basic rejoinder had always been very simple, but one which did not have the slightest effect upon Cantor. From the very beginning Veronese emphasized that he was operating upon entirely different *hypotheses*.[75]

Cantor was appalled at the very suggestion; he wrote directly to Veronese and said so. The letter still survives in its draft form and offers many valuable insights into Cantor's reasons for regarding mathematics as he did.[76] He began his retort to Veronese by denying any use of hypotheses in his arithmetical studies. He turned directly to finite arithmetic for support of his contention that hypotheses were entirely inappropriate and had no place whatsoever in such mathematics, except insofar as attempts to establish their existence in nature were concerned.

Unlike Riemann and Helmholtz ("meta-geometers" as Cantor called them), Cantor insisted it was absolutely impossible to set out hypotheses for arithmetic.[77] The basic laws had been recognized since the beginning of time, recorded in the sequence 1, 2, 3, . . . , and were as immutable as Cantor believed the laws of transfinite arithmetic to be. Any hypotheses which violated the basic truths of arithmetic had to be regarded as contradictory or false, just as $2+2=5$ was manifestly impossible. It was as incongruous with the nature of finite arithmetic as was the idea of a square circle in geometry. The best evidence Cantor felt he could offer for the impropriety of Veronese's variant hypothesis was the completely contradictory existence of infinitesimals it permitted.[78] But the roots of their difference reached much further than the question of hypotheses alone. Veronese's belief that Cantor's was not the only definition of equality possible among transfinite ordinals, for example, was founded upon his contention that definitions could be given arbitrarily, just as hypotheses could be changed and alternate theories produced. Cantor denied that either was allowable in arithmetic. Once more he drew support from his view of the nature of numbers. Nothing was variable or arbitrary in their definition, everything depended upon the nature of sets, and from these in a natural, yet absolute way, all of arithmetic followed.[79] It was as simple as that. There *could* be no hypotheses, because there were no alternatives. Finite arithmetic could not be changed, and Cantor's transfinite set theory, with its corresponding transfinite arithmetic, was similarly absolute.

Hypotheses were allowable in metaphysics, where one had to face the problem of ascribing a reality in nature, in the phenomenological world, to mathematical concepts. But even here Veronese was to have no luck in winning even the slightest recognition for his "hypotheses" in the *Fondamenti di Geometria*. Was he completely unaware, asked Cantor, that he had implicitly formulated purely *arithmetical* hypotheses with the creation of his *geometrical* ones? Veronese's attempt to provide an entirely new foundation for geometry was doomed to failure, because Cantor could only permit geometrical hypotheses that were consistent with arithmetical ones.[80] Veronese had failed on both counts, and thus even where hypotheses might have been allowed, Veronese's were still unacceptable. They contradicted prior and more basic mathematical principles.

Cantor always regarded Veronese's entire program as an elaborate "fantasy" and returned on various occasions to the subject in his correspondence. The fundamental questions at issue were always the same. Infinitesimals generally and Veronese's transfinite numbers in particular were inconsistent; they could have no real mathematical existence and should be taken as nothing but meaningless marks on paper. The paper, Cantor added, belonged in the wastebasket.[81]

When Cantor wrote, at the very beginning of the *Beiträge,* "Hypotheses non fingo," it was more than a rejoinder to Veronese and the infinitesimalists. At its most profound, it was indicative of his entire philosophy of mathematics. In three words he characterized what he believed to be true not just of the results compiled in the *Beiträge* but of his entire life's work. He had produced a mathematical theory of everlasting value, and its character in all essentials was natural, necessary, and absolute.

Cantor's declaration that he framed no hypotheses in presenting his theory of transfinite sets was only one aspect of the *Beiträge* he meant to illuminate in terms of his three aphorisms. The second was a line chosen from Francis Bacon's *Scripta in naturali et universali philosophia:* "For we do not arbitrarily give laws to the intellect or to other things, but as faithful scribes we receive and copy them from the revealed voice of Nature."[82] This too fit securely into place as an expression of Cantor's overall view of both his mathematics and his philosophy. It was intimately connected with his selection of Newton's disclaimer: "Hypotheses non fingo." In fact, Bacon's words served as the caption to Cantor's long letter to Vivanti of 1893 in which he condemned the doctrine of infinitesimals.[83] Thus Cantor directly linked Bacon's words with Newton's pronouncement rejecting hypotheses. The laws of thought, the laws of mathematics, were not subject to the arbitrary whim of individual caprice; on the contrary, they were inherently established and uniquely understandable.

The reference to nature in Bacon's pronouncement is also suggestive. By nature Cantor meant the realm of the possible, and he once noted parenthetically that this was exactly the meaning he attributed to nature in its widest sense.[84] If it is therefore permissible to substitute "the possible" for nature in Bacon's statement, and if we recall that Cantor regarded the reality of the possible as guaranteed by its consistency, by its necessary existence as an eternal truth in the *Intellectus Divinus,* then it is possible to suggest an even deeper meaning of the aphorism for Cantor.

Where God was concerned, it was impossible to entertain hypotheses. There were no alternatives to be considered. The principles of mathematics, of set theory, and the transfinite numbers, followed directly from Nature. Cantor saw his own role, as mathematician, in terms of a faithful secretary, receiving and describing what had been revealed to him by God. Thus the *Beiträge* was meant to be a faithful record of the mathematics of Nature. Cantor was merely the intermediary through whom these great, immutable truths had been communicated.[85]

The last of Cantor's three quotations was a familiar passage from the Bible and is best understood in conjunction with his quotation from Bacon: "The time will come when these things which are now hidden from you will be brought into the light."[86] As Nature's faithful scribe, Cantor in the *Beiträge* was bringing a new understanding, hitherto hidden from mathematicians, into the light. The passage, drawn from First Corinthians, similarly suggests that Cantor saw himself as the intermediary, serving as the means of revelation. It may also have been meant to express his belief that despite prevailing resistance to his work, it would one day enjoy recognition and praise from mathematicians everywhere.

In the meantime, Cantor could remain secure in the knowledge that he had been the one to show the way. He was convinced that his work eventually would have to be acknowledged as providing the only possible arithmetic embracing both the finite and the infinite. God was the source of his inspiration and the ultimate guarantor of the necessary truth of his research. Therefore it could not fail to be absolutely correct, and time would eventually vindicate everything he had done. Cantor reinforced such confidence through his belief that he had been specially chosen, something he may even have recognized as a young man. In 1862 he had written to his father (who had just consented to his son's pursuing a career in mathematics) in order to explain that "My soul, my entire being lives in my calling; whatever one wants and is able to do, whatever it is towards which an unknown, secret voice calls him, *that* he will carry through to success."[87]

CHAPTER 11

The Paradoxes and Problems of Post-Cantorian Set Theory

Solatium miseris, socios habuisse malorum,
—Gottlob Frege, 1902.

Felix Bernstein, a mathematician of the generation that began in earnest to develop Cantor's work, once remarked that set theory had suffered the fate of all new discoveries in that its early development had been haphazard and unsystematic.[1] But whatever its initial imperfections, Bernstein was certain that Cantor's theory of transfinite sets could be given rigorous and satisfactory treatment.[2] Gottlob Frege had expressed a similar opinion when he wrote that despite his many objections to what he termed Cantor's "psychologism", he nevertheless believed that the foundations of set theory could be repaired, and that Cantor's results would therefore remain valid.[3] Such optimism rested upon Frege's belief in the compelling necessity of mathematical logic, faith in which had led him to defy anyone to find fault with his own reduction of arithmetic to purely logical principles. As he was bold to say, "It is *prima facie* improbable that such a structure could be erected on a base that was uncertain or defective." Thus he challenged his opponents to show that such principles could ever lead to manifestly false conclusions and added confidently, "But no one will be able to do that."[4]

Bertrand Russell did. In a letter written to Frege shortly before Volume II of his *Grundgesetze der Arithmetik* was published in 1903, Russell described a purely logical contradiction arising from the concept of a class which did not belong to itself.[5] As Frege said, nothing could be worse than discovering that a basic premise of a completed work was faulty. In Frege's formulation, Russell's paradox cast doubt upon the meaning and legitimacy of use of the extensions of concepts, that is, of classes. Translated into Cantor's terminology, discovery of the paradoxes of set theory suggested that something was wrong with Cantor's definition of set. The problem was clearly a much deeper one than Frege's earlier critique of Cantor's definition had implied, for instead of merely requiring a more careful description to make the idea of set precise in a purely logical, nonpsychological form, it suddenly seemed as though the concept of set (as both Cantor and Frege had alternately expressed it) was *inherently* faulty. The more rigorous the logical formulation, in fact, the clearer seemed the contradictions.

Frege quipped that if any comfort was to be had, he drew it from the fact that misery loved company. As he put it, he was not alone at least in having to face the dilemma prompted by the antinomies. Anyone who had used ideas similar to Frege's extensions of concepts, classes, or sets was equally affected by Russell's paradox. The question was not one of whether a given particular foundation was adequate for arithmetic, but whether *any* logical foundation was possible at all. When applied in the context of a carefully developed logical program like Frege's *Grundgesetze,* the devastating effect of the paradoxes was evident: they threatened the entire structure. At first the difficulties of Cantor's theory seemed rather easily remedied, but as the years passed, early solutions provoked closer scrutiny and soon proved to be inadequate. The foundations of set theory, it appeared, required more systematic treatment than Cantor had ever felt necessary. The nineteenth century, often dubbed the age of rigor in mathematics, had to make way for a new era in which rigor would assume an even more demanding sense. For set theory this meant an end to the naive approach taken by Cantor. In its place emerged a more refined axiomatic approach designed to deal successfully with the problem of mathematical consistency. If mathematics could not be made certain, then where could anyone turn and hope for absolute knowledge about anything? By the turn of the century it was clear that important matters of epistemology affecting the foundations and future of mathematics would be decided along with the fate of Cantor's transfinite set theory.

THE DISCOVERY OF INCONSISTENT SETS: CANTOR'S CORRESPONDENCE WITH DEDEKIND

It is not a little ironic in light of subsequent history that the first mathematician to discover the antinomies of set theory was Cantor himself. Though Burali-Forti was the first to publish an account of the inherent inconsistency of the set of all ordinal numbers, Cantor had already anticipated the problem and by 1895 he was at work trying to remedy the paradoxes with a minimum of damage to his system of transfinite numbers.

Two major problems of set theory had been left unanswered in Part I of the *Beiträge*. The continuum hypothesis was still unresolved, as was the problem of the comparability of all transfinite cardinal numbers. Cantor had implied in Part I that he was in a position to show that among any two cardinals $\mathfrak{a}$ and $\mathfrak{b}$, exactly one of the relations $\mathfrak{a}<\mathfrak{b}$, $\mathfrak{a}=\mathfrak{b}$, $\mathfrak{a}>\mathfrak{b}$ had to be true.[6] Despite the claim, he never actually produced a proof.

Until this last question was settled, it was conceivable that there might be infinitely large nonequivalent sets of which neither one was equivalent to a proper subset of the other. Then, although $\mathfrak{a}\neq\mathfrak{b}$, it would be impossible to say that either $\mathfrak{a}<\mathfrak{b}$ or $\mathfrak{b}<\mathfrak{a}$. Unless it could be shown that every set could be well-ordered, thereby associating every set with an aleph as cardinal number, there was no rigorous means of ensuring that every cardinal number was also an aleph. Since he could not even confirm for the special case of the continuum that it could be well-ordered, it was entirely possible that it might *not* be, and that it would then have no aleph as its corresponding cardinal number. Thus the continuum hypothesis itself remained as recalcitrant as ever.

All these matters turned on the question of well-ordering. Central to this problem was the question of whether any infinite sets might exist that were not equal in power to one of Cantor's transfinite alephs. Such sets would be noncomparable in terms of their cardinal numbers, and a natural candidate was the set of *all* sets. If it contained any noncomparable set, then it too could not be expected to find its power among the alephs. This was exactly what Cantor discovered. From his paper of 1891, he could argue that the set of all sets had to give rise to a set of larger cardinality; the set of all its subsets.[7] But since this set had to be a member of the set of *all* sets, the paradoxical conclusion was inevitable that a set of lower cardinality actually contained a set of higher cardinality. This violated the most elementary of Cantor's conclusions about the comparability of sets. But what was the flaw in his thinking that produced the paradox?

Certainly there was an obvious circularity in trying to think of the set of all sets. It was impossible to regard it as a completed, self-contained object. Cantor's proof in 1891 had made it clear that whatever the cardinality of a given set might be, it was always possible to produce one of greater cardinality. Hence there was *no* end to the succession of transfinite alephs.[8] To posit a largest aleph was as impossible as trying to imagine the set of all sets as constituting a well-determined, bounded unity. As early as the *Grundlagen* he had recognized the fact that there could be no largest transfinite number, just as there could be no largest finite number.[9] But the natural numbers could be considered as a whole without contradiction. The first transfinite number ω was then the least of all infinite numbers. But there was no similar counterpart for the set of *all* transfinite numbers. The idea was self-contradictory, as he had discovered.

To a mathematician like J. Thomae, this was a shattering discovery. As a result of the paradoxes, he felt that nothing had become so uncertain as mathematics.[10] Cantor, on the contrary, regarded the antinomies as positive results which fully complemented the advance of his research. It was Bertrand

Russell who finally made it unmistakably clear that such was not the case at all. But before considering the paradoxes, what they meant to other mathematicians, and the subsequent development of set theory and mathematics generally, the question arises as to how Cantor could possibly have regarded the contradictions as beneficial? How could they have seemed constructive additions rather than early warnings that the entire theory might be in imminent peril of total collapse?

Cantor explained his interpretation of the paradoxes in an exchange of letters with Dedekind during the summer of 1899. His first letter of August 3 reiterated what he had known since the *Grundlagen*: the construction process of alephs and of the corresponding number classes was absolutely limitless.[11] But apparently he had never considered the formal, logical contradictions a system like the collection Ω of all order types engendered. When he first thought about such matters in 1895, he recognized that as a well-ordered set, Ω should have a corresponding order type. If one assumed Ω to be of type δ, it clearly had to be larger than any order type in Ω. But since Ω contained *all* order types, one was faced with the dilemma of having to say that $\delta > \delta$. Thus there was something inherently illegitimate in regarding Ω as a set, as a consistent collection, and Cantor expressed himself to this effect in a theorem:

> *Theorem A:* The system Ω of all (ordinal) numbers is an absolutely infinite, inconsistent collection.[12]

The same could then be argued in a similar fashion for the system ת (taw) of all transfinite cardinal numbers:

> *Theorem B:* The system ת of all alephs:
>
> $$\aleph_0, \aleph_1, \ldots, \aleph_\nu, \ldots, \aleph_{\omega_0}, \aleph_{\omega_0+1}, \ldots, \aleph_{\omega_1}, \aleph_{\omega_1+1}, \ldots,$$
>
> comprises in its order of magnitude a system similar to Ω, and thus is likewise an absolutely infinite, inconsistent succession.[13]

It was but a short step to showing how such inconsistencies actually led to highly constructive and far-reaching results which Cantor had been unable to obtain in any other way. In terms of Theorem B, for example, he could now settle the question of whether any cardinal numbers might exist that were not listed in Ω. In other words, could there be sets whose powers were not alephs? Cantor answered with an emphatic no and explained to Dedekind how this could be shown by recognizing the very inconsistency of the sets Ω and ת. If one were to assume that some collection V did exist which was not equal in power to any aleph, then Cantor claimed it had to be an inconsistent set.[14] He based this conclusion on the assumption that, because of the nature of V, it must be possible to project the entire system Ω into V. Since there was consequently some subcollection $V' \subset V$ which was similar to Ω, the entire collection V had likewise to be inconsistent, as was Ω. Cantor therefore concluded that every consistent set had to be power-equivalent to a definite aleph:

> *Theorem C:* The system ת of all alephs is nothing other than the system of all transfinite cardinal numbers.[15]

Above all, Cantor could conclude that the cardinality of the continuum must be an aleph indexed in ת. But he could go further. He could settle at last the question left unanswered in the *Beiträge* concerning the comparability of all cardinal numbers. Since Theorem C ensured that all transfinite cardinals were necessarily alephs and since all alephs were strictly comparable, then it followed that all cardinals were comparable. The case of nonequivalent infinite sets where neither was equivalent to a subset of the other could not occur. In one theorem, based on the inconsistency of the system of all transfinite numbers, Cantor had succeeded in resolving several perplexing and long-standing problems of set theory.

Had it been otherwise, had the system ת been consistent, Cantor's difficulties would have been considerably magnified. For then the system ת, as a consistent set, would have had a corresponding "largest" cardinal, an aleph, while even larger sets beyond the limits of the "largest" cardinal were conceivable. This would have left the continuum hypothesis even less certain while rendering proof of the comparability theorem impossible. The fact that such was not the case depended upon establishing the necessary inconsistency of Ω and ת. But if Cantor was unaware of the unwarranted assumption in his argument (his claim that there must exist some subcollection $V' \subset V$ similar to Ω), he was nevertheless fully cognizant that his proof was not the best one possible. He hoped that a more direct demonstration might yet be found.

In the only surviving reply to Cantor's letters of 1899, Dedekind came to Cantor's assistance by offering a version of the Bernstein-Schroeder Theorem, which dealt with the comparability of sets in terms of the "chains" he had used in his short book *Was sind und was sollen die Zahlen*.[16] Cantor replied immediately, thanking Dedekind for his interest and noting, quite unabashedly, that it would be wonderful if Dedekind could do the same for the *Beiträge's* Theorem A.[17] This was the major comparability result for all cardinal numbers, and considering the various combinations possible between two sets and their subsets, three of four cases had already been decided. Cantor added that no one had yet succeeded in showing that the last of these, involving two sets neither of which was equivalent to a part of the other, was impossible where transfinite sets were concerned. He admitted his own inability to prove the claim by any means as simple as the ones Dedekind had used and added that only by the comparability result based upon the inconsistency of the system ת had he even managed an indirect proof.[18] He hoped that Dedekind might be able to do better, but no breakthroughs were forthcoming. If Dedekind ever responded to Cantor's request, his reply has been lost. Cantor's letter to Dedekind of August

30, 1899, was the last of their correspondence to survive, and their exchange of ideas on the subject seems to have ended there.

Though the technical grounds for Cantor's easy acceptance of the paradoxes are understandable, he had even deeper, more abstruse reasons as well. He alluded to such reasons in his second letter to Dedekind, where he noted that:

> A collection [*Vielheit*] can be so constituted that the assumption of a "unification " of *all* its elements into a whole leads to a contradiction, so that it is impossible to conceive of the collection as a unity, as a "completed object." Such collections I call *absolute infinite* or *inconsistent* collections.[19]

The terminology of "absolute infinite sets" was not new to Cantor's set theory. It had been introduced, although only in a footnote to the *Grundlagen,* where he tried to make precise the meaning of the word set.[20] Cantor explained that by sets he meant to include only well-ordered collections that could be joined by some rule into a whole. Sets therefore were unities, and unless a given collection could be regarded as a *Ding für sich,*[21] (like the set of all natural numbers, or the set of all denumerably transfinite numbers of the class $Z(\aleph_0)$), then it was necessarily inadmissible as a set.

Did Cantor already foresee the paradoxes of set theory which the absence of such restrictions would allow? Probably not. There is a simpler explanation, one that has already been suggested: unless a set were unifiable, one could not produce concepts of ordinal or cardinal numbers by Cantor's method of abstraction. Although he does not seem to have consciously or immediately recognized the inconsistencies of set theory, he always emphasized the impossibility of a largest cardinal number, which he originally interpreted as meaning that the set of all cardinal numbers was an *incomprehensible* one.

The set of all transfinite numbers, like the absolute itself, could be acknowledged, but it could never be completely understood. He had even found a passage in Albrecht von Haller on eternity that expressed his idea exactly: "I abstract it (the monstrous number) and you (eternity) lie entirely before me."[22] The mystical-religious implications were all a part of Cantor's conceptualization of the infinite, and of his *Transfinitum* in particular. He always regarded the absolutely infinite succession of transfinite numbers as a thoroughly appropriate symbol for the Absolute. That all of this was intimately related to his understanding of God was finally made clear in the *Mittheilungen* of 1887.

For example, there were a number of different ways in which Cantor felt the concept of infinity could be regarded. One of these was the form it assumed "in Deo extramundano alterno omnipotenti sive natura naturante," or the Absolute.[23] When the infinite served in this capacity he regarded it as capable of no change or increase, and he said that it was therefore to be thought of as mathematically indeterminable. Were it determinable, then it would have been limited in some manner. This may again seem to share very close affinities with the paradox of the largest aleph, but Cantor still saw nothing inconsistent about the idea of the unending succession of all transfinite numbers.

In letters to Cardinal Franzelin Cantor was explicit in relating God and the Absolute by saying that the Absolute as the "Infinitum aeternum increatum sive Absolutum" concerned God and his attributes.[24] Furthermore it was God who ensured, through the infinite perfection of his nature, the existence of the entire *Transfinitum ordinatum*.[25] In the end, Cantor regarded the transfinite numbers as leading directly to the Absolute, to the one "true infinity," whose magnitude was capable of neither increase nor decrease but could only be described as an absolute maximum that was incomprehensible within the bounds of man's understanding.

By recognizing the connections Cantor drew between his transfinite numbers and the Absolute, it is easier to understand why the paradoxes of set theory did not upset him as they did so many mathematicians at the turn of the century. Essentially, he had recognized the impossibility of subjecting the entire succession of transfinite numbers to exact mathematical analysis. The nature of their existence as a unity in the mind of God constituted a different sort of perfection, and Cantor was not disturbed that it was beyond his means to comprehend it precisely. In fact, the inaccessibility of the Absolute to any maximal transfinite determination seemed both fitting and appropriate. It was also clear from a remark in 1886 that Cantor did not believe that any of these matters compromised the consistency of his own set theory. He was perfectly willing to assert that his thoery was completely free of contradictions, and this he regarded as an established fact.[26]

At no time before 1895 had Cantor ever dwelt upon the question of sets, nor had he ever explicitly referred to any paradoxes or contradictions that might be encountered if certain limitations were not observed. This in itself marks a very noticeable difference between the *Grundlagen* and the *Beiträge*. There was nothing particularly suspect about his claim in 1883 that the first and second principles of generation, taken together, meant that there was no such thing as the largest cardinal number. The entire sequence of transfinite numbers existed as an absolute entity in the unchanging, incomprehensible mind of God. But as Cantor's viewpoint shifted and as sets moved to the forefront of his analysis, the study of order types and powers presented immediate difficulties. The set of all ordinals Ω represented a type of order δ that must have been greater than all ordinals in Ω, but since Ω contained all ordinals, it must also contain δ. Hence $\delta > \delta$. The same problem occurred in positing the cardinality, or power of the sequence ת, or of the set of all sets. Each of these difficulties was neatly eliminated by Cantor's characterization of the differences between consistent and inconsistent sets and by his restriction of set theory to the provenance of consistent sets alone.

The recognition of inconsistent sets like ת even seemed to bring a new hope that the continuum hypothesis was not an empty combination of meaningless symbols. He had shown, indirectly, that all cardinals were alephs and therefore it seemed certain that the continuum, as a well-defined and self-contained set of elements, was also a consistent set and thus it too was certain to be equivalent in power to some transfinite aleph. That much settled, only one final step was missing: proof that $2^{\aleph_0} = \aleph_1$. Unfortunately, this equality was by no means evident. Indeed, one of the greatest shocks Cantor ever experienced was the proof J.C. König presented at the Third International Congress of Mathematicians held at Heidelberg in 1904. His startling news was the discovery that the power of the continuum could not possibly be an aleph.

KÖNIG'S PROOF AND THE HEIDELBERG CONGRESS OF 1904

By the time that the Third International Congress took place, Cantor's fame and importance were almost universally acknowledged. Even before the turn of the century his research had begun to enjoy extensive recognition and enthusiastic praise. One of the first and most influential tributes was paid by Hurwitz,[27] who addressed the inaugural session of the First International Congress for Mathematicians held in Zürich in 1897. In surveying the most recent advances in the general theory of analytic functions, Hurwitz illustrated the enormous contribution Cantor's set theory had made. Three years later the significance of Cantor's work was further emphasized by Hilbert in his address to the Second International Congress held at the Paris World Exposition of 1900.[28] Reviewing the major unsolved problems confronting mathematicians at the beginning of the twentieth century, Hilbert placed Cantor's continuum hypothesis at the head of his list. It was as though the hope which Cantor had expressed to Felix Klein had been fulfilled; the *Beiträge* had succeeded in exciting interest among mathematicians in the transfinite numbers and in the value of set theory generally.[29]

Indicative of this growing interest, books soon began to appear which employed the basic ideas of Cantor's set theory in new ways. Here the French were particularly active, at least initially, and Couturat, Baire, Borel, and Lebesgue were representative of those who were influenced by, or who began to develop, certain aspects of Cantor's work. Schoenflies' report, published in the *Jahresbericht der Deutschen Mathematiker-Vereinigung* for 1899, outlined the basic elements of the theory of point sets.[30] In the decade following 1900, countless articles which either applied or advanced Cantor's work appeared in Germany, many written by Bernstein, Hausdorff, Schoenflies, and Schroeder. Even before Schoenflies' report, Vivanti's surprisingly early history of set theory (published in 1892), reflected the great interest among Italians in Cantor's work, as did papers by such mathematicians as Burali-Forti, Dini, Peano, and Veronese.[31] England too produced much literature concerning Cantor's set theory, particularly the paradoxes, and among the first important treatments of the subject were Bertrand Russell's *Principles of Mathematics* and the volume by W.H. and G.C. Young: *The Theory of Sets of Points*.[32] Moreover, from London to Krakow there were mathematical societies which elected Cantor to corresponding or to honorary membership. He was lauded with distinguished degrees from several European universities, and in 1904 he was awarded the highest honor the Royal Society of London can confer, the Sylvester medal.[33]

But the sweetness of international accolades and of growing recognition was mixed with concern over the paradoxes and scrutiny of the foundations upon which set theory had been built. Bertrand Russell had convinced Frege in 1902 (and in turn everyone concerned with the logical consistency of formalized mathematics) that there were certain antinomies inherent in set theory as it was then understood. Somewhat earlier Burali-Forti had attempted to account for his paradox of the largest ordinal by suggesting that from the conclusion $\delta > \delta$, one could only agree to dispense with any hope of strict comparability between transfinite numbers.[34] Following Burali-Forti and Russell, others joined in questioning the soundness of Cantor's work. Doubts voiced in France finally reached such proportions that Cantor felt obliged in 1903 to make a direct reply, and at the meeting of the Deutsche Mathematiker-Vereinigung in Kassel that year he launched a counteroffensive designed to reinforce the basic tenets of set theory.[35] The debates in Kassel were symptomatic of the split which was beginning to appear between mathematicians of differing philosophical persuasions. And just as Frege had predicted, the conflict was being waged in terms of Cantor's theory of the infinite. In the following year, Cantor found himself face-to-face with an even more upsetting challenge. Jules König, from Budapest, appeared in Heidelberg for the Third International Congress and read a paper which claimed that the power of Cantor's continuum was not an aleph at all.[36]

As Schoenflies once put it, belief that the power of the continuum was equal to $\aleph_1$ was a basic dogma with Cantor.[37] Not only did König's proof challenge that dogma, but it further implied that the continuum could not be well-ordered. This rendered doubtful another article of Cantorian faith—that *every* set could be well-ordered. To the participants of the congress it was apparent that two of the most fundamental doctrines of transfinite set theory had been neatly felled. Cantor was shocked and infuriated, apparently more for having been brought to such humiliation before his colleagues than for any other reason. His work and reputation were in the balance. Gerhard Kowalewsky, who was present at the session where König read his paper, recalled in his memoirs that Cantor had somehow felt that God should never have allowed his errors to be revealed in such a way.[38] The certainty Cantor had always attributed to his set theory, his faith in its absolute perfection as virtually the word of God, seemed abruptly shattered. But the disillusionment, though devastating, was only temporary.

König's proof was not all that it seemed to be. He had begun his paper with an analysis of the continuum problem and various general remarks about powers and well-ordered sets. Then he launched his proof[39] that the power of the continuum was not an aleph by employing a general theorem Bernstein[40] had presented in his Inaugural-Dissertation in 1901:

$$\aleph_x^{\aleph_0} = \aleph_x 2^{\aleph_0}.$$

Though pure mathematics rarely makes a ripple in the daily press today, the local papers then were full of reports describing König's sensational discovery. And if one account is accurate,[41] interest in Germany was so great that the Grossherzog of Baden had Felix Klein explain the entire matter to him personally.

Cantor, however, was unimpressed, and remained skeptical that König could possibly be right. He refused to accept the proof as conclusive, and though he could discern no gaps in König's argument, he steadfastly believed that his continuum hypothesis had not been refuted. Perhaps Cantor continued to draw upon his deep convictions that set theory was necessarily and absolutely true and therefore could not possibly be in error. Indeed, he was not sufficiently shaken to have lost entirely his sense of humor. After König's paper had been read, Cantor quipped that whatever had been done to produce the alleged proof, he suspected the king less than the king's ministers.[42]

If Kowalewski's memory was correct, Cantor's reservations were vindicated within twenty-four hours.[43] He had been certain from the first reading that there was something faulty at bottom of König's analysis, and his references to the king's ministers suggests that he doubted the lemma used to produce the result more than the logic König had applied. König had a high reputation among his colleagues for being extremely sharp and reliable, and therefore it was likely that the flaw in his analysis lay in an unwarranted assumption. Less than a day later in fact, Zermelo demonstrated that König's use of Bernstein's result had been improper. The general form of the theorem which claimed that $\aleph_x^{\aleph_0}=\aleph_x 2^{\aleph_0}$ was false. As Cantor had predicted, the king's ministers were clearly at fault.

But Zermelo's discovery did little to calm Cantor's irritation, and he continued to worry about the doubts König's paper had raised. There was always the chance that eventually someone would find another way to undermine or to refute the continuum hypothesis. Until Cantor could establish more conclusively that the power of the continuum was necessarily an aleph, his work would remain open to challenge. Others were equally concerned, and following the Heidelberg Congress, a small group of mathematicians including Hausdorff, Hensel, Hilbert, and Schoenflies gathered for a brief vacation in Wengen where they continued to discuss the implications of König's paper and the future of Cantor's set theory.[44] Cantor soon found his way into the group, and the intensity of his preoccupation in the aftermath of the congress was readily apparent. As Schoenflies told the story, Cantor even appeared one morning at the hotel where he and Hilbert were staying, long before breakfast, in order to greet them (and everyone else in the dining room) with yet another excited analysis of where König had gone astray.[45] But such was the nature of Cantor's mind. Once he had grasped a problem, he was not satisfied until a solution had been formulated to his complete satisfaction.

Despite Cantor's eagerness to produce the complete and absolute refutation of König's results, he was unable to do so. Even before the congress was over everyone was fully aware that König's proof had failed, but that did not dispel the possibility that at any moment he might find a new way to repair his proof. The only conclusive answer to König would be the proof that the continuum could be well-ordered or that every transfinite cardinal number was in fact an aleph. But these were results which Cantor was unable to establish himself. Zermelo, however, having found the error in König's argument, shortly thereafter found the most significant rejoinder to König's proof. Before the year was out, Zermelo had devised a highly controversial proof which established one of Cantor's most basic tenets: every set could be well-ordered.

ZERMELO, WELL-ORDERING, AND THE AXIOM OF CHOICE

Within a month of the Heidelberg Congress Zermelo wrote to Hilbert and told him the news of a discovery which added a wholly new element to the controversy over Cantor's set theory.[46] Having already disposed of König's alleged proof, he subsequently managed to settle the question entirely. König could never hope to salvage his proof because Zermelo had at last succeeded in proving that every set could be well-ordered. The letter he sent to Hilbert was published immediately in *Mathematische Annalen*. Ironically, far from making Cantor's set theory secure and certain, Zermelo's brief paper was really but a shortlived calm before the storm.

Following the Third International Congress, Zermelo had continued to worry about the implications of König's paper. Although he had shown that König's use of Bernstein's theorem was incorrect in the general form in which König had used it, there were only two alternatives to save Cantor's set theory from future mishap or ruin. In order to show that every cardinal number was one of Cantor's alephs and that no set, including the continuum, could be of power not listed in the collection ת, it was necessary to demonstrate that every set could be well-ordered.

Zermelo discussed the problem with his friend Eberhard Schmidt, and together they devised a proof that was to raise as many difficulties as it professed to have settled.[47] By September 24, 1904, Zermelo had finished a final version. Basically, it showed how to produce an ordering for any given set in terms of a well-ordering of its subsets. The only sets newly constructed in the proof itself, given M, were the subsets of M; the set of all subsets of M, the power set $U(M)$; and the subsets of $U(M)$. The proof began by associating with every nonempty subset N of M a "distinguished element" n from N, denoted $n = \varphi(N)$. This correspondence φ Zermelo called a "covering" of the set $U(M)$, and once determined, the covering remained fixed for the entire proof. According to Zermelo, the idea for producing the well-ordering of M on the basis of *any* covering γ of $U(M)$ was Schmidt's contribution; his own concerned the amalgamation of the various possible γ-sets, specifically the well-ordered segments resulting from the ordering principle, in order to show that it was possible to find a specific covering γ from which a definite well-ordering of the original set M could be given.

Central to the entire proof was the covering γ. Zermelo described it in his own words: "For every subset M' imagine a corresponding element m_1', which is itself a member of M' and may be called the 'distinguished' element of M'."[48] The words were carefully chosen, in particular the phrase "imagine a corresponding element m_1'." Though seemingly innocuous, this form of the Axiom of Choice was one of great ontological moment. But Zermelo's basic idea, once mathematicians began to study its consequences, unleashed a flood of protest and commentary. Its vestiges can still be seen among the splintered factions of various philosophies and attitudes toward the foundations of mathematics.

Without going into every detail, Zermelo's proof can be described in outline. Special sets M_γ were defined as γ-sets if for every element a of M_γ and the segment A of M_γ determined by a (A the set of all elements $x < a$ of M_γ), the element a was always the "distinguished" element of $M - A$. Such γ-sets were the means by which Zermelo was able to "amalgamate" well-ordered subsets of M to advance beyond such special cases to produce a well-ordering for all of M. By his definition of γ-sets, Zermelo was in a position to give a specific covering γ based upon the elements a given with the sections A of the sets M_γ.

Denoting as γ-elements all elements in M that were also elements of any γ-set, Zermelo confirmed his claim that "the totality L_γ of all γ-elements can be so ordered, that it is itself a γ-set, and included all elements of the original set M." The direct conclusion was the fact that M, too, was well-ordered by L_γ.

The beauty of Zermelo's proof lay in its directness and simplicity. The mathematical argument was clear at every point, a feature in which he took some pride four years later when he offered a revised version of his theorem. Despite the numerous objections and criticisms which appeared following publication of the proof in its original form, Zermelo was pleased to note in the second proof that the careful examination of his detractors had failed to discover any mathematical errors.[49] Believing that the principles upon which his theorem was based were sound, Zermelo concluded that, eventually, with the passage of time, sufficient explanations would be produced to overcome all objections. There were a number of mathematicians, however, who were adamantly skeptical. They never granted Zermelo's proof any hope of respectability.

Zermelo emphasized the most significant conclusion of his theorem in 1904: since every set could be well-ordered, the power of every set had to be an aleph. Equally important to Cantor's set theory was the fact that the well-ordering theorem produced direct verification of the comparability of cardinals. Given any two cardinal numbers $\mathfrak{a}$ and $\mathfrak{b}$, then exactly *one* of the relations $\mathfrak{a} < \mathfrak{b}$, $\mathfrak{a} = \mathfrak{b}$, $\mathfrak{a} > \mathfrak{b}$ must hold.

This result, coupled with the fact that all cardinal numbers had to be alephs, comprised the major focus of Cantor's letters to Dedekind of 1899. But there was a significant difference between Cantor's attempt to prove the fact that all cardinals were necessarily listed in ת and Zermelo's well-ordering theorem, which led to the same result. All along, though without ever offering even the slightest hint of a proof, Cantor had assumed that every set could be well-ordered. There was no reason why this should not have seemed plausible, since it was possible to imagine a means of well-ordering any given set M by a transfinite selection of successive elements. Picking m_0 as the first element, m_1 as the second element from $M - \{m_1\}$, then a third from $M - \{m_1, m_2\}$, if the procedure eventually terminated, then M was directly well-ordered. On the other hand, should the procedure end and there be elements left yet to be ordered, then it was always possible to pick another from the remainder of M and proceed further. Thus the process could only come to an end when the successive string of choices had exhausted every element of M.

But no one would have accepted such an analysis as logically satisfying. Such reliance upon psychological methods (involving successive choices in the attempt to count or to order a given transfinite set) had no place in strictly rigorous mathematics, as Frege would have been quick to note. Nevertheless, Cantor had employed exactly this sort of argument in trying to show that were a set V to exist which was not of any power equivalent to some aleph, then it was necessarily inconsistent. To do this Cantor had assumed that V was of such character that one could project the inconsistent system Ω of all ordinal numbers into V by showing it to be similar to some subset $V' \subset V$. His reasons for believing this was possible hinged on a procedure very much like the one just described for showing that every set could be well-ordered.[50] Given any set V, Cantor claimed Ω could be corresponded with V in such a way that, if the successive correspondence between the two were ever terminated, then the cardinality of V clearly had to be an aleph. Otherwise, if the process never reached an end, then V had to contain the entire system Ω, and could therefore, as an inconsistent set itself, have no aleph as cardinal number.

Cantor was only able to draw such conclusions indirectly. Zermelo's great merit was to have established the well-ordering theorem directly, without relying upon projections or upon the psychological procedures of Cantor's more intuitive analysis. Moreover, Zermelo had avoided the "successive" choice of elements from any given set by having chosen at the beginning an element from every nonempty subset of M. One of the great problems of Zermelo's formulation was the unexplained way in which the initial covering was supposed to be produced. Of the major objections to Zermelo's entire proof, none was more obvious than the complaint that no guarantee could be given to suppose that such simultaneous selections could always be made.

These doubts in turn rested upon Zermelo's explicit use of the Axiom of Choice.

Perhaps the greatest contribution of Zermelo's paper of 1904 was its explicit recognition that the Axiom of Choice was an irreducible principle.[51] He regarded the axiom as self-evident, and turned for support to the fact that it was used everywhere in mathematics without hesitation. Even the simple theorem that the number of individual parts could not exceed the total number of elements in any given set could not be proven without the assumption that every part could be identified with one of its elements. But such arguments failed to establish the respectability of the Axiom of Choice, and after 1904, if the principle was invoked, mathematicians were more careful than ever to explain its use and to provide whenever possible alternative methods of proof which did not rely upon the axiom at all.

Zermelo's proof of the well-ordering theorem was purely existential. It asserted that for any set, a well-ordering of the set must exist, but no definite example of such an ordering was indicated. For the continuum this meant that, although one could conclude on the basis of Zermelo's arguments that it must be capable of being well-ordered, there was no constructive way to actually carry out the well-ordering itself. Unlike the simply-ordered sets of rational and algebraic numbers for which Cantor had produced well-orderings and the corresponding cardinal number $\aleph_0$, the continuum remained elusive. For those, like König, who doubted that the continuum *could* be well-ordered, there was nothing to do but to doubt the veracity of Zermelo's basic assumption concerning the admissible use of the Axiom of Choice. König's objections, like those of the axiom's other detractors, rested upon the question of how far a mathematical procedure could be carried before it was either questionable or unacceptable.[52] In particular, were there limits to the Axiom of Choice beyond which it was foolish to venture? Was the continuum perhaps of such magnitude that any connection with the Axiom of Choice might be meaningless? Some mathematicians argued that if the certainty of any proof was to be unimpeachable, then only finitely generated concepts and proofs could be allowed. Others insisted an infinite number of logical steps might be permitted, but only denumerably many. Some were more liberal and went further, beyond the second number class $Z(\aleph_0)$, while other mathematicians seemed willing to accept anything short of the totality of all numbers, Ω or ת, taken as a whole. Of all such alternatives, the most restrictive course followed the lines of Kronecker's early advocation of finitism. Among the most virulent and persistant of the twentieth-century Kroneckerians were the French.

THE REACTION IN FRANCE

The first issue of *Mathematische Annalen* for 1905 included an international sampling of comment and criticism directed toward Zermelo's proof published in the preceding issue. The English were represented by P. E. B. Jourdain[53] (who sent in his own proof that every set could be well-ordered), and the Germans by Bernstein[54] and Schoenflies.[55] The editors of the journal had specially solicited the opinion of the French mathematician Émile Borel.[56] Judging from the analysis of Cantorian set theory which Borel had given in his *Leçons sur la théorie des fonctions,*[57] they could expect lively opposition.

Borel admitted that Zermelo had attempted to solve one of the most important problems concerning a given set M:

A: Given any M, put it in the form of a well-ordered set.[58]

Borel noted that this problem actually reduced to another whose solution would lead directly to a confirmation of A:

B: Given any subset M' of M, choose from M' in a *determined* (but otherwise arbitrary) manner one element m', to which the name "distinguished element of M'" is given; this choice must be made for *all* the subsets M' of M.[59]

Certainly every solution for A produced a specific solution for B, but Borel did not think the converse was at all evident. One only had the equivalence of A and B on Zermelo's authority. Borel rejected such authority entirely and refused to accept Zermelo's contention that B actually supplied a solution for A. His major objection was Zermelo's failure to provide even a theoretical means of determining the distinguished elements for all subsets M' of M. With the continuum in mind, Borel suggested the task would be extremely arduous, if not impossible. Thus, he flatly denied Zermelo's claim that it was possible to choose, ad libitum, *the distinguished element* m' from a particular M', and that this could be done for every M'.

Borel did not regard Zermelo's procedure as being any more rigorous than attempts to establish the well-ordering theorem by picking elements one at a time from a given set until it was entirely exhausted, even if one had to proceed a transfinite number of times.[60] No mathematician could regard such an argument as a valid source of proof, and Borel claimed that his objection held whenever an arbitrary choice was supposed to be made a nondenumerably infinite number of times. Such reasoning, he added, lay completely outside the province of mathematics.

Borel's objections in turn stimulated much controversy among mathematicians in France, and in the second edition of his *Leçons sur la théorie des fonctions* he included five letters (written in 1904) indicating the diversity of opinion which Zermelo's proof had provoked among Borel, Baire, Hadamard, and Lebesgue.[61] Of the four mathematicians, only Hadamard seemed willing to support the position which Cantor and Zermelo had taken.

Hadamard disagreed with virtually everything Borel had written in his note for the editors of *Mathematische Annalen*.[62] He refused to admit any comparability between the independent and simultaneous choices of distinguished elements Zermelo had posited and the inadmissible example Borel had raised of a transfinite number of *successive* choices where each depended upon all the preceding choices. Moreover, Hadamard saw no reason why Borel should have permitted a denumerably infinite number of choices, but no more. The difference would have been plain had the choices been made successively. Since Zermelo made them simultaneously and independently, he admitted no grounds for Borel's objection.

What Hadamard *did* stress was the fact that Zermelo had given no means of effectively carrying out the choice of distinguished elements, and he doubted that anyone could.[63] Hadamard was emphatic in saying that the problem of actually determining the correspondences in question was a separate, quite distinct one from the question of whether such correspondences existed. Zermelo had only considered the latter, and Borel was in fact demanding something entirely different from anything Zermelo had either professed to show or even had in mind.

In 1897 Jules Tannery had already distinguished mappings which could be *defined* from those which could be *described*.[64] Hadamard took this distinction and noted that many important questions in mathematics changed their sense entirely if the one was substituted for the other. Ultimately he believed it was all simply a matter of opinion, because the idea of a correspondence "which could be described" was, in Borel's words, "outside of mathematics." In Hadamard's own words:

> It is moreover a pure question of sentiment. But the idea of correspondences "which can be described," to use your [Borel's] expression, is "outside of mathematics"; it depends upon the realm of psychology and is related to a property of our mind [*esprit*]. The question of knowing whether the correspondence used by Zermelo could ever be indicated *in fact* is a question of this sort.[65]

Did this mean that Zermelo's assumption (that simultaneous choices were always possible) was improper? Borel seemed to think so, and until a means for making every choice was actually produced, he felt that no one could legitimately assume otherwise. Hadamard disagreed sharply, and replied that the possibility of choosing a single element from any given set was just as certain as two propositions equivalent to Borel's:

> A: Given a number x, there exist numbers y which are not bound [*liés*] to x by any algebraic equation of integer coefficients.[66]

and

> B: There exist functions y of x such that, for any value of x, y has neither an algebraic value, nor a value bound to x by any algebraic equation of integer coefficients.[67]

It was clear that such functions could always be formed, but Hadamard insisted that none was necessary to establish the correctness of Theorem B. Likewise, in terms of Zermelo's theorem, there was no need to produce an actual correspondence naming the distinguished elements for all subsets M' in M, so long as the pure existence of such correspondences could be established. Hadamard believed that Zermelo had accomplished this, and for him that was enough.

Borel passed the contents of Hadamard's letter on to Baire,[68] who promptly agreed with Borel. Baire complained of Zermelo's frequent references to "a given set," since he could not guess what that might mean. In fact, he suspected that the phrase was meaningless. He professed to be more conservative than Borel in believing that even denumerable infinities should not be permitted in rigorous mathematics.[69] If Baire were to have his way, progress in mathematics would be measured by confining all discussion to only definable domains. In the end, everything should be returned to the realm of the finite. It was as though Cantorian set theory were once more faced with the objections of Leopold Kronecker. Certainly Baire, Borel, and Lebesgue had reopened all of the finitist doubts to which Cantor had been subjected at the outset of his career. Of all who shared these doubts, Lebesgue spoke most persuasively, if somewhat more moderately than Borel or Baire.

Lebesgue clarified what Borel and Baire had expressed only vaguely.[70] In terms of Cantor's formulation, to define a set M was to give a property P pertaining to certain elements of a previously defined set N. But one knew nothing about the elements of M except, as Lebesgue put it, that they possessed all the unknown properties of N and were the only elements which satisfied the *unknown* property P. There was nothing to differentiate between two elements in M, let alone one, as Zermelo had required in order to produce alternative formulations, like B:

> B: Correspond to every subset M' of elements of M one particular element m' of M'.[71]

Unfortunately, Lebesgue did not see how any proof of Theorem B was possible except for sets that were already known to be well-ordered. This did not seem to happen very often.

Lebesgue also felt that Hadamard had been much more faithful to Zermelo's interpretation of the well-ordering proof than had Borel. Zermelo's theorem did not pretend to establish anything beyond the necessary existence of a solution. Lebesgue wondered if it was possible to affirm the real question which seemed to underlie the entire debate: "Can one demonstrate the existence of a mathematical object [*être*] without defining it?"[72] Lebesgue was certain that merely defining something did not necessarily imply its existence. This was largely Kronecker's view of the transcendental and transfinite numbers: one could offer as many definitions as one liked, but there was no reason to think that the entities involved actually existed. In this sense Lebesgue insisted it was impossible to distinguish between Borel's proposition A and his own formulation C:

C: Can every set be well-ordered?[73]

If the convention had been universal that the existence of mathematical objects could be established without defining them, Lebesgue would have said nothing more. But he was representing the view of many mathematicians when he wrote that the word existence was frequently used in quite another sense.[74] Cantor's work even provided Lebesgue with an example. His classic proof establishing the existence of a nondenumerable infinity of numbers failed to provide any means of identifying or naming such an infinity. What Cantor had actually shown was that whenever one had a denumerably infinite collection of numbers, one could always define a number that was not contained in the original collection. Here Lebesgue explained that define was taken to mean that one named a characteristic *property* of the elements defined.

Zermelo had not done this to Lebesgue's satisfaction. Only if one took the existence of M' in the Kroneckerian sense did Lebesgue feel the existence of distinguished elements m' could be certain, since then to say M' existed was to affirm that one knew how to name its elements. Otherwise, speaking as generally as Zermelo did, how could the set of γ-coverings Γ be said to exist (as Hadamard for one believed)? Could one name or choose even a single element of Γ? Lebesgue denied that one could; otherwise it would have been possible to produce a definite solution of B for the set M.

Could Zermelo's interpretation of "to choose" as meaning "to think of" instead of Lebesgue's stricter "to name" be enough to ensure the existence of a well-ordering for any given set? Again, Lebesgue objected that this could not be allowed since Zermelo never made clear what one was "to think *of*."[75] Yet this was imperative to the sense of his proof, which required that one think of a single, determined correspondence at the outset which remained unchanged for the entire proof. Worse yet, sets M in question could be of such magnitude and complexity that it was doubtful an initial covering could ever be chosen.

Lebesgue concluded that Zermelo's proof, like so many others in set theory, was too general and failed to be "Kroneckerian enough."[76] He was willing to challenge Borel's comparison of Zermelo's method with the successive counting of a set and sided with Hadamard in thinking Zermelo's simultaneous and independent choice of initial distinguished elements might have been an advance. But it could also have been an elusive one, for as Lebesgue reflected on the matter, Zermelo had assumed a covering to exist for all *subsets* of a given set M. Thus a denumerably infinite case of ordered choices might have been replaced by independent choices, but ones necessary for an even greater power's worth of elements.[77] The nonordered covering of choices did not seem to diminish the problem anyway, since the real point concerned the absence of a law by which the choices were to be made, whether the infinity in question was denumerable or not. To make a choice, Lebesgue believed one had to be able to write or to name the element chosen. Some means had to be given for showing how a choice could be carried out. Thus Lebesgue was willing to agree with Hadamard that the question of whether or not the choices were denumerably infinite seemed irrelevant. What was imperative was some rule by which the distinguished elements were to be selected.

This did not mean Lebesgue was ignoring the essential difference between the two sorts of infinity. He hastened to reassure Borel that whenever he encountered any case involving a transfinite number of choices, he had always been suspicious.[78] He was familiar with the problem of making denumerably many choices; transfinitely many were beyond his experience. Nevertheless, there could be just as many doubts in the realm of the denumerably infinite. Lebesgue in fact refused to accept Cantor's well-known argument that every set of power greater than $\aleph_0$ necessarily contained a set equivalent in power to the second number class $Z(\aleph_0)$. Nor did he believe that because no set had ever been named which was neither finite nor infinite, that the impossibility of such sets had been demonstrated.

Thus the matter at issue among the French mathematicians was reflected in Lebesgue's basic question: could one demonstrate the existence of a mathematical entity without defining it? Borel, Baire, and Lebesgue thought not—Hadamard was the only exception among the four. Thus he preferred to ask (as did Zermelo) whether an ordering was *possible*, while the opposition, instead, preferred to ask whether, given any set, one could actually produce a well-ordered version. Whether *one* could do so seemed an entirely subjective matter to Hadamard; the answer involved psychology and the capacity of the human mind, not mathematics.[79]

Writing in the *Revue générale des Sciences* for March 30, 1905, Hadamard summarized his conclusions in an interesting way. He regarded the controversy over Zermelo's proof as similar to the debate waged nearly half a century earlier over Riemann's view of function. The essence of progress in mathematics, he suggested, rested upon flexibility and innovation; at every step in its history advances were made possible in mathematics by those who posited the utility or existence of ideas which mathematicians at the time had refused to recognize.[80]

Hadamard's position hardly differed from the one adopted in the *Grundlagen* more than twenty years earlier, when Cantor had also been concerned with defending his views from Kronecker's finitism. Cantor insisted then that if the finitists were to have their way, progress in mathematics would come to an end. Their position might ensure certainty, but it was simultaneously sterile. If mathematics were to be enriched by new ideas, then confining limitations could only inhibit the very freedom which Cantor believed to be the essence of mathematics. Hadamard agreed. Despite the stature of his opponents, he preferred to stand alone, like Cantor, in defense of the position he believed to be correct.

Equally at issue were the sets which led to the antinomies of set theory. Borel objected, for example, to the "set of all sets" because he believed that it involved concepts that were not properly defined.[81] Hadamard, by contrast, treated the inconsistent system W of all ordinal numbers in much the same way as Cantor had in his correspondence with Hilbert and Dedekind. The attempt to regard the entire collection W as a completed unity created the contradiction; in such cases the word set was improperly applied. Correctly speaking, sets could only be formed from preestablished and definite objects, ones that already existed. Self-generating sets like W and ת, or the set of all sets, did not conform to such requirements and thus could not legitimately be regarded as sets.[82] This was not the only means, however, of dealing with the antinomies of set theory. Burali-Forti, the Italian mathematician, interpreted the contradictions somewhat differently and approached the entire matter from quite another perspective.

THE ANTINOMIES AND THEIR EARLY INTERPRETATION

Burali-Forti, in 1897, was the first mathematician to make public the paradoxes of transfinite set theory.[83] By considering the entire succession of *all* ordinal numbers W, he observed that as a well-ordered set it had to have a corresponding ordinal number β greater than the totality of ordinals represented by W. Since W contained *all* ordinal numbers, it could not fail to include β, and one was consequently forced to conclude the impossible, that $\beta < \beta$. However, Burali-Forti did not suggest, as Cantor had, that the system W of all order types was *inconsistent*. Instead, he rejected the *comparability* of ordinal numbers and supposed that given two ordinal numbers α and β, then it was not always true that at least one of the relations $\alpha < \beta$, $\alpha = \beta$, $\alpha > \beta$ must necessarily hold.[84] The corresponding theorem for alephs was simultaneously lost.

Had Burali-Forti's conclusion been the only alternative, then the damage to Cantor's entire program would have been severe. Fortunately, there were several others. Bertrand Russell noted in his *Principles of Mathematics* that Burali-Forti's suggestion was directly at odds with a theorem Cantor had established in the *Beiträge*, where he had proven that for any two well-ordered sets, their corresponding ordinal numbers *were* necessarily and strictly comparable.[85] Cantor had also shown that given the whole series of ordinal numbers W, every *segment* of W was well-ordered. Russell preferred to interpret Burali-Forti's paradox of the largest ordinal as suggesting that the entire collection W was not well-ordered.[86] Since the well-ordered character of W could not be established from the fact that all its segments could be well-ordered, Russell believed his own interpretation of Burali-Forti's paradox to be preferable. Though Russell's solution did not affect as much of Cantorian set theory as did Burali-Forti's alternative, it still violated the basic reasonableness of the belief that W *was* well-ordered by virtue of the nature of its generation, despite the conclusion Russell offered.

If Russell was correct, Cantorian set theory still stood to lose a great deal. In particular, comparability of all cardinals and the theorem that every set could be well-ordered were both impossible. Furthermore, one could no longer assert that every cardinal number was indeed an aleph. To Philip Jourdain, however, it seemed entirely possible that one could prove that the system of all ordinal numbers was well-ordered.[87] In the process, of course, Burali-Forti's paradox returned and had to be dealt with anew.

Jourdain therefore proposed a solution very similar to Cantor's original conclusion that W and ת, though well-ordered, were to be regarded as inconsistent systems. As such neither W nor ת could be assigned any transfinite ordinal or cardinal number. Likewise, Jourdain echoed Cantor's characterization in writing that "we may attempt to define an inconsistent aggregate as an aggregate of which it is impossible to think as a whole without contradiction."[88] Thus Jourdain, like Cantor, concluded that a system was inconsistent if it contained any part equivalent to W.

Felix Bernstein, on the other hand, did not believe that Jourdain could regard either W or ת as well-ordered and then fail to admit the inevitable conclusion that as well-ordered sets they must, by the very terms of set theory, have both a definite order type and a corresponding transfinite cardinal number.[89] Rather than solve the inherent contradictions by facing them squarely, it seemed that such a solution only stopped just short of disaster; the logical next step would bring back the old difficulties. Bernstein believed he could resolve the paradoxes much more easily. He merely pointed out that the system W was defined as the collection of all order types of sections of well-ordered sets but that W itself was not a segment of any well-ordered set.[90] As a result, there could be no element e which followed all elements of W. If such were the case, then W could be regarded as the segment of (W,e). Bernstein thus tried to resolve Burali-Forti's paradox by saying the paradox itself was inadmissible. The set (W,β) was inherently contradictory in terms of the definition Burali-Forti had given it, which was not at all the same as saying W itself was inconsistent. In this respect Bernstein's solution was similar to Russell's, who denied that W could be regarded as well-ordered just because each of its segments was well-ordered. But Bernstein went a step further and doubted the rigor of Zermelo's well-ordering proof on the basis of his own view that W was noncontinuable.[91] Zermelo had introduced the maximal γ-set, L_γ, which established the well-ordering of M in terms of $M = L_\gamma$. Bernstein worried that if L_γ were similar to W, Zermelo would have introduced the ordered set (W,m'), which was not allowed under Bernstein's convention concerning the noncontinuable nature of the entire succession W.

Zermelo responded to these criticisms in 1908 by stating that, whereas Jourdain had always regarded "inconsistent" sets with some doubt, Bernstein had made them the objects of a dogmatic theory.[92] Since the inconsistency of W became apparent only when an element was added to produce (W,e), Bernstein believed he had solved the problem by saying that the continuation (W,e) contradicted the very *nature* of W. But Zermelo protested that Bernstein was merely playing with words. If W were taken to exist as a well-ordered set, as Bernstein thought it did, then there was little point in arguing about whether it could have a corresponding order type or not. It must, from the very definition of order types drawn from the concept of well-ordered sets, and if W was augmented by an element β coming after *all* elements of W, then the step to (W, β) was perfectly legitimate, even if it produced an upsetting paradox. Zermelo argued that W was in fact continuable, regardless of Bernstein's definition and all prohibitions to the contrary.[93] Zermelo rejected even more emphatically the solution Bernstein had offered to resolve this state of affairs. Since the Burali-Forti paradox arose only when a new element was added, producing (W, β), where β was assumed to follow all elements of W, Bernstein suggested that if one merely considered the *addition* of β to W, without determining any *ordinal* relation between W and β, then the system $(W; \beta)$ gave rise to no contradiction. The new element β was added to W, but it was not taken to *follow* every element of W. Zermelo protested that Bernstein had only pretended to solve the paradox by *ignoring* it. If it were so simple, merely a matter of the word "ordinal relation" or of the notation $\alpha < \beta$, then objective mathematical fact might be abrogated by merely avoiding the paradox-provoking words.[94] But Zermelo refused to allow that mathematics could be regarded so capriciously. He insisted that Bernstein could not define away so easily the deeply rooted paradoxes which were inherent in Cantor's largely *intuitive* formulation of set theory. Zermelo feared that if Bernstein's method of dealing with contradictions were ever adopted, mathematics too would cease to enjoy the certainty that logic had always conferred upon it.

MATHEMATICAL LOGIC AND AXIOMATICS

Mathematical logic, in fact, seemed a promising domain in which a satisfactory solution to the paradoxes might reasonably be sought. One of the most tireless of the early proponents of this view was Bertrand Russell.[95] Despite the fact that Frege's *Grundgesetze der Arithmetic* had suffered an irreparable setback in the face of Russell's antinomy, Russell believed that the methods Frege had outlined were the proper ones to follow. If Frege had committed any error, it was to be found somewhere among his basic laws, since from these the rest proceeded in clear and irrefutable steps expressed in Frege's *Begriffsschrift* itself. The very fact that the damage done by Russell's paradox was so plainly recognizable in Frege's system argued the advantages of his method. What was needed, then, was a more careful study of the fundamental logical premises from which arithmetic and all of mathematics could be safely derived.

Had the antinomies discovered by Cantor and Burali-Forti been limited to the systems of all transfinite ordinal and cardinal numbers, they could have been treated in various ways and would probably never have affected arithmetic in the form Frege had developed it. But Russell was able to show that the paradoxes of transfinite set theory were only special cases of much more general logical paradoxes which *did* affect any kind of reasoning. Russell had been led to this discovery by his study of Cantor's paradox of the largest cardinal number.[96] Cantor's proof that any given set could generate another set greater in power than its parent set depended upon the power-set formulation: $2^{\aleph} > \aleph$, whatever the particular aleph in question might be. Russell wondered in particular about the set of *all* entities. Cantor's theorem ensured the existence of elements that could not be included in any one-to-one correspondence between the parent and its power set. Russell wanted to know what sort of elements would be omitted from such a correlation and he decided to apply the mappings used in proving the Bernstein-Schroeder Theorem to decide the matter.[97] What emerged and served as the basis of his antinomy was Russell's discovery that sets determined by the proposition "x is not a class which is a member of itself," were among those omitted from the Bernstein-Schroeder correlation. Russell therefore concluded that such a property as "x is not a class which is a member of itself" could not define a set at all. The upsetting consequence, however, was the fact that Russell had established, without any consideration involving the nature of cardinal numbers or any considerations belonging to arithmetic, the existence of at least one clearly determined property which did not define a set. Worse yet, Russell saw that he had found a formula by which similar paradoxes could be manufactured in abundance.[98] In all cases the difficulties seemed to arise by assuming that certain properties could be used to determine sets when in fact they could not. This discovery led Russell to distinguish between "predicative norms or properties," those which did determine legitimate sets, and "nonpredicative norms," which did not, though it still remained to determine which norms were predicative. In answer to the paradoxes of set theory and logic, he later advanced his theory of types designed expressly to cope with the comprehensive sort of contradictions he had discovered.[99]

But this is not the place for a lengthy discussion of the approach Russell undertook in conjunction with his friend and senior partner A.N. Whitehead in producing their *Principia Mathematica*. For now it is sufficient to recognize that in conjunction with an axiomatic treatment, Russell hoped to show that mathematics could be reduced to the province of rigorous symbolic logic. In the process the major results of Cantorian set theory, short of the paradoxes and higher transfinite numbers like $\aleph_\omega$ and ω_ω, would be established within the limits of a perfectly consistent framework.

However, it soon became apparent to Russell and Whitehead that their goal could not be reached in as straightforward a manner as they had hoped. As the demands of increasingly advanced and complicated details of the *Principia* began to interfere with the ease and naturalness of the project as a whole, the additional restrictions required to exclude the paradoxes made it seem increasingly artificial.[100] To most mathematicians of Cantor's generation the *Principia* must have seemed a dry and desolate undertaking, and to younger mathematicians in Germany another approach seemed preferable, certainly more in keeping with the traditional character of mathematics. David Hilbert led the school which regarded axiomatic formalism as the surest road to recovery and sought to cure the ills of mathematics brought on by the antinomies of set theory by methods he had applied with success to the axiomatization of geometry. Such an approach had enabled him to show the consistency of that branch of mathematics, and he saw no reason why the same could not be done for set theory. Consequently, the German formalists produced an impressive axiomatization of set theory which they expected would rehabilitate Cantorian principles and end the debate over whether certainty in mathematics was possible or not.

THE AXIOMATICS OF HILBERT AND ZERMELO

The promise of a new kind of certainty in mathematics, despite the momentary problems which set theory seemed to create, was reflected in the optimism of the younger generation of German mathematicians prior to the First World War. At no time in the history of Western thought had the impressive gains of science generally and of mathematics in particular seemed so plain. There were few who thought that there could be any limitations to what mathematics could accomplish. Among the more promising approaches coming into prominence at the turn of the century, formal axiomatics seemed an impressive aid in helping verify that a given theory was actually a solid, certain structure rather than a fragile, elusive house of cards.

Hilbert had made clear the benefits to be counted from formal axiomatics in his *Grundlagen der Geometrie* of 1899, and in 1900 he returned to the same theme in discussing the concept of number.[101] He was particularly anxious to give a rigorous proof that the real numbers constituted a consistent set, borrowing Cantor's use of the term.[102] His interest might have seemed natural at the time, since Hilbert had been the first to whom Cantor had communicated his discovery of the paradoxes of set theory. But there was more involved than a simple desire to help Cantor determine whether the power of the set of real numbers was among the alephs or not. Hilbert had good reasons of his own for wanting to show the consistency of the continuum hypothesis. Though he had managed to apply the axiomatic method with success in showing the completeness and consistency of geometry, Hilbert had only been able to do so in relative terms, assuming the consistency of the real numbers. Any doubt, therefore, as to their consistency cast an immediate shadow on the conclusions drawn in his *Grundlagen der Geometrie*. Cantor's discovery of inconsistent systems coupled with his inability to show that the continuum (or any set in general) could be well-ordered was certain to make Hilbert uneasy. Establishing the consistency of the real numbers was therefore a matter of considerable priority.

Hilbert's program reduced essentially to setting out primitive concepts called numbers (analogous to the primitive elements point and line in Euclidean geometry) and to determining various relations between the primitive concepts by introducing certain axioms.[103] Hilbert's axioms included those for the familiar operations of arithmetic, commutative and associative laws, order relations, an axiom of continuity (the Archimedean Axiom), and a completeness axiom, which stated that from the axioms listed no other numbers could be generated in addition to the primitive elements, or numbers, assumed at the outset. From these it was then necessary to establish that the system of primitive concepts and axioms were together consistent and complete, that no contradictions could result from any combination of axioms, and that the system of axioms was sufficient to prove all theorems possible concerning the real numbers. This, in basic outline, was the aim of Hilbert's axiomatic method.

Confirmation of the consistency of arithmetic and the healthy status of the real numbers was certainly the goal of Hilbert's axiomatization. But there was an equally positive subsidiary question which he also meant to resolve. Hilbert hoped that his own philosophy of mathematics, his formalism which equated mathematical existence with self-consistency, would serve to neutralize the annoying finitist objections to so much of higher mathematics.[104] He challenged such conservatism by facing the intuitionist view of number directly.[105] Beginning from the sequence of natural numbers generated successively from the unit, the integers 1, 2, 3, . . . , eventually led to negative numbers, fractions, and ultimately to real numbers by successively extending the simple concept starting with the positive integers. But despite the great pedagogical and heuristic value of this "genetic method," as he called it, Hilbert believed that it could never ensure complete logical certainty. To satisfy the demands of absolutely certain knowledge, only an axiomatic method would suffice. Thus Hilbert was not interested in the real numbers as defined by the entirety of fundamental sequences; rather, he was concerned with a system of objects whose connections were described by a finite and closed system of axioms from which new propositions might subsequently be derived in only a finite number of steps. It was then merely a matter of showing that the number system which then emerged axiomatically did in fact correspond with the familiar system of real numbers.

The advantage of the axiomatic formulation was the procedure it made possible for establishing the *consistency* of the system it generated. By axiomatizing the theory of real numbers, Hilbert believed one could not only prove the *existence* of the set of real numbers but simultaneously show that the real numbers constituted (in Cantor's sense of the word) a *consistent* set.[106] This did not mean that Hilbert had established any conclusion about powers or cardinal numbers in general. If the axiomatic method were applied in a similar attempt to prove the existence of all the Cantorian alephs, Hilbert warned that the attempt could only fail, since such a system did not exist.[107] He was referring directly to Cantor's conclusion that the system of all powers was inconsistent. It was an unfinished set.

Hilbert was followed in this attempt to provide a system of axioms for the real numbers by Zermelo, who offered a system of axioms for set theory. Inspired by his belief that Cantor's research was of indispensable value to mathematics, he was determined to reaffirm the soundness of Cantor's basic principles by showing how the contradictions could be eliminated completely. Russell's antinomy suggested that there were fundamental problems of a logical nature which had to be overcome. Despite temporary measures to stem the tide of criticism by distinguishing between consistent and inconsistent sets, as Cantor had suggested, there was no guarantee that such a dichotomy would serve to cover all paradoxes that might later be discovered. Proof was needed to establish conclusively that systems were possible which preserved all that was essential of set theory but which strictly eliminated any possibility of contradiction. This Zermelo set out to do.

In 1908 he not only published his second, revised proof of the controversial well-ordering theorem, but in the same issue of *Mathematische Annalen* Zermelo also published his "Untersuchungen über die Grundlagen der Mengenlehre. I."[108] Both papers had been written during the summer of 1907 and were finished within two weeks of each other. The significance he attached to set theory was apparent from the language of his opening sentence, in which he tried to prepare his readers for the importance of the axiomatization he was about to introduce. Set theory, he wrote, was that branch of mathematics which concerned the basic concepts of number, or order, and of functions, and as such could only be regarded as providing the logical foundations for all of arithmetic and analysis.[109] Zermelo's sweeping characterization of the importance of set theory was grandly summarized in his claim that it was an indispensable component of all mathematical sciences. The paradoxes might seem to endanger nearly all of set theory, but Zermelo was prepared to show how the threat they represented could be minimized. Axiomatization, as Zermelo hoped to show, was a powerful antidote to the antinomies.

Unfortunately, Zermelo was only able to indicate the directions in which a completely successful axiomatization of set theory ought to move. He had made initial advances toward the ultimate goal, but there was a good deal of ground still to be covered. As a first step, he had managed to isolate the fundamental principles of Cantorian set theory that were requisite for a systematic foundation, one that would eventually produce the certainty and rigor of axiomatic set theory. Zermelo believed he had succeeded in restricting the basic elements sufficiently to ensure that no contradictions could ever be generated from his axioms; he was similarly confident that they were not so narrow as to preclude any of the essential and most valuable parts of the original theory. But there were nevertheless difficulties which he had to admit. Despite the limited number of definitions and no more than seven axioms, he could only say that the axioms *appeared* to be independent. Moreover, he confessed that the consistency of the entire system was a very complex matter which had yet to be established.[110] But at least he could recommend his results as they stood insofar as they excluded all of the antinomies known at that time.

THE SPLINTERING OF MATHEMATICS

Frege's fatal prediction of 1892 concerning the future of mathematics had come true in little more than a decade. The question of infinity had brought mathematicians to the edge of uncertainty. This confrontation, as Frege had foreseen, was the inevitable product of irreconcilably opposing views concerning the nature and legitimacy of Cantor's transfinite numbers. The emergence of dissident factions was immediately apparent in the debate over Zermelo's proof that every set could be well-ordered. It was decisive in determing how one might choose to regard set theory itself. Everyone was anxious to resolve the paradoxes and to reestablish the former faith in the certainty and exactness of mathematics, but the ways in which individuals chose to accomplish this were as diverse as some were drastic.

Intuitionists like Poincaré argued that most of the ideas of Cantorian set theory should be banished from mathematics once and for all.[111] He believed that knowledge concerning mathematical entities and propositions originated in the intuition and were thus in Kant's sense real *synthetic apriori* judgments. His running debate with Russell made it clear that he opposed any attempt to reduce mathematics to logic.[112] If the logicians were correct, then Poincaré asserted that mathematics should be nothing more than a complex system of tautologies. Instead, he was convinced that mathematics was much richer than it could ever have been were it simply the product of numerous variations on a tautological theme. At the same time he thought that mathematical certainty was limited to the strict bounds of finite argumentation. Transfinite set theory, Cantor's great contribution to mathematics, involved nothing in Poincaré's view but contradictory and therefore meaningless concepts. The paradoxes of set theory were direct evidence that Cantor's ideas were a grave disease that seemed to infect all mathematics. Poincaré's medicine was hard stuff indeed: his prescription called for the elimination of virtually every aspect of Cantor's work from respectable, permissible, and finite mathematics.[113]

Not all intuitionists were so strict. Borel, despite his own strong bias in favor of a finite mathematics, was nevertheless capable of allowing Cantor's work a modicum of acceptance. Unlike the more extreme of his colleagues, Borel thought that if Cantor's set theory were really destitute of any value, it would never have led to anything, since it would have been nothing but mere assemblages of words void of any sense.[114] Borel preferred to think of transfinite set theory as something more akin to mathematical physics. In this sense Cantor's set theory was *not* taken to express any sort of reality. Instead, it served as a guide which might be used to discover new results that would then have to be verified by acceptable methods. For Borel, it seemed as though set theory was to be regarded as a mathematical divining rod. One should neither believe it nor take it too seriously, but if it should prove helpful in leading one to new ideas, then it might be allowed to enjoy a silent partnership in an enterprise that should ultimately be given an acceptably finite public face. But as Frege had warned, the infinite could not be expected to remain content like a well-behaved child whose presence was neither to be seen nor heard. Cantor had already done too much to make the infinite mathematically respectable, and no one could hope to deny his transfinite numbers their rightful place in arithmetic without provoking a divisive conflict. All of mathematics was witness, for example, to the series of rejoinders which passed between Poincaré, the arch-intuitionist, and Bertrand Russell, then a staunch logician, each fighting for his cause in the pages of the *Revue de Métaphysique et de Morale*.[115]

Russell, in fact, was firmly opposed to the intuitionists. Basically, he affirmed the ultimate reduction of all mathematics to purely logical principles. No view of reality, no consideration of one's perception or intuition came into play. Mathematics, for the logician, concerned only the proper use of symbols. Their definition and the rules by which they could be manipulated and related were perfectly arbitrary. It was only a matter of ensuring that no contradictions could be produced within a given logical system before one could legitimately claim its existence. But many found mathematics in this new guise rather barren, little more than a pale reflection of the very real content they all associated with rational numbers, elliptical functions, π and e, partial derivatives, proper integrals, even sets, and the transfinite numbers. The formality of logic seemed to remove the very life from mathematics.

Thomae, for one, found the position of logicians like Frege and Russell unacceptable, and he expressed his feeling that pure logicism alone was greatly lacking in any sense of what numbers were or ought to be. He regarded someone like Frege as interested only in taking what he felt was needed from the number concept, perhaps reducing the ideas in question to their minimum essentials, but boiling away the most vital ingredients of a dynamic mathematics in the process.[116] Rather than strip away the personality and character of their discipline, Thomae and others felt that difficulties like the paradoxes of set theory could be effectively diagnosed and remedied within the discipline itself. This was clearly the view of Gerhard Hessenberg, who surveyed the state of mathematics in 1908 and concluded that productive set theory had at last recovered from the paradoxes.[117] He likened the contradictions to the old sophisms against the possibility of change which Zeno had once promulgated with such provocative success several thousand years ago. History had shown how the idea of continuity was eventually understood to answer the riddles of motion, and he hoped it would not take as long to dispatch the paradoxes of set theory. In the meantime, Hessenberg opposed any retreat into conservatism, he shunned the timidity of intuitionists who were too restrictive in what they took to be mathematically allowable, and he added his fervent support to Schoenflies' war cry, as he called it: "Down with all resignation and down with all Scholasticism."[118] In fact, Hessenberg took courage in the positive character of mathematical opinion he noted in the literature. This change seemed to reflect the steady decline of skepticism in dealing with such matters as set theory and transfinite numbers.

None were more faithful to the original content and spirit of Cantor's set theory than those mathematicians who sought to axiomatize it systematically, thereby hoping to establish its consistency and mathematical legitimacy. Zermelo felt he detected a pervasive prejudice against certain basic tenets of Cantorian set theory (particularly the belief that every set could be well-ordered) in the abundance of criticism which his paper of 1904 had met.[119] His goal had been to reduce the fundamental principles of set theory to a minimum number of self-evident axioms which would yield all the results and richness of traditional set theory but which would exclude as well the possibility of paradoxes. Others like Fraenkel, Hausdorff, von Neumann, Bernays, Skolem, and Gödel also joined the attempt to bring Hilbert's original dream of a consistently formalized interpretation of mathematics into being.

It was Gödel, however, who published a major and dismaying discovery in 1931.[120] Gödel's famous Incompleteness Theorem showed that it would never be possible to establish the consistency of set theory by working entirely within the limits prescribed by any axiomatization. Consistency could only be proven by an appeal to external or higher principles. In particular, Gödel showed that in any system rich enough to contain elementary arithmetic, there were always theorems that could neither be proven nor disproven. They were undecidable, and it seemed quite possible that Cantor's continuum hypothesis might be a prominent example of such an undecidable proposition.

On the other hand, though axiomatic set theory, as Gödel had treated it, might never be capable of deciding the power of the continuum, it could be shown that Cantor's continuum hypothesis could not give rise to any paradoxes if added to an axiom system that was itself consistent. This Gödel managed to do in 1936, when he succeeded in arguing the consistency of both the continuum hypothesis and the Axiom of Choice.[121] Finally, in 1963, Paul Cohen established what Gödel's Incompleteness Theorem had indeed suggested was possible. Cohen showed that neither the continuum hypothesis nor the Axiom of Choice could be proven from the axiomatic system of Zermelo-Fraenkel set theory.[122]

Two years later, while lecturing at Harvard University, Cohen tried to make some predictions about the conclusions mathematicians might some day reach in light of his independence theorems.[123] Although Cantor would have found the prediction upsetting, Cohen suggested the likelihood that mathematicans might one day come to the conclusion that the continuum hypothesis was obviously false, and his reasons harkened back in part to opinions expressed during the controversy over Zermelo's well-ordering theorem.[124] In Cantor's famous claim, $2^{\aleph_0} = \aleph_1$, $\aleph_0$ represented the power of the continuum while $\aleph_1$ represented the cardinality of the set of all ordinal numbers of the second number class $Z(\aleph_0)$. These were two very different things. The class $Z(\aleph_0)$ contained nothing but *denumerable* ordinal numbers of the type ω. As such, $Z(\aleph_0)$ was generated by the simplest of all means possible for producing sets of increasingly higher cardinality. By means of the first and second principles of generation, the very elementary procedures of successively adding units and of determining the limits of fundamental sequences had been used to produce the first nondenumerable cardinal $\aleph_1$. By contrast, $2^{\aleph_0}$ represented a much more sophisticated concept, the idea of the power set. In terms of coverings, Cantor had managed to show how transfinite exponentiation could guarantee that any set could give rise to sets of necessarily distinct and higher cardinality. Why, then, asked Cohen, should so rich a concept as the continuum, as represented by $2^{\aleph_0}$, ever be equivalent in power to anything so simple as the class $Z(\aleph_0)$ determined by only the first and second principles of generation? Cohen conjectured, in fact, that $2^{\aleph_0}$ might well turn out to be larger than any transfinite aleph, surpassing all elements like $\aleph_n$, $\aleph_\omega$, $\aleph_{\aleph_\omega}$, . . . , of the succession of alephs ת. In Cohen's view, the continuum was clearly an incredibly rich set, one produced by a bold new axiom which could never be approached by any piecemeal process of construction.[125]

This interpretation was by no means a recent innovation. As early as 1905 René Baire was thinking along similar lines when he suggested that Cantor's continuum hypothesis assumed the identifiability of two concepts that were intrinsically different and of noncomparable orders of magnitude.[126] In a letter to Borel, Baire argued that if one wanted to consider the continuum, it was certainly a concept which lay well beyond the powers of any class of denumerably infinite order types to describe. Borel later chose to use an excerpt from Baire's letter in concluding his own article of 1905 for the *Mathematische Annalen*.[127] His words served as a fitting capstone to Borel's critique of Zermelo's well-ordering theorem because Baire refused to believe the continuum could ever be well-ordered. The two ideas were inherently antithetical: the nature of the continuum, regarded as the collection of all infinite sequences of integers, was something totally different. It was useless to ever speak of well-ordering such sequences used to define elements of the continuum.

Just as Frege had predicted, Cantor's set theory had brought mathematicians to a frightening and perilous precipice. Cantor's infinite had shaken the traditional faith in mathematics' everlasting certitude, and although Cantor had never doubted the ultimate validity of his transfinite numbers, others of a more conservative persuasion refused to follow his lead into the uncertain regions which he had chartered in the *Beiträge*. The final irony, confirmed by the results of Cohen's research, was the futility of Cantor's greatest hope: the verification of his continuum hypothesis. Though he had given the world a tremendously powerful and useful theory, his transfinite numbers were flawed from the outset. Though he knew that his life's work was incomplete, he could not have known that, despite his own inability to come to satisfactory terms with the continuum hypothesis, no one else would succeed either. The solution remained an enigma.

CHAPTER 12

Epilogue: The Significance of Cantor's Personality

> Ce qui est interéssant pour le psychologue, ce n'est pas le théorème, ce sont les circonstances.
>
> —Henri Poincaré

This book began with the premise that a study of the early history of set theory should illuminate more than the evolution of Cantor's work, its creation and intellectual development. By tracing the course of Cantor's research from his papers on trigonometric series in the early 1870s to his discovery of the paradoxes of set theory at the end of the century, one might hope to understand more fully the nature of progress in mathematics and of scientific creativity generally. But to appreciate the way in which set theory developed during its formative years, it is also necessary to consider the complex kind of genius Cantor represented as a creative individual. Therefore, in bringing this book to a close, it is appropriate to consider certain details of his biography and to assess the influence of his personality, including psychological factors, upon the development of transfinite set theory.

Historians of mathematics are generally accustomed to discussing ideas rather than individuals. Consequently, a mathematician's biography and his mathematics are frequently regarded as wholly separate. His biography may provide human interest, but his mathematics is taken to form the heart of the matter. Even so, the analysis of personality, in particular when creative individuals are concerned, can reveal a great deal about the nature of intellectual discovery. In Cantor's case this is especially true. If one were to consult only the published record of his research, the factors influencing his discovery and subsequent development of set theory and the transfinite numbers would remain obscure. At best, one could hope for only a partial, and probably inaccurate, view of what Cantor accomplished. Why should he have been the mathematician most likely to defend transfinite set theory, despite tremendous criticism from mathematicians, philosophers, and even theologians? How did his character and scientific temperament shape the earliest development of set theory? Satisfying answers to such questions cannot be expected without going beyond the record of Cantor's publications. In letters, and in the memoirs and reminiscences of those who knew him, some hints can be found. Unfortunately, much of the evidence needed to produce a full biography of Georg Cantor has been lost. In his literary estate now in the care of the Akademie der Wissenschaften in Göttingen, some family letters have survived, and these have been used along with other documents described in this chapter, to produce as complete a sketch of Cantor's personal life as possible. One cannot help but wish that more of the family history were preserved, and yet what is known proves to be extremely interesting and very useful in permitting a fresh assessment of Cantor as a mathematician. In fact, without some consideration of the larger context of Cantor's family history, including certain aspects of his personal and medical history, it would be impossible to appreciate the reasons why set theory, and transfinite set theory in particular, persisted despite continuing opposition to thrive in the hands of its creator, Georg Cantor.

GEORG CANTOR: BIOGRAPHY, GENEALOGY, AND MATHEMATICS

What follows is not an attempt to prepare the way for any detailed psychoanalysis of Cantor's personality. Still, it must be said that much of Cantor's personality was shaped under the very strong influence of his father. To understand why Cantor approached certain problems and harbored specific anxieties as he did, it is necessary to survey what little is known of the family history and to add some relevant new facts from unpublished sources.

Although the family history is unclear, there is agreement that Cantor's grandfather, Jakob, lived in Copenhagen. Exactly when Cantor's father, Georg Woldemar Cantor was born is uncertain, though at the time of his death in Heidelberg his birth (in Copenhagen) was officially registered as falling on March 24, 1814. But this does not accord at all with traditional family history, which places Cantor's family in Copenhagen during the English bombardment of 1807.[1] According to this story, the family lost everything in the attack and moved to St. Petersburg, Russia, where Georg Woldemar's mother apparently had relatives. There is, however, evidence for an earlier date of birth: a Danish passport issued to Georg Woldemar Cantor in August 1833. The passport recorded his age then as twenty-four, placing his birth in March 1809. This fits more accurately with the English siege of Copenhagen, but not exactly. Nevertheless, following the family's move to St. Petersburg, the upbringing

Cantor in 1917, a few months before his death.
In the possession of Helga Schneider and Sigrid Lange.

and education of the child Georg Woldemar was entrusted to an evangelical Lutheran mission.[2] What became of his parents at this time is unknown, though his mother's family, by the name of Meier, was reportedly a respected and successful family in St. Petersburg. A sister was married to Josef Grimm (a Roman Catholic), who was a chamber musician of the royal court, and a nephew was established as a professor of law at Kasan. He was apparently instrumental in promoting the legal apparatus involved in freeing the Russian serfs in 1861, and as the family liked to recall, Tolstoy had once been his student. As for Georg Woldemar's parents, nothing more is known except for the fact (mentioned in a report of the Danish Genealogical Institute) that Jakob Cantor was still alive in 1841, when he sent congratulations to his son on the occasion of his engagement.

Georg Woldemar Cantor married Maria Anna Böhm the following year in St. Petersburg, on April 21, 1842.[3] She had been born in St. Petersburg and baptized a Roman Catholic. The entire family was gifted musically, and it included several classical violinists, the most noted being Cantor's great uncle Joseph Böhm, director of the Vienna Conservatory and founder of the great school for violinists which produced many a virtuoso at the time, including Joseph Joachim and Joseph Hellmesberger. The fact that Georg Woldemar and Maria Anna Böhm were married in the Evangelical Lutheran Church in St. Petersburg again reflects the close links that Cantor's father felt to the mission where he had been raised. The couple's six children, of which Georg was the eldest, were baptized as Lutherans and given strict instruction by their father in matters of religion. The seriousness and pervasive consciousness of religion in Georg Woldemar's daily life was strongly reflected in nearly every letter he addressed to his son—letters the young Cantor saved during his school days in Darmstadt and later at the Polytechnic Institute in Zürich. One letter in particular deserves special mention. It was written on the occasion of Cantor's confirmation, and was a letter he never forgot; he kept it with him always. It deserves to be cited almost entirely, for in so many ways it is an uncanny prediction of the successes and failures of Cantor's personal life and public career.

> Dearest Georg:
>
> Through the goodness of the Almighty, the Creator of the universe and Father of all living creatures, may this day be of blessed influence upon your entire future life. May you constantly and unremittingly keep before your eyes the virtuous resolutions which you have no doubt made today in silence as a solemn vow! . . . The future of one's *life* and the fate of the individual lie hidden from him in the most profound darkness. And it is good that it is so. No one knows beforehand into what unbelievably difficult conditions and occupational circumstances he will fall by chance, against what unforeseen and *unforeseeable* calamities and difficulties he will have to fight in the various situations of life.

How often the most promising individuals are defeated after a tenuous, weak resistance in their first serious struggle following their entry into practical affairs. Their courage broken, they atrophy completely thereafter, and even in the *best* case they will still be nothing more than a so-called ruined genius! ... Indeed, it is *not seldom* that young men come to such an end, even those who are *seemingly* endowed with the most promising features of mind and body, and whose prospects for the future through ability and family connections in their youth were also the rosiest in the world!

But they *lacked* that *steady heart* [*feste Kern*], upon which *everything* depends! Now, my dear son! *Believe me,* your *sincerest, truest,* and *most experienced* friend—this sure heart, *which must live in us,* is: a *truly* religious spirit [*Gemüt*]! This reveals *itself to us* through a *sincere, humble feeling of the most grateful reverence for God,* from which feeling grows as well the *victorious, unshakable, enduring faith in God,* and which keeps and maintains us throughout our entire life in that silent, undoubting communion with our heavenly Father! ...

But in order to prevent as well *all those other* hardships and difficulties which inevitably rise against us [*entgegentürmen*] through jealousy and slander of open or secret enemies in our eager aspiration for success in the activity of our *own specialty or business;* in order to combat these *with success* one needs above all to acquire and to appropriate the greatest amount possible of the most basic, diverse technical knowledge and skills. Nowadays these are an *absolute necessity* if the industrious and ambitious man does not want to see himself pushed aside by his enemies and forced to stand in the second or third rank.

To the procurement of diverse, thorough, scientific and practical knowledge; to the *perfect* acquisition of foreign languages and literatures; to the many-sided development of the mind in many humanistic disciplines—and of this you must always be thoroughly conscious!—*to all this* the *second period* of your life, your *youth,* now just beginning, is *destined,* in order first to *equip yourself with dignity* by means of all this for *those struggles* yet to come. Whatever one neglects in this period or through premature extravagance of his best strength, health and time, he *squanders* [*verludert*], so to speak, that is irretrievably and *irreplacably* lost *for ever;* just as innocence, *once lost,* is for ever and eternally, irretrievably lost ...

I close with these words: Your father, or rather your parents and all other members of the family both in Germany and in Russia and in Denmark have their eyes on *you* as the eldest, and expect you to be nothing *less* than a Theodor Schaeffer and, God willing, later perhaps a *shining star* on the horizon of science.

May God give you *strength,* persistence, health, sound character, and His best blessings! And therefore *you* should follow only his ways. Amen!

Your father [4]

All the major features of Cantor's later personality seem to have found a place in his father's letter: the religious faith meant to help one through the darkest hours and the most troubled times, the certainty that those who failed in life lacked a strong spirit, and the belief that a truly religious spirit had to reside in the individual. In addition, to face the slanders of obvious or of secret enemies and to meet them successfully, one had to be prepared fully and grounded completely in the expertise of one's own profession. The call to devoted and thorough study was driven home with a reminder that the entire family, in Denmark, Germany, and in Russia, expected Cantor to be as distinguished as his teacher, Theodor Schaeffer, and possibly even a renowned man of science.

In large measure Georg Woldemar Cantor's advice to his fifteen year old son reflected his own experience. Raised and educated in a Lutheran mission, the father had obviously been deeply impressed by the religious values he had been taught and he was determined to pass them on to his children. As a young man Georg Woldemar had gone into business in St. Petersburg and had established a wholesaling firm in 1834 under the title of Cantor and Company. The firm's dealings were international, with interests in Hamburg, Copenhagen, and London; transactions reached as far as New York, Rio de Janiero, and Bahia. But for reasons now unknown, Cantor and Company failed in December 1838, and by December 1839 liquidation was complete.[5] Nevertheless, Georg Woldemar Cantor still had the "inner resources and determination" to make a success of his business career and turned his talents to good use as a broker on the stock exchange in St. Petersburg. Shortly before leaving Russia for Germany (owing to ill health), he drafted his will in May 1854 and identified himself as: "Wholesale trader from Wildmanstrand and temporary [*temporairer*] merchant of Petersburg's Third Guild."[6]

Georg Woldemar's success in business was clearly reflected in his estate. When he finally died of tuberculosis in June 1863, he left his family nearly half a million marks. But more than money, he left his eldest son with a compulsive desire to succeed professionally and with a belief in his own inner resources, with which he could always be sure of overcoming even the most distressing episodes of life. Cantor appears to have followed his father's advice with great seriousness, and this, as much as any other aspect of his personal life, helps explain why he was always determined to prevail and why he fought with such tenacity and confidence in support of his theory of sets and transfinite numbers.

After leaving Russia, Cantor's family lived shortly in Wiesbaden before settling permanently in Frankfurt. Young Georg boarded at the Realschule in

Darmstadt, from which he graduated with the report that "his diligence and fervor are exemplary; his knowledge in basic mathematics including trigonometry is very good; performance praiseworthy."[7] A year later he was at the Höheren Gewerbeschule (trade school) and again upon leaving in 1862 was given the highest praise: "He showed himself to be very gifted and a highly industrious pupil."[8] Much to his father's satisfaction, George was working with great industry, and he seemed to have won special recognition for his mathematical abilities. Cantor's father took a special interest in his son's education and was careful to direct his personal and his intellectual development, offering a good deal of encouragement. The closeness of their relationship was particularly noticeable in the letters Georg Woldemar sent to his son during the years he was away at school. The careful, emphatic admonitions to work diligently and thereby to ensure success had not been lost on young Georg.

The fact that Cantor spent two years in the Darmstädter Gewerbeschule suggests that his father's earliest hopes for his son's future did not include the devotion to pure science that his son was finding increasingly hard to resist. But by the spring of 1862 the young Cantor had come to a decision: he had decided to devote his life to mathematics. Despite his initial uncertainty as to whether this was the correct decision, he was wholly supported in his choice by his father. Cantor's joy at his father's approval was echoed in every word of the following letter:

> My Dear Papa!
>
> You can imagine how very happy your letter made me; it determines my future. The last few days have left me in doubt and uncertainty. I could reach no decision. My sense of duty and my own wishes fought continuously one against the other. Now I am happy when I see that it will no longer distress you if I follow my own feelings in this decision. I hope that you will still be proud of me one day, dear Father, for my soul, my entire being lives in my calling; whatever one wants and is able to do, whatever it is toward which an unknown, secret voice calls him, *that* he will carry through to success![9]

From the very beginning, apparently, Cantor had felt some inner compulsion to study mathematics. Georg Woldemar's original doubts and disappointment over his son's wish to study pure science were set aside, but this also left Cantor with a sense of having to prove that his father would not be disappointed. Characteristically, Cantor reinforced the validity of his decision by saying that it was not his alone—an unknown, secret voice was making its impression even then.

Cantor's father followed his son's career with enthusiasm. In August 1862 Cantor took the *Reifeprüfung,* which qualified him for advanced study in the sciences. Georg Woldemar sent his best wishes and again stressed a philosophy to which his son would later turn in periods of doubt and depression. His father's words were a firm call to success, but as Cantor sat down to write his examination on August 18, 1862, he had more than his father's prayer that God might grant him blessings and luck. He also had his father's reminder:

> May God give you His blessing and good fortune for the examinations beginning today .... In critical moments of one's life an unwavering and joyous trust in God and a profound, moving prayer to the Almighty Giver of all goodness before the start of the day gives stability, courage, and self-confidence. And thus commit yourself to God! Be fresh, joyous and cheerful in work![10]

By the end of August Cantor's examination had been read and was "recognized as proof of first-level preparation [*Maturitätszeugnis Nr. I*] for the study of the natural sciences."[11] Apparently Cantor had done less well in subjects like geography and history, because his father was prompted to write, stressing in a letter the importance of a diverse and thorough study of all branches of knowledge.[12] Before Cantor had even left Darmstadt, his father had counseled him not to concentrate on a single subject, like botany, but suggested that he study everything from languages to music as thoroughly as possible. Thus he was delighted to hear of his son's interest in hearing a series of lectures by the renowned Friedrich Theodor von Vischer on Shakespeare, of whom his father wrote enviously that he had never heard anything but the highest praise. When Cantor formed his own string quartet, his family was delighted, and his father invited the entire group to spend the Christmas holidays in Frankfurt. Perhaps the most graphic indication, however, of the degree of paternal pride Georg Woldemar could take in his son's achievements is a proclamation he wrote to commemorate a drawing (see page 279) his son had sent home as evidence of the progress he was making at the Polytechnicum in Zürich:

> Whereas Georg Ferd. Louis Phil. Cantor has not spent years in the study of drawing according to the ancient models; and whereas this is his first work and in this difficult art a perfected technique is only achieved after great diligence; and whereas, furthermore, until now he has greatly neglected this beautiful art; the thanks of the nation—I mean of the family—is unanimously voted to him for this first effort, which already shows great promise.[13]

Contrary to the claims made by E.T. Bell that Georg Woldemar had a thoroughly deleterious and ruinous effect upon his son's psychological health, the surviving evidence of their relationship indicates quite the opposite.[14] Cantor's father was a sensitive and gifted man, who loved his children deeply and wanted them to live happy, successful, and rewarding lives. As a child, young Georg was expected to be a credit to his family, and as he approached a career in mathematics, he wanted nothing more than to fulfill the hopes of his

Pencil drawing by Cantor of a dog under a tree.
Signed G[eorg] C[antor] 1862, in the bottom right-hand corner.
It is the only surviving example of Cantor's considerable artistic gifts.
In the possession of Egbert Schneider.

parents. Rather than disappoint them, he worked diligently and proved to be an energetic student. But underlying even his earliest decisions was a sense of destiny, a feeling that unknown forces were at work and that they were propelling him toward a career he felt he could not deny.

PERSONALITY, PSYCHOLOGY, AND DEPRESSION

Cantor's first serious mental breakdown occurred shortly after his thirty-ninth birthday in the spring of 1884. The episode has always figured prominently in the accounts of his life given by sensational popularizers, and of these none has treated his case with less reliability than E.T. Bell.[15] Not content with the view advanced in 1927 by Schoenflies that Cantor's breakdown was the result of conflict with Kronecker and his inability to solve the continuum hypothesis, Bell turned to the popular psychology of his day and produced a predictable, Freudian analysis.[16] He traced the roots of Cantor's mental instability back to childhood and blamed Cantor's father for having left his son with deeply planted anxieties. These eventually led to the breakdown which required hospitalization in 1884. But unless Bell had access to material unknown to historians before or since, there seems to be little justification for the exaggerated scenario he describes. On the contrary, the actual details of Cantor's illness, as it is currently understood, are considerably more interesting than the analysis manufactured by Bell, and in fact they make a great deal of sense when related to other features of Cantor's personality.

When Cantor suddenly suffered his first breakdown in May 1884 he had just returned from an apparently successful, quite enjoyable, trip to Paris.[17] He had met a number of French mathematicians, including Hermite, Picard, and Appell, and was delighted to report to Mittag-Leffler that he had liked Poincaré very much and was happy to see that the Frenchman understood transfinite set theory and its applications in functional analysis. Since other French mathematicians were away from Paris on vacation, Cantor had some time free to visit galleries and museums, to indulge his love of music at the Paris Opera, and to spend an evening at the Comédie Française. But after eight days, he was unexpectedly called back to Frankfurt to deal with some unknown family matters, and shortly thereafter his breakdown occurred. It is not possible to conjecture the immediate cause, though the tradition made popular by Schoenflies blamed Kronecker's persistent criticism and Cantor's inability to confirm his continuum hypothesis.[18] Cantor's bitterness, Schoenflies argued, was the product of tremendous opposition to his work, coupled with his failure to produce the most important result of all. This evidently drove him to the heights of doubt and distress and even caused him to despair of ever making a success of his mathematics. His efforts to solve the continuum conjecture reached a fervent pitch, and determination drove him on. Because of the strain, the breakdown inevitably followed, and he lapsed into the abyss of depression which left him weak and unsure of himself and of his transfinite set theory.

Schoenflies interpreted Cantor's attempted reconciliation with Kronecker later that summer as proof that his reconstruction of the causes and events leading up to the crucial moment of collapse was essentially correct.[19] Cantor himself recalled the entire episode in similar terms when he wrote to his friend Mittag-Leffler a few months later, on August 18.[20] Cantor was in the summer resort of Friedrichroda at the time, where he was trying to relax and to recuperate from the entire unhappy experience. He felt that it had all been his own fault, brought on by the problems of his research. The breakdown had not been caused by overwork, he insisted, but by irritations and frictions with Kronecker. He wrote that he should never have let Kronecker upset him to such a degree, and consequently, he decided that the best cure was to face Kronecker squarely and to attempt a reconciliation. On the same day that he wrote to Mittag-Leffler, Cantor posted a letter to Kronecker and tried to explain his unhappiness over the state of their rivalry.[21] Above all he hoped it was still possible to gain some semblance of equanimity. Whatever might come of the attempt, Cantor knew that writing to his rival would at least serve to lighten his heart. By the end of August, Kronecker had written a polite, even friendly response, reminding Cantor of how much closer they had been when he was a student in Berlin and registering some surprise at Cantor's sudden claims to any hostility or animosity between them.[22] Cantor was greatly reassured and wrote immediately to Mittag-Leffler to say how happy he was, that at last he could work again without the dread of personal animosity.[23]

In retrospect, the first major breakdown was relatively short, and could not have lasted more than a month. His letter to Klein of May 10 was the last before the onset of the illness; by June 21 he was writing again to Mittag-Leffler saying he hadn't felt so "fresh" lately and couldn't be sure when he'd get back to his research.[24] Despite the speed with which the depression seemed to pass, it left the entire family shaken. His eldest daughter Else, only nine at the time, was bewildered by the incomprehensible change in her father, the swiftness with which his entire manner had been transformed.[25] Cantor himself emerged from his attack emotionally enervated, and he spent the following months trying to rebuild his strength and to regain his intellectual spontaneity.

Although he returned to mathematics in the autumn and attempted to work on the continuum hypothesis again, his attitudes generally had undergone substantial alteration. He began to emphasize other interests. The amount of time he devoted to various literary-historical problems increased steadily, and he read the history and documents of the Elizabethans with great attentiveness in hopes of proving that Francis Bacon was the true author of Shakespeare's plays. As time progressed, he also began to intensify his study of the scriptures and of the Church fathers, and he developed new interests in freemasonry, Rosicrucianism, and theosophy.[26] Perhaps he felt that his earlier exclusive devotion to mathematics had been too concentrated, too painful, without the rewards he had expected to make it all worthwhile. Perhaps he hoped to dilute the intensity of his earlier devotion to mathematics, thereby avoiding the chances of a relapse of his illness. Nothing reflects more directly Cantor's shift in attitude following his first breakdown than does his sudden desire to teach philosophy in Halle instead of mathematics.[27] It was as though he were striving for a certain balance as he began to take his extra-mathematical interests more and more seriously.

This strategy was apparently successful, at least for a time, but eventually the breakdowns began anew; they grew more intense, lasted longer, and occurred with greater frequency. In 1899 (the first year since 1884 in which any record survives of his hospitalization for reasons of mental instability) Cantor was again faced with a series of dilemmas, both professional and personal.[28] During the summer the antinomies of set theory were on his mind, as we know from the letters he sent to his friend Dedekind.[29] Cantor was certainly unhappy at the unsatisfactory impass he had reached where the old problems of the continuum hypothesis, of well-ordered sets and the comparability of the transfinite numbers were concerned. He applied for a leave of absence from teaching for the fall term, and his request was granted. But in November he sent an extraordinary letter to the Ministry of Culture, saying he was anxious to abandon completely his professorship in Halle.[30] So long as his salary was not diminished (he asked for no increase from his recompense at university of 6,000 marks annually), he would be content with a modest position in some library. Nor was a prominent title necessary; he was ready to accept virtually any alternative that might release him from the confines of the German university. Whether the Ministry could coöperate or not, he was determined to break away "by all means." He emphasized his qualifications, his knowledge of history and literature, his publications on the Bacon-Shakespeare question, and even added the provocative news that he had come upon certain information in the course of his research concerning the first king of England, "which will not fail to terrify the English government as soon as the matter is published."[31] Cantor tried to inject a note of urgency into the matter by asking the ministry to send him their reply within the next two days, for if they could offer him no alternative to teaching, then as a born Russian he would apply to the Russian diplomatic corps in hopes that he might be of service to Tzar Nicholas II.[32]

Nothing seems to have come of Cantor's request, nor did he enter the ranks of the Russian diplomatic service. The entire episode fit the pattern of his continuing fears of persecution and opposition, which previously, in 1884, had prompted him to think seriously about giving up mathematics in order to teach philosophy. But he was no more successful in 1899 than he had been in 1884. Finding neither a post as librarian nor a position in diplomatic service in Berlin, Cantor remained at university in Halle. For part of 1899 he was again hospitalized, but before the year was out a final tragic event occurred which produced a surpassing disillusionment and much pain for the entire family.

Cantor's mother, born in St. Petersburg in 1819, died in Berlin more than thirty years after the death of her husband, in 1896. Three years later her son Constantin, Cantor's younger brother, died in Capri where he had retired from military life as an officer in a Hessian Dragoon Regiment to marry an Italian baronessa.[33] Then, on December 16, 1899, while Cantor was delivering a special lecture on the Bacon-Shakespeare question in Leipzig, his youngest son died suddenly. When Cantor returned home later that evening, he was told the terrible news: Rudolf had died toward mid-afternoon. In another four days the child would have been thirteen.

Cantor described his son Rudolf in a letter to Felix Klein a week later, and the tragedy was borne in every word he wrote.[34] The boy had been frail in childhood but then began to grow stronger. His spirit correspondingly strengthened, he was so loving, so amiable, that he was everyone's favorite. Rudolf had been gifted musically; his father hoped he might follow in the family tradition and become a famous violinist. In writing to Klein, Cantor was prompted to recall his own early study of the violin and to rekindle his doubts about the wisdom of having given up his inherited talents and love of music to become a mathematician. His early enthusiasm for science had faded from memory; he could not even remember *why* he had ever given up music for mathematics.[35] That early voice calling him to a new and challenging life in 1863 seemed entirely forgotten and buried under the weight of subsequent events. Disappointed and unhappy over his own choice of profession, Cantor wanted something better for his son. As Cantor confided to Klein, Rudolf represented a means through which he might still live and enjoy a way of life he now regretted having abandoned. But now, even that hope was gone.

Despite his hospitalization and the tragic events of 1899, Cantor managed to maintain his equilibrium for another three years, until he was again hospitalized and relieved of his teaching duties for the winter term 1902–03.[36] Barely a year later yet another shock sent him to the hospital again, and thereafter he was frequently in and out of the Halle Nervenklinik. The dramatic events of König's paper read during the Third International Congress for Mathematicians in Heidelberg greatly upset him.[37] He was there with his two daughters, Else and Anna-Marie, and was outraged at the humiliation he felt he had been made to suffer. He did not seem so upset at the suggestion that his theory was somehow wrong for he was certain that it was infallibly correct, despite König's proof. Instead, he was angered at the public way in which his work had been brought to question. As the Congress drew to a close, he was still in a state of agitation, and Schoenflies' story of how Cantor rushed into breakfast a week later to describe in excited terms the error in König's proof, shows that his mind was racing to find ways of dispelling the doubts that lingered in the aftermath of the Congress.[38] It is no surprise that the strain was too much for him; he was soon hospitalized, and given a leave from teaching for the winter term 1904–05. Thereafter he spent increasingly longer periods at the Nervenklinik in Halle, from October 22, 1907, to June 15, 1908, and from September 28, 1911, to

June 18, 1912, when he was then moved to yet another sanitorium.[39] He was admitted to the Halle clinic for the last time on May 11, 1917. He did not want to go, and he wrote continuously to his family asking that they come to take him home. One of his last letters to his wife included a moving poem he had written to express his love, his feelings for her, and it showed a tender and most sensitive side of the man. The lines speak for themselves, and reflect the resignation he must have felt:

That was a winter cold and wild,
Like none one can recall;
But then I had it good and mild,
Was my luck yours at all?

The love you gave me my good wife,
You cared for me so well;
However gray might seem this life,
Your cooking would excell!

And now that spring is come full-fledged,
The Holy Child is gone;
And you in greatest haste are pledged,
To visit sites unknown.

But had you then remained at home,
Cold would my winter've been;
To suffer gladly, pen a poem,
To escape the world I'm in.[40]

Summer passed to autumn, but Cantor was not allowed to return home. World War I was still raging. Food was scarce, and the lack of nourishment shows in a surviving photograph: Cantor's face is gaunt and tired, and though his eyes are still a sparkling blue, they are no longer piercing, no longer full of questions; instead they seem more passive, reflective. On January 6, 1918, Cantor died, apparently of heart failure. But as Edmund Landau wrote to Cantor's wife as soon as he had heard the news, Cantor and all that he represented would never die. One had to be thankful, he observed, that mankind had been given a Georg Cantor, from whose works later generations would learn: "Never will anyone remain more alive."[41]

THE SIGNIFICANCE OF CANTOR'S NERVOUS BREAKDOWNS

With the above details of Cantor's biography and his medical history in mind, what can be said of their significance? Above all, there is reason to suspect that much more is involved in appreciating the nature and causes underlying Cantor's periods of euphoria and depression than Schoenflies' conventional interpretation might suggest.[42] After all, the account which Schoenflies published was concerned exclusively with Cantor's first major breakdown in 1884, and it was not difficult for him to draw explicit lines between Cantor's illness and specific anxieties which the climate of Cantor's research had produced. But what of the later episodes? Though at points, particularly in 1904, there seem to be links between periods of emotional upset and difficult impasses with his mathematics, it is by no means clear that the troubles with research and professional rivalries were the sole causes of his breakdowns. The traditional account, passed by word of mouth from mathematicians of Cantor's own generation to the next, was eventually codified in Schoenflies' article devoted to Cantor's "critical years." This interpretation was elaborated and given great currency through the popular writings of E.T. Bell.[43] A final, even more unfortunate stamp of authenticity, considering the source, came from Bertrand Russell, who shortly before his own death wrote in somewhat piquant terms that "Georg Cantor, the subject of the following letter, was in my opinion, one of the greatest intellects of the nineteenth century . . . After reading the following letter, no one will be surprised to learn that he spent a large part of his life in a lunatic asylum."[44]

Fortunately, it is possible to assess the nature and consequence of Cantor's nervous breakdowns with somewhat more objectivity than that shown by either Bell or Russell. Ivor Grattan-Guinness was the first to consider Cantor's illness from a clinical point of view by reviewing the records which survive in the Halle Nervenklinik and by reconstructing the outline of a case history: "The attacks all began suddenly, usually in an autumn season, and exhibited phases of excitement and exaltation: they ended suddenly in the following spring or summer, and were sometimes followed by what we now understand to be the depressive phase. In Cantor's day it was seen as a cure, and he would be sent home to sit silent and motionless for hours on end." He then concluded, with the advice of a psychologist who examined Cantor's file in the Halle Nervenklinik, that "Cantor's illness was basically *endogenous,* and probably showed some form of manic depression: exogenous factors, such as the difficulties of his researches and the controversies in Halle University, are likely to have played only a small part in the genesis of his attacks, little more than the clap that starts the avalanche. Thus he would have suffered his attacks if he had pursued only an ordinary mundane career."[45]

Moreover, there is good evidence of a direct nature which tends to confirm that this diagnosis is more realistic and in accord with the known facts of Cantor's history than any previous attempts to explain his mental illness. Georg Cantor was known by those who met him to be energetic, forceful, and volatile. Gerhard Kowalewski, in his memoirs, recalled his student days in Leipzig when there was a coffee group that met regularly with some of the Halle mathematicians. It was at such meetings that he first met Cantor, who to

Kowalewski's eye was, "as one immediately saw, a man of surpassing significance, of very unusual character."[46] Kowalewski admired the variety and depth of Cantor's many interests, and looked forward to visiting his home on Handelstrasse, where he delighted in the highly intellectual atmosphere which prevailed. Like all great men, Cantor had his fancy, and the Bacon-Shakespeare question was his most passionate interest, as Kowalewski explained. Though he had heard a number of Cantor's lectures on the subject, Kowalewski said that he was never convinced by the Baconian arguments despite the "washbasket full of literature" Cantor always seemed to produce on such occasions![47] The most interesting information from Kowalewski's memoirs, however, concerns the way in which all of this affected Cantor. Kowalewski was being extremely generous in saying that like everyone possessed by such ideas, Cantor felt he was persecuted by many people. There were those, Cantor imagined, who feared his research, since he believed that his conclusions carried with them results of worldwide political significance. One has only to recall Cantor's letter of 1899, sent to the Ministry of Culture in Berlin, to know that he was entirely serious about such matters.[48] The letter mentioned information that would greatly distress the English, Cantor said, were it to become known. Consequently, he was wary of alleged enemies who he claimed were interested in "making him silent" because of the secrets he might otherwise divulge.[49]

These recollections of Cantor's personality and sometimes erratic behavior coincide with the reports of other mathematicians who also had known Cantor personally. Albert Wangerin, his colleague in Halle since 1881, described Cantor warmly as a man of imposing stature, witty, spirited, amiable, and in his early years particularly lively and stimulating in conversation.[50] He was also portrayed as a man of loud and intense character, true to friends, helpful when needed, a noble and generous man. Schoenflies drew a similar picture in remembering Cantor as one of the most stimulating of Germany's mathematical personalities: his presence at any congress or meeting was always an enticing attraction; his mind was as imaginative and sparkling as it was temperamental and explosive.[51] Nothing illustrates these latter qualities more vividly than does a startling episode described by H.A. Schwarz in a letter to his colleague E.R. Noevius in 1888.[52]

Weierstrass had invited a number of friends to visit him in Wernigerode, where he was vacationing, and among the mathematicians present were Gösta Mittag-Leffler, Sophie Kowalevsky, Georg Hettner, and Georg Cantor. Adding Schwarz to the group, particularly when it already included Cantor, could not have been expected to promote a cordial atmosphere, since the two had been unfriendly rivals for some time (despite their close friendship during Cantor's student days in Berlin). At one point in the conversation, apparently without warning, Cantor exploded into a rage over the fact that he had not been called to fill the position at Göttingen given to Felix Klein in November of 1885.

Cantor's fury over the matter, coming three years after the fact, shows how deeply he resented the refusal of German mathematicians to recognize his work and advance him to a major mathematical center. Schwarz was part of the alleged conspiracy to isolate Cantor in Halle and to help consign his work to oblivion. How else, he asked, could one explain that his qualifications as a mathematician had been ignored in favor of candidates of notably inferior abilities? In this instance, the resentment Cantor carried with him had not been diluted by the passage of time, and the sudden confrontation with a disagreeable situation produced a swift and violent outburst that his colleagues had learned was typical of his personality. In the later years, such periods of volatility were generally followed by longer and more serious periods of depression.

The first unmistakable indication of such conflicts leading to a breakdown occurred in 1884. But there were definite signs well before then that Cantor was prone to periods of depression, and such evidence helps support the view that the origins of Cantor's mental illness were not determined by his anxieties over mathematical doubts or rivalries. In 1863 Cantor was beginning his studies at the Polytechnicum in Zürich, when something happened to promote serious misgivings about his future, and to a drastic degree. His father sent off a worried letter to his son on January 21 in an attempt to help him out of the doldrums. The letter began with a little quatrain, frivolous to be sure, but designed to coax Cantor from his moodiness.[53] Then followed a number of exhortations that might just as easily have applied to Cantor's later bouts of depression. Above all, Cantor's father wanted to dispell young Georg's melancholy. If only he were home, his father was sure he could break his son's depression, cheer his state of mind to such a degree that he would be immune to any relapse for three years, at least. Was this an implication that such moods had happened before? Or was his father only saying that he could liven Cantor's spirits for some time were he not so far away? Whichever was true, Cantor's father realized that his son's moodiness was the result of some unfounded anxiety; it caused the young student to work long into the night, endangering his health, but even more, it prefigured the patterns associated with his later periods of fatigue and emotional disturbance. Apparently Cantor was anxious about the costs of his education, and he doubted his ability to succeed were he forced to hurry his studies and face examinations before he was fully prepared.[54] Such anxieties compelled him to study without interruption. He even went without sleep, if necessary, which left him tired and eventually so despondent that his father was worried about his condition. Georg Woldemar wrote reassuringly and made it clear that means were not lacking to allow his son to take as much time as might be necessary to complete his studies. But within five months his father would be dead of tuberculosis, and Cantor would have to leave Zürich. Moreover, he would then be without the positive and understanding support his father had always given. Cantor had promised to make a success of his career in mathematics, but after June of 1863, with the death of his father, he was left to do so alone.

CANTOR'S MUSE: THE INNER VOICE AND DIVINE INSPIRATION

Too little information has been preserved to allow any detailed assessment of Cantor's personality, which leaves the historian to say either nothing on the subject, or to conjecture as best he can. With this caveat in mind, the analysis which follows can only plead reasonableness by virtue of its consonance with the known facts of Cantor's history and makes no claims to being the only reading possible. Nevertheless, I feel a certain justification, based upon my familiarity with Cantor's life and writings, in briefly sketching the following conclusions, even if I cannot always document them with the detailed evidence I have tried to use in making judgments elsewhere in this book.

When Georg Woldemar wrote to his son at great length on the occasion of his confirmation, he was careful to emphasize values and standards of conduct which were prominently reflected in Cantor's later personality. The two most significant items his father decided to underscore included the importance of success: Cantor was expected to be no less than another Theodor Schaeffer, and later "a shining star on the horizon of science." Second, he made it clear that Cantor should always turn to God for strength in facing the difficult moments in life, for it was upon God that one could always rely for guidance and inspiration.

Similar values were reflected in the jubilant letter Cantor penned when his father consented to his study of higher mathematics. He admitted the anxiety he had suffered in wanting to devote himself to science while feeling a duty to honor his father's wish that he do something more practical. The letter went on to commit young Georg to fulfilling his father's expectations by promising to be a success in mathematics. In fact, the need to succeed was greatly reinforced, if only because Cantor now had to prove that his choice had been the proper one. Thus he insisted, despite his change of vocation, that his father "would still be proud of his son" and added that he knew he would succeed because an unknown, secret voice had compelled him to study mathematics and continued to fire his enthusiasm.[55]

With colleagues this "will to succeed" was manifest in his attempts to dominate conversation and in his delight at being the center of attention.[56] He liked nothing better than to find himself surrounded by a group of listeners, and he delighted in leading discussion on a wide variety of subjects punctuated by outspoken pronouncements. But with his family he apparently made little effort to dominate at all.[57] In fact, during meals he would often listen quietly, letting his children develop the subject at hand, and then, before leaving the table, he would make a point of turning to his wife, thanking her for the meal, and asking "Are you content with me and do you then also love me?"[58]

This dichotomy of conduct between Cantor's personal and professional lives is very revealing. With his wife and children he felt no need to prove himself, no need to establish a successful, professional image. But he did show a great need to be loved and to find approval, from his father, for example, and from his wife. Similarly, he sought approval from his colleagues. In professional meetings he loved to dominate, and his book-lined study in Halle, as Kowalewski remembered, reflected the aura of a highly intellectual life. There his father's imprint was unmistakable; the effect of his encouragement upon his son was reflected throughout. The will to succeed was combined with the best advice from his father to cultivate a variety of interests and abilities and to apply his expertise to any situation, however adverse it might seem. By following God's plan for one's life, by listening to His advice and trusting in the strength of one's own faith, Cantor believed that success would eventually be won.

If it is possible to trace in this way the origins of Cantor's desire to prove himself a success in the profession he had chosen, it is no presumption to link that same desire with Cantor's later anxieties. His inability to deal with Kronecker's opposition to set theory, or his failure to solve the continuum hypothesis could easily trigger the mental breakdowns he in fact experienced—by blocking most effectively the fulfillment of his need to be successful. But as I have already suggested, this is not the whole story. There is an even deeper, more fascinating dimension to Cantor's bouts of depression, one similarly keyed to his father and directly related to Cantor's intensively religious character.

On September 17, 1904, Cantor was admitted to the Halle Nervenklinik, where he remained until March 1, 1905. When he was released, he did something quite remarkable, something which I believe throws an unsuspected light upon his periods of mental breakdown. At least in Cantor's own mind his periods of depression and seclusion served a unique and generative purpose, providing rest and quiet during which great progress might be made in many facets of study by mere reflection. The most surprising declaration Cantor made upon his release from the hospital in 1905 was that he had had an "inspiration from above, which suggested to me a renewed study of our *Bible* with opened eyes and with banishment of all previous preconceptions."[59] He was writing to the English mathematician P.E.B. Jourdain and explained how the "captivity and solitude" of his confinement had produced unexpected insights from which Cantor produced his pamphlet *Ex Oriente Lux*.[60] It was written as a series of dialogues between a master and his pupil, and the major point of the piece was to argue that Christ was the natural son of Joseph of Arimathea.

The point of all this is not to show that Cantor was interested in somewhat peculiar aspects of "documentary Christianity," as he referred to the pamphlet *Ex Oriente Lux*, but rather to show how he regarded his periods of hospitalization.[61] During the long months of seclusion his mind was left free to ponder many things, and in the silence he could perceive the workings of a divine muse, he could hear a secret voice from above which brought him both inspiration and enlightenment. Moreover, these convictions also provided a similar link between Cantor's periods of introvertive contemplation, the long silences, and his mathematics.

Following another long period of hospitalization at the Halle Nervenklinik in

1908, Cantor wrote to his friend in Göttingen, the English mathematician Grace Chisholm Young. Apparently, Cantor's inner voice knew more than the details of Christian history; his Muse was also a mathematician. Here it is best to let Cantor speak for himself:

> A peculiar fate, which thank goodness has in no way broken me, but in fact has made me stronger inwardly, happier and more expectantly joyful than I have been in a couple of years, has kept me far from home—I can also say far from the world—since October 22, 1907, until June 15, 1908, thus until last Monday . . . In my lengthy isolation neither mathematics nor in particular the theory of transfinite numbers has slept or lain fallow in me; the first publication in years which I shall have to make in this area is designated for the "Proceedings of the London Mathematical Society," to which, for the great honor of having named me to its membership, I shall always remain grateful as well as to the Royal Society for conferring upon me the Sylvester Medal three and one-half years ago.[62]

At the close of this same letter he lashed out against Poincaré's remarks condemning Cantorism at the International Congress in Rome, 1908. In particular, Cantor made reference to the impossibility of any sort of Genus supremum embracing *all* transfinite numbers. In fact, he was adamant in underscoring that:

> I have never proceeded from any "Genus supremum" of the actual infinite. Quite the contrary, I have rigorously proven that there is absolutely no "Genus supremum" of the actual infinite. What surpasses all that is finite and transfinite is no "Genus"; it is the single, completely individual unity in which everything is included, which includes the "Absolute," incomprehensible to the human understanding. This is the "Actus Purissimus" which by many is called "God."[63]

There can be no mistake about Cantor's identification of his mathematics with some greater absolute unity in God. This also paralleled his identification of transfinite set theory with divine inspiration. Even *before* his first nervous breakdown of 1884, Cantor had told Mittag-Leffler that his transfinite numbers had been communicated to him from a "more powerful energy"; that he was only the means by which set theory might be made known. Just as he had been inspired to write *Ex Oriente Lux,* "Dialogues of a master with his student . . . reported by the student himself," so too his mathematics could be inspired through a muse from above. Thus the periods of isolation in the hospital could be regarded as periods during which, as he told Mrs. Young, the transfinite numbers lay neither fallow nor forgotten, but might be further ellucidated by the grace of God sent to inspire new lines of research. All this was very much in keeping with the principles Cantor had inherited from his father and from his religious upbringing.

Georg Woldemar had been careful to instill in his son a reverence and faith in one's ability to succeed by hard work and faith in God. Without such confidence in his own abilities, Cantor might never have had the courage to face the relentless opposition he encountered to his work almost from the start. Had he not been able to cast himself in the role of God's messenger, inspired from some higher source of inspiration, he might never have asserted the unquestionable, indubitable correctness of his research. The religious dimension which Cantor attributed to the *Transfinitum* should not be discounted as merely an aberration. Nor should it be forgotten or separated from his life as a mathematician. The theological side of Cantor's set theory, though perhaps irrelevant for understanding its mathematical content, is nevertheless essential for the full understanding of his theory and the development he gave it. Cantor believed that God endowed the transfinite numbers with a reality making them very special. Despite all the opposition and misgivings of mathematicians in Germany and elsewhere, he would never be persuaded that his results could be imperfect. This belief in the absolute and necessary truth of his theory was doubtless an asset, but it also constituted for Cantor an imperative of sorts. He could not allow the likes of Kronecker to beat him down, to quiet him forever. He felt a duty to keep on, in the face of all adversity, to bring the insights he had been given as God's messenger to mathematicians everywhere.

CANTOR'S WORLD HYPOTHESES

Cantor's insights were by no means limited to the abstract, purely theoretical side of transfinite set theory. The *application* of his ideas in other sciences, particularly in mathematical physics, had been an ever-present stimulus to his research. In 1883, while addressing the mathematicians meeting with the GDNA in Freiburg, he claimed that:

> One of the most important problems of set theory, the major part of which I believe to have solved in my paper "Grundlagen einer allgemeinen Mannigfaltigkeitslehre," consists of the challenge to determine the various valences or powers of sets present in all of Nature (in so far as we can know them). I have arrived at this by developing the general concept of the numbering [*Anzahl*] of well-ordered sets or, in other words, of the concept of ordinal numbers.[64]

But how did he expect set theory to clarify the conditions of nature or to aid in the progress of other sciences generally? This was doubtless the question Mittag-Leffler had in mind when he encouraged Cantor to end his series of papers devoted to the *Punktmannigfaltigkeitslehre* with some indication of the theory's utility in other branches of science.[65] As a result, Cantor agreed to add a half dozen or so paragraphs to his "Zweite Mittheilung" for the *Acta*

Mathematica, in which he would explain the kind of applications he foresaw (see Chapter 6).

Cantor had never been satisfied with the hypotheses underlying most theoretical research in the physical and biological sciences.[66] He was uncomfortable about the vague and imprecise assumptions made concerning the ultimate constitution of matter—either its nature was left entirely uncertain, or atoms were assumed to exist which were very small, though material and extended in space. But Cantor was convinced that to provide satisfactory explanations for natural phenomena, it was necessary to assume that the number of elementary particles of matter be absolutely infinite, and that with respect to space they be absolutely extensionless, actual points. Once these assumptions had been made, the connection with his set theory was then immediate. Cantor said that he had been confirmed in this view of the ultimate forms of matter by reading such authors of similar persuasion as Faraday, Ampère, Weber, and Cauchy.[67]

But more than any of these, Leibniz was the source of greatest inspiration, as Cantor made clear by his choice of terminology: he called the ultimate components of matter which he hypothesized "monads," extensionless unities wholly comparable to the mathematical point. From this assumption it was a short step to identifying two different varieties of monads, corporeal monads and aetherial monads. Cantor added that this dichotomy between matter and aether was perfectly consonant with contemporary principles of physics. Prevailing opinion taught that the two were perfectly adequate to explain all phenomena of nature known at the time. But what of the powers of these two sets of distinct monads? Until his own research, no one could have asked the question, but Cantor was sure that the answer would solve numerous mysteries concerning the universe.[68] He had long fancied the hypothesis that the set of corporeal monads constituted a denumerably infinite set of cardinality $\aleph_0$. The set of aetherial monads, on the contrary, he conjectured to be equivalent in power to $\aleph_1$. This, Cantor's "First World Hypothesis,"[69] must have gained in strength when he proved that among point sets there were only two infinite powers possible, each corresponding to the two cardinal numbers $\aleph_0$ and $\aleph_1$. Thus, given any material body in space, it could be conceptualized as a conglomerate of monads determining a point set P. P was necessarily of the first power by virtue of Cantor's hypothesis, while the aether monads occupying the same space constituted a set Q of the second power. Based upon his studies concerning the inherence of point sets (Chapter 7), he could then determine:

$$P = Pr + Pi_1.$$

Since P was of the first power, there were no other inherences, but for the set of greater cardinality:

$$Q = Qr + Qi_1 + Qi_2.$$

Various subdivisions of Pr and Qr were also possible, and Cantor suggested that with the inherences Pi_1, Qi_1, Qi_2, these might suffice to explain either separately or in various combinations distinct properties of matter, including differences in composition, chemical properties, and such phenomena as light, heat, electricity, and magnetism.[70] The first step, however, was a thorough understanding of the nature of point sets. Until such complex and more refined distinctions like the adherence and inherence of a set were fully developed, one could not hope to have all the resources one might need to describe fully the intricacies of matter and aether together. As Cantor said, "In order to be able to demonstrate this basic idea, general studies of *point sets* seemed prerequisite to me."[71]

Even in the *Grundlagen* Cantor hinted that underlying his entire program to define, to explain, and to justify the transfinite numbers was an ultimate belief that they would lead to an improved understanding of nature and of natural phenomena. In a note to Section 4 he said that once the transfinite number classes (I), (II), (III), . . . , had been determined, it was then necessary to investigate not only their mathematical properties, "but to establish and to pursue as well (the number classes) wherever they occur in nature."[72] What he expected to find once this had been seriously accomplished was suggested in the *Grundlagen's* Section 5:

> For in addition to or in place of the mechanical explanation of nature, (which has all of the aids and advantages of mathematical analysis at its disposal, and yet the one-sidedness and inadequacy of which has been exposed so well by Kant), previously there has not been even one attempt to pursue this beginning [i.e., Kant's] armed with the same mathematical rigor for the purpose of reaching beyond that far-reaching *organic* explanation of nature.[73]

Cantor had never been content with a purely mechanical physics. Above all else, it seemed destined to promote materialism which he attributed in large measure to Isaac Newton and to Newton's faulty metaphysics.[74] As an antidote, Cantor sought what he called an organic theory of nature determined with the same mathematical precision as Newton had brought to his materialistic, mechanical philosophy. The key to this new organic approach was to be found, Cantor believed, in his transfinite numbers:

> The elements of the set M in question are to be thought of as separate; in the intellectual copy of $\overline{M}$, which I call its order type, the units are nevertheless joined into an organism. In a certain sense these order types may be regarded as a composite of *matter* and *form;* the conceptually distinct units contained therein comprise the *matter,* while the *order* of these elements corresponds to their *form*.[75]

Cantor frequently chose a simple example to demonstrate just how powerful a unifying concept order types could be. The most diverse things could be rendered mathematically comparable with great exactness. For the sake of illustration he liked to compare a painting with a symphonic score.[76] Every point in the painting could be ordered according to a horizontal and a vertical axis, by the wave length of color and by color intensity at a given point. The symphony could also be quadruply ordered, every instant arranged in sequence by time elapsed from beginning to end, by the duration of a given tone, by its pitch and intensity. Though improbable, it might even turn out that such heterogeneous objects as a Rembrandt portrait and a Beethoven symphony might have similar order types. The possibility alone seemed enough to inspire Cantor to expect that type theory might reveal unsuspected unity among the diversity of natural phenomena, and for a time he reflected this optimism by referring to the transfinite order types as *ideal* numbers.[77]

Cantor gave specific emphasis to the universal significance of his type theory in correspondence with Mittag-Leffler, in which he mentioned not only the possibility of applications in chemistry and optics but reiterated his hope that type theory would help explain the purely organic world. In fact, he said that such expectations had inspired him *originally* to develop the principles of his theory of transfinite order types.[78] As a result, in trying to grasp the full meaning and content of type theory, in striving to understand the absolute, forever ascending sequence of transfinite ordinal numbers, he was compelled to wonder about nature and its incomprehensible immensity, or as Cantor put it, its "*unermesslichen Grösse*."[79]

THE CRITICAL DIMENSIONS OF SET THEORY: ITS VALUE TO PHILOSOPHY AND THEOLOGY

While trying to defend his transfinite set theory from criticism of virtually every sort, Cantor came to see how his work could directly aid the development of Christian philosophy. His earliest concerted effort to combat all conceivable opposition to actual, completed infinities produced the early sections of the *Grundlagen*. Shortly thereafter, Cardinal Franzelin raised the specter of pantheism and the dangerous position which any affirmation of transfinite numbers seemed to suggest.[80] Cantor was able to satisfy Franzelin that, to the contrary, his theory was in no way conducive to pantheism, but could actually be used to destroy such doctrines (see Chapter 6).

Cantor drew a special kind of strength from his defense of the transfinite numbers by regarding them as eternally existing realities in the mind of God. As part of the *Intellectus Divinus*, they followed God-given laws; moreover, it was possible to argue their existence in *natura naturata*, based upon God's perfection and power. Like the seventeenth-century natural philosophers, Cantor argued that it would have diminished God's absolute perfection had his creation been only finite—thus the *Transfinitum* was brought into the bounds of the concrete, and Cantor could then safely apply his set theory to the study of nature.[81] But as Franzelin had already warned, this seemed dangerously near to pantheism, and equally distressing, just as near to materialism. Cantor himself on several occasions cited the example of no less a figure than Isaac Newton to illustrate the dangers one had to avoid. Newton's *Principia,* despite his "subjective" appeals to religion (as Cantor described them), opened the doors to materialism and positivism. This was a lesson to be taken seriously because it demonstrated without question:

> . . . that the greatest achievement of a genius (like Newton) despite the subjective religiosity of its author, when it is now united with a true philosophical and historical spirit, leads to effects (and, I assert, necessarily does so) by which it seems highly questionable whether the good in them simultaneously conveyed to mankind won't be significantly surpassed by the bad. And the most detrimental (of these effects) it seems to me are the errors of the "positivism" of Newton, Kant, Comte and others upon which modern scepticism depends.[82]

The problem of relating God, his creation, and science in a way that avoided heterodox implications had been an early concern of Cantor's. This is revealed in a most fascinating way by an episode connected with publication of the "Zweite Mittheilung" in the *Acta Mathematica.* Originally, when Cantor added the paragraphs at the end, at Mittag-Leffler's request, he had been anxious to make it absolutely clear that his work was no refuge for materialist, positivist, or pantheist thinking. In bringing the application of set theory into the phenomenological, into the "real world," Cantor was careful to stress that despite the actually infinite nature of the universe, and the reasonableness of his conjecture that corporeal and aetherial monads were related to each other as powers equivalent to the transfinite cardinals $\aleph_0$ and $\aleph_1$, this did not mean that God had necessarily had to create the world in this way. As Cantor put it: "It ought not to be said that *other* (kinds of) matter could not have been created (or even that they have not been created) by the Creator, but only that those two *substrates* seem to be sufficient to *explain* all the *perceived appearances thus far observed* (in Nature)."[83]

This sentence was *omitted,* however, from the final published version of Cantor's "Zweite Mittheilung." Mittag-Leffler's assistant, Gustav Eneström, in editing Cantor's manuscript before sending it to the printer, read the lines that were later omitted, clearly had doubts, and subsequently wrote to Cantor about the possibility of cancelling the entire passage.[84] Eneström recalled Laplace's famous quip that in writing the *Mécanique céleste* he had needed no reference to the creator, thereby explaining to Napoleon the absence of any mention of God. Leibniz had feared that this was just the sort of fate to which Newton's mechanical philosophy would lead, and the same view was equally shared by Cantor (as a letter to Valson made clear).[85] He had originally included reference to the Creator in his "Zweite Mittheilung" to make it absolutely certain that he had no intention of forcing any necessary conditions upon God's creation. But at Eneström's suggestion, he agreed to omit the passage entirely, even though it accurately reflected a major issue in Cantor's mind at the time.

Materialism and determinism, Cantor was certain, could be effectively combatted by a thorough understanding of transfinite set theory. He was particularly appalled by the steady march of such ideas late in the nineteenth century and was shocked by the realization that he was the only *in*determinist at Halle.[86] The factors encouraging the revival of such aberrant philosophies, which he described as "pathological", were directly attributable to a growing belief in the eternity of time and in the correspondingly eternal persistence of matter, for which God's existence and role in a Newtonianly mechanical universe were at best superfluous.[87] Orthodox Roman Catholic doctrine, as Cantor summarized it, had always denied the possibility of any eternal existence of matter, space, or time by virtue of the professed impossibility of actually infinite numbers.[88] But having discovered a consistent mathematical theory of actual infinities, Cantor insisted his research rendered that traditional line of argument useless. But this did not mean the door was then open to assert the real existence of eternity. On the contrary, he argued that his theory of the transfinite numbers offered a rigorous means of establishing quite the opposite. After all, despite attempts of mathematicians like Benno Kerry to pull infinitesimals from his transfinite numbers, he had succeeded in showing that his view of numbers made it possible to *prove* the nonexistence of actual infinitesimals. Cantor suggested that his theory of transfinite numbers could also be used to demonstrate the absolute impossibility of the eternity of time, space, and matter.[89]

Moreover, Cantor found that set theory could be useful to theology in an even more important capacity. As he began to consider the ways in which he might confront and resolve philosophical objections to his work, he also began to see that his research could be applied directly to the *perfection* of Christian philosophy. His efforts to do so were most arduously promoted in the mid 1890s, especially during the years immediately following the appearance of Part I of the *Beiträge*. In October of 1895 he promised Jeiler he would prepare a substantial document "in scholastic form" to show how set theory could be used to correct religious dogma.[90] In particular, he hoped his efforts would help to promote the same spirit Pope Leo XIII seemed to encourage in urging a revival of neo-Thomism. Cantor was determined to help as best he could. There were important lacunae, imprecisions, even errors in the system of St. Thomas, but Cantor was certain they could all be eliminated through study and application of his new theory of the infinite.[91] He reminded Esser that the Church ought to profit from the advance of science, a view that he supported by direct reference to Pope Leo XIII's encyclical *Aeterni Patris*.[92]

Thus the motives encouraging the early development of transfinite set theory were far from simple. Similarly, the purposes to which Cantor believed his work might be put and the ends toward which he seemed to direct so much of his energy were likewise diverse. But one last question has yet to be answered.

How did these myriad features of Cantor's personality, his theoretical, philosophical, and religious interests actually affect the development of his research? Apart from all the technical considerations one would expect, it remains to be shown how his personality and nonmathematical preconceptions also played their part in shaping the nature and determining the fortunes of set theory.

THE EARLY DEVELOPMENT OF TRANSFINITE SET THEORY: THE EFFECT AND INFLUENCE OF CANTOR'S CONTRIBUTION

One has only to read the opening sections of Cantor's *Grundlagen* to see how his interest in philosophy affected his presentation of set theory. We know that he went into great detail concerning the philosophy of the infinite because he wanted to confront and dispel the long-standing objections traditionally levied against the idea of actual infinities. Cantor knew his work was revolutionary; he too had found it difficult to overcome long-standing prejudices against completed infinities.[93] But gradually he came to see the power of his new ideas, and he found that it was possible to dispel even the objections of philosophers and theologians. Nevertheless, his heavy philosophical encasement of set theory in the *Grundlagen* was a disadvantage to its widespread acceptance. This was suggested by Klein's appeal and was mentioned specifically in Mittag-Leffler's plea that Cantor remove all such nonmathematical material from his publications on the subject.[94] Cantor, however, continued to believe that his mathematics and his metaphysics went hand in hand. As a result, many thought of set theory as belonging more to philosophy than to serious mathematics, and for many years journals and mathematical indices alike continued to list works in set theory under the rubric "Philosophy."[95] Only after publication of the *Beiträge,* a monograph which was completely mathematical and void of any overtly philosophical content, did his theory of transfinite numbers begin to receive its full share of study and recognition.

If philosophy affected the form and early acceptability of Cantor's set theory, his deeply religious convictions played a rather different role. His personality, as we have considered in some detail, was such that success was very important to him. He was determined to show to his parents that his decision to study mathematics had been the proper one. But as the years passed, theoretical difficulties with the continuum hypothesis and mounting external criticism from opponents like Kronecker conspired to raise doubts and to feed Cantor's anxieties about the acceptability and success of his work. As a result of his breakdown in 1884, he was forced to deal with the problems that had provoked his illness. He attempted a reconciliation with Kronecker, but that was short-lived. He worked less fervidly at his mathematics, and for a time seemed rather detached from any interest in the success or promotion of set theory. What seemed to keep his spirit alive, however, was a growing faith in the absolute certainty of his theory of transfinite sets. Perhaps a weaker man would never have recovered from the experiences of 1884–85. It is conceivable that others might have given up so seemingly hopeless a struggle against criticism of such power and influence as Kronecker's, but Cantor refused to be intimidated. He could count a special resource of his own that was immensely significant to the fortunes of his mathematics. Because he believed that set theory, having been divinely inspired from God, was therefore absolutely and necessarily true, nothing could succeed in shaking his confidence that it could never be flawed. It was exactly this point that he once made in a letter of 1888 to Heman: he knew his theory was correct because he had taken care to investigate over the years all objections that could conceivably be raised against it. In every case Cantor was satisfied that he could defeat any opposition:

> My theory stands as firm as a rock; every arrow directed against it will return quickly to its archer. How do I know this? Because I have studied it from all sides for many years; because I have examined all objections which have ever been made against the infinite numbers; and above all, because I have followed its roots, so to speak, to the first infallible cause of all created things.[96]

Such unwavering, infallible convictions were Cantor's greatest allies in periods of stress and difficulty. As his father had predicted, faith in God and a certainty in the capacity of one's own abilities to meet any problem, however great, would always ensure that one could weather any storm. Had it not been for the force of his own faith, supported by convictions entirely nonmathematical in nature, his theory of transfinite numbers might never have survived the criticism of finitists and intuitionists alike. Cantor did much more for set theory than merely discover its basic principles. He not only shaped its early character and formulated all its most essential elements virtually single-handedly, but he also ensured that once mathematicians were ready to consider the significance of the transfinite numbers, the entire theory would be ready to stand on the foundations which he had given it.

Though distasteful, and ultimately injurious to Cantor's health and well-being, the continuing conflict with Kronecker had its positive results. It forced Cantor to look more carefully at the foundational aspects of the new theory he was building. It caused him to search for more direct, mathematically acceptable, rigorous means by which to introduce, to define, to explain his transfinite numbers. It might well have prompted him to take all philosophical objections more seriously. And in the end, it stimulated his concern for the Deutsche Mathematiker-Vereinigung, and his earnest desire to promote not only a German forum for mathematicians but to further as well the organization of mathematical congresses on an international scale. Finally, it may be that without the keen and relentless opposition of Kronecker, Cantor might never have been driven to produce so detailed and carefully argued an exposition of his theory as the *Beiträge* of 1895 and 1897.

Later generations might forget the philosophy, smile at the abundant references to St. Thomas and the Church fathers, overlook his metaphysical pronouncements and miss entirely the deep religious roots of Cantor's later faith in the veracity of his work. But these all contributed to Cantor's resolve not to abandon his transfinite numbers for less controversial and more acceptable interests. Instead, his determination seems actually to have been strengthened in the face of opposition. His forbearance, as much as anything else he might have contributed, ensured that set theory would survive the early years of doubt and denunciation to flourish eventually as a vigorous, revolutionary force in scientific thought of the twentieth century.

APPENDIXES

NOTES

BIBLIOGRAPHY

INDEX

Appendixes

Correspondence is divided into groups by archive for easy identification and reference. In the case of Sophie Kowalevski's letter of May 21, 1885 to Mittag-Leffler, the transcription from the original Swedish is followed by a translation. For help with the translation I am grateful to Lennart Carleson of the Institut Mittag-Leffler, Djursholm, Sweden.

Appendix A: Letter from Cantor to Goldscheider, preserved in the archives of the Niedersächsische Staats- und Universitätsbibliothek, Göttingen.

Cantor to Goldscheider, October 11, 1886.

Appendix B: Letters from Heine to Schwarz, currently in the possession of Professor Herbert Meschkowski, Berlin.

Heine to Schwarz, March 8, 1870.

Heine to Schwarz, May 26, 1870.

Appendix C: Letters from the archives of the Institut Mittag-Leffler, Djursholm, Sweden.

Cantor to Kowalevski, December 7, 1884.

Kowalevski to Mittag-Leffler, May 21, 1885.

Appendix D: Letter from Kronecker to Schwarz, from the archives of the Akademie der Wissenschaften der DDR, Berlin.

Kronecker to Schwarz, December 28, 1884.

APPENDIX A

Georg Cantor to Franz Goldscheider
October 11, 1886
Universitätsbibliothek, Göttingen

1. Ist α irgend eine Zahl ≥ 1 der ersten oder zweiten Z.cl. und bildet man:

   $\alpha_1 = \omega^{\alpha}$, $\alpha_2 = \omega^{\alpha_1}$, $\alpha_3 = \omega^{\alpha_2}$, ...

   so ist, wie leicht zu sehen die Zahl:

   $$\text{Li}\ (\alpha, \alpha_1, \alpha_2, \ldots)$$

   stets ein Gigant; wir wollen ihn in seiner Abhängigkeit von α mit $\mathfrak{g}(\alpha)$ bezeichnen.
2. Ist nun γ irgend eine Gigant, γ' der auf ihn nächstfolgende Gigant und η irgend ein Zahl zwischen beiden, so dass:

   $$\gamma < \eta < \gamma', \text{ so ist immer:}$$

   $$\mathfrak{g}(\eta) = \gamma'.$$

   Denn man hat, wegen: $\gamma < \eta < \gamma'$ auch:

   $$\omega^{\gamma} < \omega^{\eta} < \omega^{\gamma'}$$

   und da: $\omega^{\gamma} = \gamma$ und $\omega^{\gamma'} = \gamma'$, so hat man:

   $$\gamma < \eta_1 < \gamma', \text{ wo } \eta_1 = \omega^{\eta}.$$

   Analog folgt hieraus, dass:

   $$\gamma < \eta_2 < \gamma', \text{ wo } \eta_2 = \omega^{\eta_1}, \text{ u.s.w.}$$

   Man hat also:

   $$\gamma < \text{Li}\ (\eta, \eta_1, \eta_2, \ldots) \leq \gamma', \text{ d.h.: } \gamma < \mathfrak{g}(\eta) \leq \gamma'.$$

   Da nun nach 1) $\mathfrak{g}(\eta)$ ein Gigant und γ' der auf γ nächstfolgende Gigant ist, so kann nicht $\mathfrak{g}(\eta) < \gamma'$ sein und es bleibt nur übrig, dass:

   $$\mathfrak{g}(\eta) = \gamma'.$$
3. Setzt man in 2) $\eta = \gamma + 1$, so folgt:

   $$\gamma' = \mathfrak{g}(\gamma + 1).$$

   Aber es ist auch: $\gamma' = \mathfrak{g}\ (\gamma + 2) = \mathfrak{g}\ (\gamma + 3) = \ldots = \mathfrak{g}\ (\gamma^2)$, da sich leicht zeigen *lässt,* dass die Zahlen $\underset{\leq \gamma^2}{> \gamma}$ *keine* Giganten, also kleiner als der auf γ *nächstfolgende* Gigant γ' sind.

4. Ist $\gamma, \gamma', \gamma'', \ldots$ irgend eine einfach unendliche *steigende* Folge von Giganten, so ist immer:

$$\mathrm{Li}\ (\gamma, \gamma', \gamma'', \ldots)$$

auch ein Gigant und zwar der auf alle jene Giganten *nächstfolgende* Gigant, wie leicht zu sehen.

5. In 3) u. 4) ist das gesuchte Gesetz enthalten, nach welchem die Reihe der Giganten:

$$\gamma_1, \gamma_2, \ldots, \gamma_\omega, \gamma_{\omega+1}, \ldots$$

fortschreitet. Wir können sagen: ist α eine Zahl *erster* Art, so ist: $\gamma_\alpha = \mathfrak{g}(\gamma_{\alpha-1}+1)$; ist aber α von der zweiten Art, etwa: $\alpha = \mathrm{Li}\ (\alpha_1, \alpha_2, \ldots, \alpha_\nu, \ldots)$, so hat man:

$$\gamma_\alpha = \mathrm{Li}\ (\gamma_{\alpha_1}, \gamma_{\alpha_2}, \gamma_{\alpha_3}, \ldots).$$

Somit ist jene Reihe auf *vollständiger* Induction basirt.

6. Ist γ irgend ein Gigant, so lässt sich daraus eine neuer Gigant wie folgt bilden: man setze:

$$\gamma^\gamma = \gamma^1, \quad \gamma^{\gamma^1} = \gamma^2, \quad \gamma^{\gamma^2} = \gamma^3, \ldots$$

und bilde: $\delta = \mathrm{Li}\ (\gamma, \gamma^1, \gamma^2, \ldots)$.
Es fragt sich, in wlechem Abstande von γ dieser Gigant δ steht und ich behaupte, δ ist gleich dem auf γ nächstfolgenden γ'. In der That, nach 3) ist: $\gamma' = \mathfrak{g}\ (\gamma^2)$ und wir haben also nur zu zeigen dass auch: $\delta = \mathfrak{g}(\gamma^2)$.
Beweis: $\gamma^1 = \gamma^\gamma = (\omega^\gamma)^\gamma = \omega^{\gamma^2}$
$\gamma \cdot \gamma^1 = \omega^\gamma \cdot \omega^{\gamma^2} = \omega^{\gamma+\gamma^2}$
Es ist aber: $\gamma + \gamma^2 = \omega^\gamma + \omega^{\gamma \cdot 2} = \omega^{\gamma \cdot 2} = \gamma^2$, also
$\gamma \cdot \gamma^1 = \omega^{\gamma^2} = \gamma^1$.
Daher: $\gamma^2 = \gamma^{\gamma^1} = (\omega^\gamma)^{\gamma^1} = \omega^{\gamma \cdot \gamma^1} = \omega^{\gamma^1}$.
Ferner $\gamma \cdot \gamma^2 = \omega^{\gamma+\gamma^1} = \omega^{\gamma^1} = \gamma^2$, weil $\gamma + \gamma^1 = \omega^\gamma + \omega^{\gamma^2} = \omega^{\gamma^2} = \gamma^1$.
Mithin $\gamma^3 = \gamma^{\gamma^2} = \omega^{\gamma \cdot \gamma^2} = \omega^{\gamma^2}$.
Ebenso: $\gamma^4 = \omega^{\gamma^3}$, u.s.w.
Allgemein: $\gamma^\nu = \omega^{\gamma^{\nu-1}}$, falls $\nu \geq 2$.
Und da $\gamma^1 = \omega^{\gamma^2}$, so folgt nach 1), dass:

$$\delta = \mathfrak{g}(\gamma^2), \text{ w.z.b.w.}$$

7. Mit Hülfe des Ihnen mitgetheilten Theorems, wonach jede Zahl α in der Form:

$$\alpha = \omega^{\alpha_1}\nu_1 + \ldots + \omega^{\alpha_\mu}\nu_\mu$$

eindeutig darstellbar ist, kann man leicht folgenden Satz über die Giganten beweisen. Ist α irgend eine Zahl >1 der ersten oder zweiten Z.cl., so wird die Gleichung: $\alpha^x = x$ dann und nur dann befriedigt, wenn man für x irgend einen Giganten setzt, der grösser ist als α. Somit werden die Giganten auch durch die Gleichungen $2^\gamma = \gamma$, $3^\gamma = \gamma$, ..., $\alpha^\gamma = \gamma$, falls $\alpha < \gamma_1$ ebenso characteris[irt] wie durch $\omega^\gamma = \gamma$. Sie sehen also, dass die Gleichungen $\alpha^x = x$, auch wenn α von ω verschieden ist, auf keine neuen Zahlenarten führen, da ihnen nur Giganten als Wurzeln genügen.

8. Folgende Gleichungen:

$$\omega^x = x \cdot \omega;\ \omega^x \cdot 2 = x^2;\ \omega^x = x^\omega;\ \omega^x = x^{\omega^2};$$

haben respective die folgenden Lösungen:

$$x = \gamma + 1;\ x = \gamma \cdot 2;\ x = \gamma \cdot \omega;\ x = \gamma \cdot \omega^2;$$

wo γ irgend einer Giganten bedeutet und es lässt sich auch beweisen, dass keine anderen Lösungen von ihnen existiren.

APPENDIX B

Eduard Heine to H.A. Schwarz
March 8, 1870
Nachlass Schwarz: Professor Herbert Meschkowski, Berlin.

Mein lieber Herr College,

Auf Ihren Brief vom 27/2. antworte ich erst heute, weil ich mich zunächst über die Methode genau unterrichten wollte, welcher Sie sich zum Beweise des Hülfssatzes bedienen, der wesentl[ich] zur Verallgemeinerung der Resultate beiträgt die ich gewonnen und bereits an Borchardt gesandt habe. Cantor hatte mir bereits mitgetheilt, er wolle Ihnen den Hülfssatz vorlegen, und auch Ihren ersten Brief gezeigt; ich bin Ihnen aber nichts destoweniger sehr dankbar dafür, dass Sie mich auch direct über Ihren Fortschritt unterrichteten, und habe dann Ihren Beweis des Cantorschen Hülfssatzes vom 28ten durch Cantor erfahren. Ich weiss nicht ob es auf einem Missverständnisse beruhte, wenn Sie annehmen, dass der Hülfssatz ausreiche, um die fragliche Verallgemeinerung zu gestatten, oder ob Cantor erst später bemerkte, dass noch ein sehr wesentlicher Punct zu erledigen bleibe, der bei der nicht völlig genauen Untersuchung von Riemann dem Leser allerdings leicht entgehen kann. Jedenfalls hat Cantor (er wird Ihnen selbst wohl das Nähere gelegentlich mittheilen) diesen zweiten Umstand durch eine feine Untersuchung völlig erledigt, welche mit einer Darstellung von Irrationalitäten zusammenhängt, welche er in seinen kleinen Arbeiten in Schlömilch's Journal giebt.

Trotz der fortgesetzten Untersuchungen habe ich meine Arbeit nicht zu weit gezogen. Erstens berührt ja Cantor gar nicht den Theil des Stoffes, der sich auf die Untersuchung der "im allgemeinen" in gleichem Grade convergirenden Reihen bezieht; zweitens ist meine Arbeit in sich geschlossen, und lenkt zunächst von dem Wege ab, der hier keine Resultate gab, und den schon Dirichlet für hoffnungslos bei solchen Untersuchungen zu halten scheint, ich meine von den Schlüssen aus der Beschaffenheit des Inneren auf die Convergenz vom Rande,—hin zu Riemann's $F(x)$. Ich zeige dort, dass für $f(x)=0$ in meinem Fall $F(x)=a+bx$ sein müsste. Die bedeutende Verallgemeinerung, die Cantor anstrebt als er Sie um den Beweis des Hülfssatzes anging besteht darin, dass er nachweisen will, es für $F(x)$ noch $=a+bx$ wenn nur Convergenz, statt gleichmässiger Convergenz, verlangt wird. Von da an bleibt Alles in meinem Beweise ungeändert; Satz II u[nd] III, die ich aus $F(x)=a+bx$ herleite, würden denn von selbst nur die einfache Convergenz statt der Convergenz in gleichem Grade erfordern.

Ob mein Verfahren angegriffen wird muss ich abwarten; ich selbst habe keine bedenkliche Stelle darin gefunden. Dagegen leugne ich nicht, dass Ihr Beweis des Hülfssatzes nach Bolzano-Weierstr. Principien, wie schön er auch ist, mir *nicht* völlig beweisend erscheint, u [nd] ich deshalb nicht zugeben kann, dass der Satz erledigt sei. Cantor sagt mir, dass er seine Untersuchungen,—ich glaube auch eine Mittheilung über Ihren Beweis des Hülfssatzes—Weierstrass mitgetheilt habe. . . .

E. Heine

Eduard Heine to H. A. Schwarz
May 26, 1870
Nachlass Schwarz: Professor Herbert Meschkowski, Berlin.

Mein lieber Herr College,

Meine kleine Arbeit *"Über trigonometrische Reihen"* von der ich in diesem Augenblick die Correctur in Händen habe, und die nunmehr, noch vielen Verhandlungen mit Kronecker, der mich veranlassen wollte sie zurückzuziehen (das Nähere unten) im laufenden (71 ten.) Bande des Journals S.353 erscheint hatte mir viele Freude gemacht; ich hatte sie im Februar Borchardt gesandt, bei dem sie Kr. sah, der sie sich geben liess u. behielt ohne dass ich es erfuhr, ehe ich nach Berlin kam. Dadurch wird es sehr wahrscheinlich, dass sie erst im nächsten Bande erscheinen könnte, also lange nachdem die wesentlichsten Punkte derselben schon bekannt waren. Dass sie in Thomae's Buch (sehr ungenügend) stehen ist mir übrigens jetzt ziemlich gleich-gültig, da dies wohl nicht zu sehr verbreitet werden wird; merkwürdigerweise zeigt er S.13/14 in der Note, dass eine stetige Function auch in gleichem Grade stetig ist! Es mag wahr sein, aber das Beweis passt nicht.

Ich erwähne, dass Kronecker sagte, er habe Sie veranlasst, die gleichmässige Stetigkeit nicht einzuführen; Ihre Sätze seien ganz richtig auch ohne meine Forderung, (Ich habe ihm *schriftlich,* wie ich glaube *bewiesen,* dass er sich irrt). Um so mehr konnte ich mich wohl verwundern als ich bei Ihnen sie eingeführt fand, aber nicht *expressis verbis*. Ich gebe sofort zu, dass nur sie die eigentlich Stetigkeit ist, die Stetigkeit, welche man überall stillschweigend voraussetzt; es handelt sich einfach darum, dass ich diese Voraussetzung enthällt habe. Es reimte sich für mich nicht, dass Sie, wenn Sie wie ich nach K's Äusserungen annehmen konnte, jetzt auf K's Standpunct stehen u. die gewöhnlich Definition genügend finden, doch von meiner Gebrauch beim

Beweise machen. Es war überhaupt dies Mal seltsam in Berlin. Weierstrass u. K. waren sehr viel zusammen, und die Beweise von Cantor und mir wurde viel besprochen. Kron. sagt uns (Cant. u. mir) öfter, diesen u. jenen Punct habe ich nun Weierstr. zugegeben, während Weierstr. auf Befragen nichts zugegeben haben wollte. Es war eine seltsame Erregung; dazu Kr. kränklich, und das glaube ich, war zum Theil der Anlass, aus dem kein rechtes Einverständniss nicht zu Stande kam.

APPENDIX C

Georg Cantor to Sophie Kowalevski
December 7, 1884
Institut Mittag-Leffler

Halle d. 7ten Dec. 1884

Hochgeehrte Frau.

Soeben erhalte ich die mir durch Ihre Güte zugeeigneten Abhandlungen "Ueber die Reduction einer bestimmten Klasse Abelscher Integrale" und "Ueber die Brechung des Lichtes in cristallinischen Mitteln" und ich erlaube mir, Ihnen meinen besten Dank für Beides auszusprechen. Es wird mich besonders die letztere interessiren und beschäftigen, weil ich darin willkommene Hülfsmittel für eigene Untersuchungen zu finden hoffen darf, welche mich seit Jahren beschäftigen. Momentan bin ich mit der Ausarbeitung eines *Mémoire sur la théorie des types d'ordre* in Anspruch genommen, welche in den ersten Monaten des nächsten Jahres fertig werden soll und wovon ich vorher die Principien in möglichster Kürze zusammenstelle, um sie in Ihrer Zeitschrift zu publiciren; ich habe die fünf ersten Paragraphen dieser "Principien" Herrn Professor Mittag-Leffler eingesandt; es gehören zu dieser ersten Mittheilung noch weitere mindestens fünf Paragraphen, die ich bereit haben werde, sobald ich erfahre, dass der Druck anfangen soll.

In den bisherigen Paragraphen ist nur von den Typen einfach geordneter Mengen die Rede; es gibt aber ebenso auch Typen von zweifach, dreifach, n-fach ja selbst ω-fach etc. geordneter Mengen, durch welche viel Licht auf alte und neue Fragen der Arithmetik und Kosmologie zu fallen scheint. Alles was ich Ordnungstypus nenne hat einen sowohl arithmetischen, wie auch geometrischen Character, letzteren namentlich bei den Typen mehrfach geordneter Mengen. Während die Descartes-Newton-Leibnizsche Methode das ihrige leistet in der Abgrängung der Naturphänomene, glaubte ich schon vor vielen Jahren dass es an einem entsprechenden streng-mathematischen Hülfsmittel noch fehle, durch welches man befähigt wäre, gewissermassen in die natürlichen Vorgänge mitten hineinzutreten, sich dieselben nicht nur von aussen—sondern auch von Innen genau anzusehen, um alsdann darüber genauer als bisher berichten zu können; ob meine Typentheorie dieses gesuchte Werkzeug sei, wage ich selbst nicht zu entscheiden und wird sich erst mit der Zeit herausstellen. Doch schien mir diese Theorie auch abgesehen von allen Anwendungen in rein mathematischen Hinsicht nicht ohne Interesse, daher ich mir nun die Freiheit nehmen werde, darüber zu schreiben. Wenn schon die *endlichen* Typen einen namenlosen Reiz ausüben bei jedem der für die Gesetze ewiger Wahrheiten einen empfänglichen Sinn hat, woraus die *théorie des nombres* hervorgegangen ist, so bieten die Typen *unendlicher* Mengen eine immer grössere Steigerung in der Befriedigung dieses wissenschaftlichen Interesses dar. Denn es sind hier nicht bloss die Operationen des Addirens und Multiplicirens, welche den Gesetzen Grunde liegen, sondern es kommen hier noch ein paar andere fundamentale Operationen hinzu, die bei den endlichen Typen aus dem Grunde nicht hervortreten, weil sie hier ohne Interesse sind (indem sie nämlich dabei nur auf die Null führen). Es sind dies die Operationen welche ich in einem Schreiben an Prof. Mittag-Leffler v. 20^{n} Oct. mit Differenz, Cohärenz, Adhärenz, Inhärenz, Supplement und Residuum in Zeichen mit d, c, a, i, s, r bezeichnet habe; es stellt sich nämlich heraus, dass diese Operationen *nicht nur an Punctmengen,* sondern *an Typen selbst* vornehmbar sind.

Georg Cantor

APPENDIX C

Sophie Kowalevski to Gösta Mittag-Leffler
May 21, 1885
Institut Mittag-Leffler

... Eneström har återkommit till H.för ett par dagar sedan och har varit hos mig igår.

Han berättade att Kronecker, hvilken han upsökt i Berlin, var mycket kritisk öfver funktionsteorien och riktigt bitter öfver Cantor. W. skrifver också att Kr. synes vara ledsen öfver att han icke blifvit kallad till domare i vår prisfråga. Han talar alldrig om densamma, och detta är ett tecken att han är mycket ond. Eneström har också uppsökt Cantor. Densamma har börjat under den föregående terminen att föreläsa om Leibnitz's philosophie. I början hade han 25 åhörare, men så småningom ha de smältet ihop först till 4, sedan till 3, så till 2, slutligen till en enda. Cantor höll ändå ut och fortsatte att föreläsa. Men ve! En vacker dag kom den sista mohikanern, något generad, och tackade professorn så mycket men förklarade att han hade så många andra sysselsättningar att han icke orkade följa prof's föreläsningar. Då har Cantor, till sin hustrus outsägliga fröjd, gifvit ett högtidligt löfte att aldrig föreläsa om philosophien igen!

Och nu synes det mig, snälla Herr Professor, att jag talat om för Er om allt hvad jag kunde tänka mig skulle interessera Er. Jag hoppas snart få underrätelser från Er. Ännu har jag fått endast ett enda bref från Er och från Anna Charlotte intet. Mina bästa helsningar till Signe.

Eder varmt tillgifven vän
S.K.

Sophie Kowalevski to Gösta Mittag-Leffler
May 21, 1885
Institut Mittag-Leffler, translation:

. . . Eneström returned to H[elsingfors] a few days ago, and has been with me yesterday.

He related that Kronecker, whom he visited in Berlin, was very critical of the function theory and quite bitter about Cantor. W[eierstrass] wrote also that Kronecker seems to be sorry that he had not been called as a judge in our contest. He never speaks about it, and that is a sign that he is very angry. Eneström has also visited Cantor. The latter began during the previous semester to lecture on Leibnitz's philosophy. In the beginning he had 25 students, but then little by little, melted together first to 4, then to 3, then to 2, finally to one single one. Cantor held out nevertheless and continued to lecture. But, alas! One fine day came the last of the Mohicans, somewhat troubled, and thanked the professor very much but explained that he had so many other things to do that he could not manage to follow the professor's lectures. Then Cantor, to his wife's unspeakable joy, gave a solemn promise never to lecture on philosophy again!

And now it seems to me, kind professor, that I [have] told you about everything which I could imagine would interest you. I hope soon to receive news from you. Still I have received only a single letter from you and from Anna Charlotte none. My best greetings to Signe.

Your warmly affectionate friend.
S.K.

APPENDIX D

Leopold Kronecker to H.A. Schwarz
December 28, 1884
Nachlass Schwarz: Akademie der Wissenschaften der DDR, Berlin.

. . . Wenn mir noch Jahre und Kräfte genug bleiben, werde ich selbst noch der mathematischen Welt zeigen, dass nicht bloss die Geometrie, sondern auch die Arithmetik der Analysis die Wege weisen kann—und sicher die strengeren. Kann ich's nicht mehr thun, so werden's die thun, die nach mir kommen, und sie werden auch die Unrichtigkeit aller jener Schlüsse erkennen, mit denen jetzt die sogenannte Analysis arbeitet . . . .

Notes

All citations are keyed by author and date of publication to items in the bibliography. Unless otherwise noted, all references to the works of Georg Cantor are taken from the edition by Zermelo: Cantor (1932). Nevertheless, Cantor's papers are designated by their original years of publication. Thus "Cantor (1872), 93" refers to Cantor's article of 1872 *as given in Zermelo's edition* of the *Gesammelte Abhandlungen* on page 93. For the sake of economy, the following abbreviations are used in reference to the three surviving *Briefbüchern* in which Cantor drafted his letters before writing out final copies:

Cantor (I): Cantor's letter-book for 1884 through 1888.
Cantor (II): Cantor's letter-book for 1890 through 1895.
Cantor (III): Cantor's letter-book for 1895 and 1896.

These letter-books are part of Cantor's remaining *Nachlass,* which has recently been entrusted to the Akademie der Wissenschaften, and which is now kept in the archives of the Niedersächsische Staats- und Universitätsbibliothek, Göttingen.

Introduction

1. For Kronecker's accusation, see Schoenflies (1927), 2; Russell (1967), 335.
2. Hilbert (1926), 170; Poincaré (1908), 182.
3. According to Bell's jaundiced viewpoint: "There is no more vicious academic hatred than that of one Jew for another when they disagree on purely scientific matters. When two intellectual Jews fall out they disagree all over, throw reserve to the dogs, and do everything in their power to cut one anothers' throats or stab one another in the back" Bell (1939), 562–563.
4. Grattan-Guinness (1971a), 351, though the matter is not as clear as it might be, considering a reference Cantor once made to his *israelitische* grandparents in a letter to P. Tannery of January 6, 1896: in P. Tannery (1934) 306. But neither in an orthodox rabbinical sense, since his mother was a Roman Catholic, nor in the sense of a practiced faith can it be said that Cantor was Jewish.
5. Russell (1967), 335. For a more balanced view of Cantor's mental illness, consult Grattan-Guinness (1971a), 368–369.

6. Meschkowski (1965), 503; Grattan-Guinness (1971d).
7. See Grattan-Guinness (1971a), 354–355; Dugac (1976a), 126–128.
8. Cantor (1970); Grattan-Guinness (1971a), 357–358.
9. Fraenkel (1930).
10. Bell (1937).
11. Meschkowski (1967).
12. Grattan-Guinness (1971a).
13. See Jourdain (1915) and Grattan-Guinness (1971b).
14. Koyré (1957).

1. Preludes In Analysis

1. For discussion of the broader history of trigonometric series, refer to Sachse (1879); see as well the reply to Sachse by du Bois-Reymond (1880b). Cavaillès (1962), 44–60, devotes some space to a discussion of trigonometric series, as does Jourdain (1910), 21–43. More recently attention to the trigonometric background of set theory has been given by Manheim (1964), Paplauskas (1966) and (1968), and Hawkins (1970). The lengthy report by Burkhardt (1901) is still useful, as are Burkhardt (1914), Van Vleck (1914), Hilb (1923). The best and most comprehensive surveys of the purely mathematical aspects of the theory of trigonometric series are to be found in Zygmund (1959) and Bary (1964).

2. The first source anyone interested in Dirichlet should consult is Biermann (1959b). Further remarks of a biographical nature are to be found in Borchardt (1860), Kummer (1860). See also Minkowski (1905).

3. For a detailed analysis of Fourier's part in the history of analysis and for an evaluation of the degree to which Fourier was able to appreciate methematical sophistication, refer to the studies by Grattan-Guinness (1969), (1970b), and especially (1972d).

4. For Dirichlet's evaluation of Cauchy (1823a), consult Dirichlet (1829), 157, and Dirichlet (1889), 119.

5. See Dirichlet (1829), 158; Dirichlet (1889), 120.

6. Riemann (1874), 235. For a discussion of the importance of convergence in analysis during the early nineteenth century, turn to Grattan-Guinness (1970b), 93–94.

7. Dirichlet (1837c), 48–49; Dirichlet (1889), 318–319.

8. Dirichlet (1829), 159; (1889), 121.

9. Dirichlet (1829), 168–169; (1889), 131.

10. Dirichlet (1829), 169; (1889), 132.

11. Dirichlet (1837a).

12. Dirichlet sent a letter to Gauss on February 20, 1853. See volume II of Dirichlet's collected works: Dirichlet (1897), 385–387.

13. Dirichlet (1897), 386.

14. Dirichlet (1829), 169; (1889), 131–132.

15. Gauss' opinion of Riemann's *Promotionsschrift* is quoted by Schering (1866). The article on Riemann was commissioned by the Royal Society of Göttingen, read to the public session on December 1, 1866, but not published until it appeared in Schering's collected works, specifically volume II, Schering (1909), 367–383. Biographical information concerning Riemann may be found in Schering (1866) and (1909). See also Klein

(1894), 3–18. Specific attention to Riemann's mathematical work is given in Courant (1926). Dedekind also contributed a lengthy historical evaluation of Riemann's life and work for the edition of collected works. See Dedekind (1876) and Riemann (1876b).

16. Dirichlet's characterisitc function was the familiar $\chi(x)$, where $\chi(x)=0$ for rational values of x; $\chi(x)=1$ for irrational values of x.

17. Riemann (1854). For recent discussions of Riemann's work, see Paplauskas (1966), 139–141, 203–214; and Hawkins (1970), 16–20.

18. Riemann (1854), 239.

19. Cauchy (1823b), 81.

20. Riemann (1854), 239.

21. Riemann (1854), 241.

22. Riemann (1854), 242, and Hawkins (1970), 18–19.

23. Among the most recent authors to offer this suggestion are Grattan-Guinness (1970b), 124; and Hawkins (1970), 20. See as well Courant (1926), 1267.

24. Riemann introduced his function $F(x)$ in Section 8 of his *Habilitationsschrift,* which also contained the three lemmas dealing with $F(x)$ and applied later in Section 9. See Riemann (1854), 245–251.

25. Riemann (1854), 246–251.

26. The three theorems are given in Section 9 of the *Habilitationsschrift:* Riemann (1854), 251–253.

27. Riemann (1854), 237–239.

28. Riemann devoted the last two sections of the *Habilitationsschrift,* to descontinuous functions with special properties: Riemann (1854), 257–264.

29. Riemann (1854), 259.

30. Both this example and the one which follows appear in the last section of Riemann's memoir: Riemann (1854), 260–264.

31. Riemann (1854), 263.

32. Lipschitz (1864). See the French translation by P. Montel: Lipschitz (1913). For a short biography of Lipschitz, consult Kortum (1906).

33. Lipschitz (1864), 297; Lipschitz (1913), 283.

34. These alternatives are outlined in Lipschitz (1864), 298–300; Lipschitz (1913), 283–285.

35. I have chosen to translate Lipschitz's expression *primae speciei* as "first form" in order to avoid any confusion with Cantor's sets of the first and second species [*Gattung*]. In French, the phrase is translated "première espèce." See Lipschitz (1864), 298; (1913), 283.

36. Lipschitz (1864), 300–301; (1913), 286; Dirichlet (1829), 168.

37. Lipschitz (1864), 299–300; (1913), 285.

38. Lipschitz (1864), 300–301; (1913), 286.

39. Lipschitz (1864), 301; (1913), 286–287.

40. Lipschitz (1864), 301; (1913), 287.

41. Hankel (1870). For general biographical remarks, consult von Zahn (1874), 583–590, and M. Cantor (1879), 516.

42. For Hankel's criticism of Dirichlet see Hankel (1870), 5; (1905), 49.

43. Specifically, Hankel (1870), 15–19; (1905), 61–65.

44. Hankel (1870), 19; (1905), 64.

45. Hankel's definition of linear functions: (1870), 23; (1905), 69.

46. Hankel (1870), 25–26; (1905), 72.
47. Hankel (1870), 27; (1905), 74.
48. Hankel (1870), 30; (1905), 77.
49. Hankel (1870), 37; (1905), 84–85.
50. Euler (1748), 4; Dirichlet (1837b), 152–153.
51. Hankel (1870), 39; (1905), 87.
52. Hankel (1867), 10–13.
53. Hankel (1870), 42; (1905), 90.
54. Hankel (1870), 42; (1905), 90.
55. Hankel (1870), 43; (1905), 92.
56. Hankel (1870), 43; (1905), 92.
57. Hankel (1870), 28; (1905), 74.

2. The Origins of Cantorian Set Theory

1. Cantor (1867). For details mentioned here concerning Cantor's career and for biographical material consult Jourdain (1906), (1909), (1910), (1913), and his *Introduction* to Cantor (1915); Fraenkel (1930); Meschkowski (1967); Grattan-Guinness (1971a); and Kertész (1977).

2. Cantor (1869).

3. Heine (1870), 355. The expression "in general" refers to any continuous interval with the possible exception of a finite number of points.

4. Heine (1870), 353.

5. Cantor (1870b), 80.

6. Cantor (1870b), 80–81.

7. Cantor (1870a), 71. This theorem is generally known as the Cantor-Lebesgue theorem, for although Cantor gave the proof for trigonometric series which converge over the interval (a, b), Lebesgue generalized the result for any interval or set of positive measure in the Lebesgue sense. See Lebesgue (1906), 110–111.

8. Riemann (1854), 245.

9. Though neither Schwarz nor Cantor gave the Schwarz second derivative any special designation, I shall follow current practice and use the notation $D^2F(x)$ to represent this function. See as well Cantor (1870b), 81.

10. See Schwarz (1890), II, 341–343. The paper here in question concerning the Schwarz second derivative was never published separately and appears only in Schwarz's collected works. See also Schwarz's letter to Cantor of April 1, 1870, in which he discussed his delight at the major successes of the Weierstrassian school, despite the objections of Kronecker, in Meschkowski (1967), 228–229.

11. Cantor (1870b), 83.

12. Cantor (1870b), 83.

13. Cantor (1871a).

14. Cantor (1871b), 87.

15. Cantor (1871a), 85.

16. Heine (1870), 359.

17. Riemann (1854), 248. See Chapter 1.

18. Cantor (1871a), 85.

19. Cantor (1871c).

20. Cantor (1883c), 185.

21. No account given by Weierstrass of the theory of irrational numbers survives. However, several of his students, preserving their notes of his lectures, published them in various forms. See for example Kossak (1872); Pincherle (1880); Biermann (1887). The latter claims to be developing an essentially Weierstrassian theory of the irrationals, but he devotes an entire section to derived point sets and declares his indebtedness to Cantor for the concept of the fundamental sequence. See Cantor (1883c), 186. Biermann, in fact, is clear evidence that not only were Cantor's ideas being circulated and discussed, they were also being put to good use. For Dedekind's work on the irrationals see his famous paper of 1872.

The history of the irrational numbers has been dealt with extensively. Of particular interest is the translation by Molk (1904). See also Jourdain (1910) and Cavaillès (1962), 35–44.

22. Cantor (1872), 92–93. For conciseness I shall call such sequences, as Cantor does, fundamental sequences: Cantor (1883c), 186. Cantor also noted the connection between his theory of irrationals and the fundamental sequences in Cantor (1889). It should be noted that the first published mention of Cantor's work on the irrationals referred to the fundamental sequences as *Elementarreihen*. See Heine (1872), 173–174.

23. Cantor (1872), 93.

24. This association is very important, because it allows Cantor to avoid the accusations Russell makes: "There is nothing to show that a rational can be subtracted from a real number, and hence the supposed proof is fallacious." If Cantor suggested such a subtraction, it was only symbolically and not meant to be taken in any literal sense. Cantor made all this clear in his response to Illigens, who had also objected to Cantor's theory of irrationals. See Cantor (1889), 114, and Russell (1903), 283–286.

25. Cantor (1872), 93. Here Cantor may have been relying too heavily on geometric intuition. Interpreting $b - a_n$ in the language of infinite series helps legitimate his claim. The argument also appears, however, quite circular, since b is defined, or merely associated with $\{a_n\}$. Cantor's remark in the limit reduces to the tautology $(a_n - a_n) = 0$. What Cantor probably intended to say was that since all sequences in the class of sequences associated with b differ in the limit from a_n by an arbitrarily small amount, there is a clear justification for thinking of b as the limit of $\{a_n\}$. Cantor would have avoided Russell's criticism in particular had he not relaxed his terminology and returned to more familiar notations or intuitions.

26. Cantor (1872), 93–94.

27. Cantor (1872), 95.

28. Cantor (1872), 93.

29. Cantor (1872), 94. This further emphasized Cantor's view that the elements of the set B should not be taken as actual limits. This, coupled with Cantor's earlier expression for $b + b' = b''$ as lim $(a_n + a'_n - a''_n) = 0$, further confirms my interpretation of the meaning Cantor had intended for $b - a_n$. See also note 25.

30. Cantor (1872), 95.

31. Cantor noted specifically that "the number concept, in so far as it is developed here, contains in itself the germ of a necessary and absolutely limitless extension," Cantor (1872), 95.

32. Cantor (1883c) and the section in this chapter entitled "Derived Sets of the Second Species."

33. Cantor (1872), 96.

34. Cantor (1872), 97. Dedekind noted in the preface to his *Stetigkeit und irrationale Zahlen* that his own work was sent to press after he had read Heine's development of irrationals (which drew heavily on Cantor's work, as Heine acknowledged), and immediately after he had seen Cantor's own paper of 1872. Dedekind, in describing Cantor's exposition of the irrationals, was particularly struck by the Cantorian axiom: "As I find after a hasty reading, the axiom in § 2 [of Cantor (1872)], except for the external form of its wording, agrees completely with the one which I designate as the essence of continuity in § 3 below," Dedekind (1872), 3. But the axiom by which Dedekind attributed continuity to the line was given as follows: "If all points of the line fall into two classes such that every point of the first class lies to the left of every point of the second class, then there exists one and only one point which gives rise to this division of all points into two classes, and which reproduces this cut of the line into two pieces," Dedekind (1972), 11. Though this was interpreted as equivalent to Cantor's assumption that to every real number there corresponded a unique point of the Euclidean line, the identification was by no means made explicit in Dedekind's form of the axiom. This, of course, was one of Cantor's great contributions and insights, clearly and precisely presented. The assumption that such a correspondence was possible led Cantor to the axiom that essentially underlay the complete understanding of the nature of continuity in any continuous domain.

35. Dedekind makes this same point concerning the value of geometric guides, though stressing their lack of rigor, in the preface to Dedekind (1872), 1.

36. Cantor (1872), 98. It should be stressed that the concept of limit point and the associated idea of the first derived set P' of limit points of a set P were the fundamental, basic elements of Cantor's early development of set theory. Not only did they play an obvious part in proving Cantor's uniqueness theorem as published in 1872, they also contained the seeds of important consequences. One such development included the transfinite numbers. The concept of the first derived set of P, however, was also the means by which Cantor was later to categorize sets as being, for example, closed, perfect, or dense. For a more detailed discussion of these and other more sophisticated ways in which Cantor was able to characterize sets in terms of P and P' alone, see Chapter 7, as well as the introduction by I. Grattan-Guinness to Cantor (1970), 70–71.

37. Cantor (1872), 98.

38. Cantor (1872), 98.

39. Cantor (1872), 98. Cantor, however, had not forgotten the importance of the case where $P^{(\infty)} \neq 0$. The significance of such point sets of the second species, as Cantor was later to call them, is discussed in greater detail in Chapter 2.

40. Cantor (1872), 98.

41. Meanwhile a shift in terminology has occurred. Earlier a value of the λth kind meant that the value was contained in the derived set L obtained after λ extensions from the set of rationals A. Now, in describing a *point* set P of the λth kind, Cantor meant that $P^{\lambda+1}$ did not exist.

42. Cantor (1872), 99.

43. Riemann (1854), in particular section 8 of his paper, pages 245–251.

44. Cantor (1872), 99–100. Condition 2 involved the Schwarz second derivative. Condition 3, however, involved the idea generally referred to as the smoothness of the function $F(x)$, which implies that if $F(x)$ possesses a derivative $F'(x)$ at x, then both its left derivative $F'(x^-)$ and the right derivative $F'(x^+)$ exist and are equal.

45. Cantor (1872), 100.

46. Cantor (1872), 101.

47. Cantor (1880d), 358. Grattan-Guinness (1970a), 69, where the same point is emphasized.

48. Cantor (1872), 95.

49. Cantor (1883c), 148.

50. Dedekind, from his remark in the preface to *Stetigkeit und irrationale Zahlen,* seemed to have been unable to appreciate the significance of Cantor's development of the domains of higher order: "In terms of my conception of the domain of real numbers [as being] complete in itself, I am as yet unable to see the uses which will be furnished by the distinction, if only conceptual, of real numbers of higher orders," Dedekind (1872), 3.

51. Refer in particular to pages 52–53.

52. Cantor (1874).

3. Denumerability And Dimension

1. Dini (1878); Mittag-Leffler (1884a) and (1884b).

2. Dedekind (1872). The paper appeared in the same year as Cantor's last paper on trigonometric series. For details concerning their friendship and correspondence, see Cantor/Dedekind (1937), 3–9; and Fraenkel (1930), 8–10.

3. See Cantor's letter to Dedekind, dated April 28, 1872, in Cantor/Dedekind (1937), 12; and Cantor (1883c), 185.

4. Dedekind describes all this in the preface to his monograph: Dedekind (1872). It should be noted that in America, the subject of continuity and the infinite had also been scrutinized by C.S. Peirce. Peirce, however, took an entirely different approach. His inspiration was not analysis, and his interests were not in probing the foundations of mathematics in order to provide a certain, unshakeable beginning from which function theory could proceed without difficulty. Instead, Peirce was led to study the mathematics of infinity, infinitesimals, and continuity as a result of his interests in logic and philosophy.

Moreover, in 1881 Peirce published a paper in the *American Journal of Mathematics,* "On the Logic of Number," in which (as he was later always proud to emphasize) he had characterized the difference between finite and infinite sets well before Dedekind did so in 1888. Peirce asserted that Dedekind's *Was sind und was sollen die Zahlen* was doubtless influenced by his own paper, because Peirce had sent Dedekind a copy. But the most interesting feature of Peirce's entire approach to mathematics was not the way in which it was like the research then being conducted in Europe, but in the ways it was unlike the approaches taken by Georg Cantor and Richard Dedekind to the problems of continuity and infinity. For details, see Peirce (1881), (1976), and Dauben (1977b).

5. Dedekind (1872), 1–2.

6. Dedekind (1872), 9.

7. Cantor/Dedekind (1937), 12–13.

8. Cantor/Dedekind (1937), 12.

9. Liouville (1851), 133–142.

10. Cantor to Dedekind, in a letter dated December 2, 1873, in Cantor/Dedekind (1937), 13.

11. Cantor (1874). Cavaillès (1962): "Remarques sur la formation de la théorie abstraite des ensembles," discusses the 1874 paper. It should be noted that in treating Cantor's proof of the nondenumerability of the continuum in terms of denseness and compactness, Cavaillès introduces elements which Cantor developed only later, and which were certainly foreign to the paper as it appeared in 1874. See also Medvedev (1965), 94–96.

12. As Cantor explained, algebraic numbers were those which satisfied equations with integer coefficients and finite exponents in the form $a_0\,\omega^n + a_1\,\omega^{n-1} + \ldots + a_n = 0$. Since the Fundamental Theorem of Algebra guaranteed that for an equation of degree n there were exactly n roots to the equation, he turned this fact into a basic method for enumerating the algebraic numbers. Given the equation for any algebraic number, it could be ordered in terms of the equation $N = n-1+ \mid a_0 \mid + \mid a_1 \mid + \ldots + \mid a_n \mid$. For further details turn to Cantor (1874), 115–116.

13. See the letter Cantor sent to Dedekind on December 7, 1873, in Cantor/Dedekind (1937), 14–15.

14. Consider Cantor's letter to Dedekind of November 29, 1973, Cantor/Dedekind (1937), 12–13.

15. Cantor/Dedekind (1937), 14.

16. Cantor/Dedekind (1937), 14.

17. Cantor to Dedekind, December 9, 1873, in Cantor/Dedekind (1937), 16.

18. Cantor/Dedekind (1937), 16.

19. Cantor (1874), 116.

20. Cantor/Dedekind (1937), 20.

21. Cantor/Dedekind (1937), 21.

22. Apparently Vally Guttmann had lost her parents and was living in Berlin with her three brothers when Cantor first met her. Cantor had grown accustomed to spending much of his time in Berlin, even after he left the university for Halle in 1869. Not only was his family there, but the intellectual center toward which he was continuously drawn was always Berlin. For details, see A. Fraenkel (1930), 189–266; in Cantor (1932), 452–483 (an abridged version). See also Meschkowski (1967), 1–9; and Grattan-Guinness (1971a), 351–354.

23. Cantor/Dedekind (1937), 25.

24. Cantor to Dedekind, June 29, 1877, in Cantor/Dedekind (1937), 34.

25. Cantor to Dedekind, June 25, 1877, in Cantor/Dedekind (1937), 34.

26. See Dedekind's letter to Cantor of June 22, 1877, in Cantor/Dedekind (1937), 27–28.

27. Cantor/Dedekind (1937), 28.

28. Cantor/Dedekind (1937), 29.

29. Canttr/Dedekind (1937), 34.

30. Cantor/Dedekind (1937), 37.

31. Cantor/Dedekind (1937), 38.

32. Cantor/Dedekind (1937), 38.

33. Cantor's letter was sent to Dedekind from Halle on July 4, 1877. See Cantor/Dedekind (1937), 39.

34. Cantor (1878).

35. Cantor (1878), 120.

36. Cantor (1878), 120.

37. Compare Cantor (1878), 120–121, with Cantor/Dedekind (1937), 33–34.

38. Cantor (1878), 121.

39. Cantor (1878), 121.

40. Cantor (1878), 122.

41. Cantor (1878), 123.

42. Cantor (1878), 124.

43. In 1878 Cantor used the terminology *ohne Zusammenhang* to describe the disjointed nature of such sets. However, by the time the French translation of this paper had appeared, Cantor had devised an explicit notation for the concept and therefore added: "I designate the set P by $\mathfrak{D}(M,N)$." In the above case P would of course be empty. See Cantor (1883b), 311–328, esp. 317.

44. Cantor (1878), 125.

45. Cantor (1878), 127.

46. See Cantor (1878), 127. Cavaillès also remarks that Cantor's demonstration was not analytic, but graphic, and adds that Cantor attached such importance to the "remarkable curve" that he reproduced it in his letter to Dedekind and in his article published in Crelle's *Journal*. See Cavaillès (1962), 77. For an example of analytic methods used in this context, see Meschkowski (1967), 37–38. The iage illustrated here is from Cantor's letter to Dedekind dated June 25, 1877. I am grateful to Professor Clark Kimberling for permission to reproduce this example from the original letters of the Cantor-Dedekind correspondence, now preserved in the archives of the University of Evansville, Evansville, Indiana.

47. Cantor noted the advances he had made in a letter to Dedekind dated October 23, 1877. See Cantor/Dedekind (1937), 40–41.

48. Cantor (1878), 132–133. This was Cantor's first explicit statement of his conjecture, given later in another form as his "continuum hypothesis"—namely that there is no set of elements whose power is greater than that of the set of all natural numbers but less than that of the set of all real numbers.

49. For details of Kronecker's antagonism against analysis generally, see the discussion in Mittag-Leffler (1900), 150–151; Meschkowski (1967), 67–69; Biermann (1968), 101–102; and Dugac (1973), 141–142, 145–146, 161–163.

50. Schwarz to Cantor, April 1, 1870, in Meschkowski (1967), 229.

51. I am indebted to Professor H. Meschkowski for making a number of letters in his possession available to me, in particular a collection representing nearly two dozen letters from Eduard Heine to H.A. Schwarz. Most of the letters were written in 1870, and two are to be found in Appendix B.

52. See the letter from Heine to Schwarz, dated May 26, 1870, given *in extenso* in Appendix B.

53. Cantor (1874), 117.

54. Dedekind made a note of this, dated December 7, 1873. See Cantor/Dedekind (1937), 19.

55. Cantor/Dedekind (1937), 19.

56. Cantor (1874), page 261 in the *original* paper, published in Crelle's *Journal*, volume 77.

57. Heine specifically acknowledged his debt to Cantor in his paper, Heine (1872), 173. Cantor later stressed the injustice of Kronecker's remarks criticizing Heine, in Cantor (1887), 385.

58. Kronecker (1886), 336.

59. Kronecker (1887), 337–355.

60. See Kronecker (1894), Section 7, esp. pp. 11–12. Consult as well the letters of Heine to Schwarz given in Appendix B, together with the letter from Schwarz to Cantor in Meschkowski (1967), 228–229.

61. Details of the incident are to be found in Fraenkel (1930), 10; and in Meschkowski (1967), 40.

62. Minnigerode (1871).

63. Cantor's Braunschweig lecture of 1897 recalled that Weierstrass had interceded on Cantor's behalf to urge publication of the paper. See Fraenkel's references to the notes taken by Staeckel, but now lost, in Fraenkel (1930), 10. Further details are given by Meschkowski (1967), 40. See also references by Cantor to the matter in his letters to Dedekind, esp. that of October 23, 1887: Cantor/Dedekind (1937), 40.

64. See in particular Fraenkel (1930), 10; and Meschkowski (1967), 40.

65. Meschkowski (1967), 40.

66. Kronecker made his legendary remark during a lecture to the Berliner Naturforscher-Versammlung in 1886. See Weber (1893a), 15; (1893b), 19. See also Kneser (1925), 221; and Pierpont (1928), 39.

67. Lüroth (1878). Cantor had immediately recognized such difficulties at the outset, and mentioned them specifically to Dedekind in a letter of 1877: Cantor/Dedekind (1937), 40.

68. Lüroth (1878), 191.

69. Lüroth (1878), 195.

70. Jürgens (1878).

71. Thomae (1878), 466–467.

72. Jürgens (1878), 139–140.

73. See Jürgens (1879), and note 76 below.

74. Cantor/Dedekind (1937), 47.

75. Netto (1878), 265.

76. Jürgens (1879). I am indebted to Dr. Dale Johnson of Hatfield Polytechnic, Middlesex, England, for calling to my attention a rare copy of this paper which Jürgens published in 1879, and in which a counter-example vitiating Netto's proof of the previous year was produced. Basically, Jürgens showed that no set containing interior points could be continuously corresponded by any one-to-one mapping with a second set without interior points. Since this case had not been considered by Netto in his proof, it was consequently open to question. As Jürgens concluded: "Thus the inductive proof [given by Netto] is correct only if the impossibility of a unique, continuous correspondence of a continuous space of n-dimensions with a discontinuous part of a second space of n-dimensions, i.e., with a part without inner points, is also established," 18.

Jürgens' paper was privately printed by B.G. Teubner, and was apparently given only limited circulation. This helps to explain why practically no mention is made of the work in the literature dealing with the problem of the invariance of dimension. There is, however, at least one copy of the pamphlet (amounting to no more than 18 pages) in the library of the Karl-Marx-Universität, Leipzig, Deutsche Demokratische Republik.

77. See Cantor's letter to Dedekind, December 29, 1878, in Cantor/Dedekind (1937), 42–43. See also a letter of January 17, 1879, 44.

78. Cantor/Dedekind (1937), 44. The theorem appears with slightly different notation as Theorem I in Cantor (1879a), 135.

79. Cantor (1879a), 135–136.

80. Cantor (1879a), 136.

81. Cantor/Dedekind (1937), 48.

82. Cantor/Dedekind (1937), 49.

83. Cantor (1879a), 137.

84. Cantor (1879a), 137.

85. Jürgens (1898), 53. In particular he pointed out that Cantor's assertion, which claimed that the map between $K_{\mu-1}$ and K_{n-1} was continuous, was merely an assumption still requiring proof. This was all the more difficult since multiple images of points had been allowed, i.e., the correspondence was not one-to-one.

86. Jürgens (1898), 51.

87. Jürgens (1898), 52–53.

88. Jürgens (1898), 52.

89. Jürgens (1898), 53–54. In establishing the insufficiency of Cantor's proof, Jürgens reopened the entire question of the invariance of dimension which was not satisfactorily resolved until Brouwer's famous paper of 1911. For remarks concerning the significance of that paper and Brouwer's subsequent work, consult Freudenthal (1975).

4. Cantor's Early Theory of Point Sets

1. Cantor (1879b), (1880d), (1882b), and (1883a).

2. Cantor (1883c).

3. Cantor (1879b).

4. Cantor (1879b), 139.

5. Cantor first distinguished such sets of the second species, or *Gattung,* in his paper of 1879. See Cantor (1879b), 140.

6. Cantor (1879b), 140.

7. Cantor (1879b), 141.

8. Cantor (1878), 119; and Cantor (1879b), 141.

9. Cantor (1879b), 142.

10. Murata (1971).

11. Cantor (1880d).

12. Cantor (1880d), 146.

13. Cantor (1880d), 148.

14. Cantor (1880d), 148.

15. Cantor (1883a), 160.

16. Cantor (1879b), 141.

17. Cantor (1882b).

18. Cantor (1872), 97; Dedekind (1872), 1–2.

19. Cantor (1882b), 150. For Cantor's letter to Dedekind of April 15, 1882, see Cantor/Dedekind (1937), 51–52.

20. Cantor (1882b), 150.

21. Cantor (1882b), 150.

22. Hermite (1873).

23. Lindemann (1882).

24. Cantor (1882b), 152.

25. See, for example, Cantor's later indications of a theory of matter based upon

set theoretic concepts: Cantor (1885a), 275–276, and in a letter to Mittag-Leffler of November 16, 1884, in Meschkowski (1967), 247–248. See also the letter to Sophie Kowalevski in Appendix C.

26. Cantor (1882b), 152.

27. Cantor (1882b), 152.

28. Cantor (1882b), 153.

29. Cantor (1882b), 153.

30. Cantor (1882b), 153. Originally Cantor had given the value of the volume of the sphere as simply $2^n\pi$, but he later realized that this was inaccurate for spaces of general dimension. In Mittag-Leffler's copy of the proof sheets of the original paper, Cantor inserted his revised formulation, and also recorded it at the end of his fourth paper as a correction to the earlier formulation; see the original version of the paper: *Mathematische Annalen 21* (1883), 58. In a letter to Felix Klein of June 6, 1882, Cantor explained the details of his more accurate determination of the volume of the unit sphere of dimension n in a space of dimension $n+1$. It was true that the volume was always less than or equal to $2^n\pi$. But equality was true only for $n=1$, $n=2$. For higher dimensional spheres, the correct calculation of volume was given by $\frac{2\pi^{\frac{n+1}{2}}}{\Gamma\left(\frac{n+1}{2}\right)}$, where Γ was described as an Eulerian integral of the second kind. More precisely, Cantor gave the volume as $2^{n+1}\int \frac{\partial x_1\, \partial x_2 \ldots \partial x_n}{\sqrt{1-x_1^2-x_2^2-\ldots}}$. Using a method of evaluation attributed to Dirichlet, Cantor concluded that it was possible to show that $2^n\pi \geq \frac{2\pi^{\frac{n+1}{2}}}{\Gamma\left(\frac{n+1}{2}\right)}$. See letter #408, Georg Cantor to Felix Klein, *Nachlass Klein,* Niedersächsische Staats- und Universitätsbibliothek, Göttingen, as well as Cantor (1883a), 157.

31. Cantor (1882b), 153.

32. Cantor (1882b), 154.

33. Cantor (1882b), 157.

34. Cantor (1882b), 156.

35. Cantor (1872), 97.

36. Cantor (1882b), 156.

37. The note did appear in the German paper: Cantor (1882b), 156, but was omitted in the French translation. It *should* have appeared in Cantor (1883d), 370.

38. Cantor (1872); Dedekind (1872); and Lipschitz (1877).

39. du Bois-Reymond (1875b), (1877) and (1882); and Hardy (1954). See also Vivanti (1891a), (1891b), (1899), (1901), and (1906).

40. For an idea of Cantor's opposition to actual infinitesimals, see his arguments in (1883c), 171–172; (1887), 407–409; and the analysis of Cantor's formalism in Chapter 6.

41. Cantor (1882b), 157.

42. See Cantor's letter to Dedekind of June 25, 1877: Cantor/Dedekind (1937), 34; and Cantor's remarks in Cantor (1878), 120–121.

43. Cantor (1883a).

44. Cantor (1883a), 158.
45. Cantor (1883a), 158.
46. Cantor (1882b), 153.
47. Cantor (1883a), 159.
48. Cantor (1883a), 159.
49. Cantor (1883a), 160.
50. Cantor (1883a), 160.
51. Cantor (1883a), 160.
52. Du Bois-Reymond (1875a), (1879), (1880a), and (1882); Harnack (1880), (1881), (1882a), (1882b), and (1891).
53. Cantor (1883a), 160.
54. Cantor (1883a), 160.
55. Cantor (1883a), 161.
56. Cantor (1883a), 161.
57. Cantor (1883a), 161.
58. Cantor (1883a), 161.
59. Cantor (1883a), 162. See also Hawkins (1970), 62–63; 71–74.
60. Weierstrass recognized that Riemann's definition of the integral required further generalization and that Cantor's set theory, in particular his suggestions concerning a new theory of content, would help provide a more general theory of integration. See Weierstrass's letter to Sophie Kowalevski of May 16, 1885 (Institute Mittag-Leffler), and his letter to Schwarz dated May 28, 1885 (Deutsche Akademie der Wissenschaften der DDR). The letter of May 28 is transcribed in Dugac (1973), 141.
61. Du Bois-Reymond (1879), 287; (1880a), 128.
62. Harnack (1882a), 251; du Bois-Reymond (1882), 188–189.
63. Cantor (1879b), 140.
64. Du Bois-Reymond (1880a), 127–128.
65. See du Bois-Reymond (1875a), (1879) and (1880a).
66. See Cantor's letter to Klein, dated October 14, 1882. The letter is #424 of the Klein *Nachlass,* kept in the archives of the Niedersächsische Staats- und Universitätsbibliothek, Göttingen, and is reproduced in Appendix D.
67. See Cantor's letter to Klein, #429 dated December 18, 1882, and his letter #430 dated December 20, 1882. Both are part of the collection of letters from Cantor to Klein kept in Göttingen. See note 66.
68. Specifically noted in Cantor's postcard of December 20, 1882, #430, to Klein.

5. The Mathematics of Cantor's Grundlagen

1. Cantor (1883b), (1883c).
2. Simon (1883). For an assessment of his life and mathematical work, see Lorey (1918). Though Simon (1844–1918) taught mathematics in Strassburg, he was best known for his textbooks, his publications on the history of mathematics, and his interest in pedagogy. He had been a close friend of Cantor during their student days in Berlin, and was part of a small circle of young mathematicians who would often meet together in the Rähmelsche Weinstube or in the Cafe Röche. In addition to Cantor and Simon, the

group included among others Lampe, Thomé, Henoch, and Schwarz. See Cantor's letter to Schwarz, January 22, 1913, in Meschkowski (1967), 269; and Fraenkel (1930), 194. A number of letters written by Max Simon to H.A. Schwarz may also be found in the *Nachlass Schwarz,* kept in the archives of the Akademie der Wissenschaften der DDR, Berlin.

3. Simon (1883), 643.

4. See Mittag-Leffler's letter to Cantor of March 11, 1883. The letter is part of the *Nachlass Mittag-Leffler,* kept at the Institut Mittag-Leffler in Djursholm, Sweden. This letter has not been given an archival number.

5. Cantor (1883c), 165.

6. Cantor (1883c), 166.

7. See Cantor (1883c), 165, 166.

8. Cantor (1883c), 166. In the opening sections of the *Grundlagen,* Cantor distinguished between *reellen* and *realen Zahlen*–real numbers as opposed to complex or imaginary numbers in the mathematical sense, and *real* numbers in the concrete, ontological sense, a difference that is noted by Ernst Cassirer in Chapter IV: "The Concept of Number and Its Logical Foundation," Cassirer (1950), 63–64, 73–74. For further discussion of the importance of Cantor's distinction between *reellen* and *realen* for transfinite set theory, see Chapter 6.

For the French translation of Cantor's *Grundlagen,* see Cantor (1883d); for the English translation, see Cantor (1976). The reader should also be warned that in addition to missing the distinction between real and *real* numbers in translating the *Grundlagen,* Parpart also fails to distinguish between *Zahlen* and *Anzahlen,* translating both as "number" throughout without making clear the differences crucial to Cantor's introduction of the transfinite numbers.

9. Cantor (1883c), 167, 195–196.

10. See Cantor (1883c), 195–196.

11. Cantor (1883c), 196.

12. Cantor (1883c), 197.

13. Cantor (1883c), 197.

14. Cantor first introduced his ω in Cantor (1883c). Its appearance can be dated even more accurately in terms of the proof sheets for the *Punktmannigfaltigkeitslehre* papers, which Cantor sent to Mittag-Leffler for use in preparing the translations for *Acta Mathematica.* In every case, ∞'s were changed to ω's. The change must have occurred sometime between September and October of 1882, when the first section of the *Grundlagen* was sent to Teubner for press. The proof sheets Cantor sent to Mittag-Leffler are kept in the upstairs library, in the box marked *Cantor,* of the Institut Mittag-Leffler, Djursholm, Sweden.

15. Cantor (1883c), 197–199.

16. Cantor (1883c), 199–201.

17. Note that this assertion involves the Axiom of Choice. For discussion of the nature and significance of the axiom in the early history of set theory, see Chapter 11.

18. Cantor (1883c), 200.

19. Cantor (1883c), 168.

20. See the examples in the French translation of the *Grundlagen,* Cantor (1883d), 393.

21. See Cantor (1883d), 394.

22. Cantor (1883c), 169.

23. Cantor (1883c), 174.

24. Cantor (1883c), 174.

25. See Cantor (1883c), 170.

26. For Cantor's analysis of prime numbers see Cantor (1883c), 170, 202–204.

27. Cantor outlined his analysis of subtraction in Section 14. See Cantor (1883c), 201–202.

28. For Cantor's discussion of multiplication, turn to Cantor (1883c), 202.

29. See Cantor (1883c), 203.

30. Cantor (1883c), 203.

31. Cantor (1883c), 203–204.

32. See the *Beiträge,* Cantor (1897), 343.

33. In the French version, refer to the closing section of Cantor (1883d), as well as the original presentation in Cantor (1883c), 190–194.

34. See Cantor (1883c), 190.

35. Cantor (1883c), 191.

36. Cantor (1883c), 191. "Schüchterne Naturen empfangen dabei den (heilsamen) Eindruck, als ob es sich bei dem "Kontinuum" nicht um einen *mathematisch-logischen* Begriff, sondern viel eher um ein religiöses Dogma handle." *Heilsamen* appeared in Cantor's original draft. Even the proof sheets, stamped Teubner, January 15, 1883, preserved the word which Cantor chose to omit from the final printed version. The proof sheets, sent to Mittag-Leffler as guides for the preparation of the *Acta* translation, are still preserved in the old library, upstairs, of the Institut Mittag-Leffler, Djursholm, Sweden.

37. Cantor (1883c), 191.

38. Cantor (1883c), 191–192.

39. Cantor (1883c), 192.

40. See Cantor (1883a) and the discussion in Chapter 4.

41. Cantor gave the outlines of his analysis in Section 9 of the *Grundlagen.* See Cantor (1883c), 183–190.

42. Cantor (1884a), 224. In the *Grundlagen,* Cantor advanced the incorrect decomposition $P' = R + S$, where R was described as reducible, meaning that for some number γ of the first or second number class, $R^{(\gamma)} = 0$, and where S was perfect (Cantor (1883c), 193). But by the time Cantor came to write the *second* half of the *Grundlagen,* he had discovered that the set R did not have to be reducible. Instead, it was shown that "R is a point set which is either finite or of the first power," Cantor (1884a), 224. Ivar Bendixson, by letter from Stockholm, had alerted Cantor's attention to the fact that R need not be a derived set of any other set. Consequently $R^{(\gamma)} = 0$ was not a necessary property of the set R in the decomposition of P'. Bendixson was able to determine the specific properties distinguishing R from other denumerable sets by showing that there was a smallest α of the first or second number class such that $\mathfrak{D}(R, R^{(\alpha)}) = 0$.

43. Cantor introduced his ternary set in note to the *Grundlagen:* Cantor (1883c), 207. It should be noted that the ternary set is also of measure zero. Cantor established this property of the set in Cantor (1884a), 235.

44. Cantor (1883c), 194.

45. See Zermelo's notes to his edition of Cantor's *Gesammelte Abhandlungen,* Cantor (1932), 194.

46. Cantor (1883c), 194.

47. See Bolzano (1851/1889) and Bolzano (1950), esp. Section 38; Cantor (1883c), 194.

48. Dedekind also studied the elements of continuity in Dedekind (1872); Cantor (1883c), 194.

49. The sixth paper, Cantor (1884a).

50. Cantor (1884a), 215.

51. Cantor (1884a), 218.

52. Cantor (1884a), 218.

53. Cantor (1884a), 218.

54. See Cantor (1883a), 158.

55. Cantor (1883c), 200.

56. Cantor (1884a), 221.

57. Cantor (1884a), 222.

58. Cantor (1884a), 223.

59. Cantor (1883c), 193.

60. A number of Cantor's letters to Ivar Bendixson (written in 1883 and 1884) were sent by Else Cantor in 1927 to Mittag-Leffler, and are now kept in the archives of the Institut Mittag-Leffler in Djursholm, Sweden.

61. Cantor acknowledged Bendixson's letter explicitly. See Cantor (1884a), 244.

62. Cantor (1884a), 224.

63. Cantor (1884a), 225.

64. See Cantor (1884a), 226.

65. Cantor (1884a), 226.

66. Cantor (1884a), 227–228.

67. Cantor (1884a), 228.

68. Cantor (1884a), 228. Cantor was not ready to settle the converse proposition. Could separated sets be given that were of a power greater than the first number class (I)? He finally established, in the "Zweite Mittheilung," that all separated sets were necessarily of the first power. See Cantor (1885a), 262 and 268.

69. See Cantor's analysis in Cantor (1884a), 229.

70. Cantor used the terminology *Inhalt*. See Cantor (1884a), 229. For an analysis of the historical development of the theories of integration and content in the nineteenth and early twentieth centuries, see Hawkins (1970).

71. See Weierstrass' letter to Sophie Kowalevski of May 16, 1885 (Institut Mittag-Leffler), and his letter to Schwarz dated May 28, 1885 (Akademie der Wissenschaften der DDR, Berlin).

72. Cantor (1884a), 231.

73. Cantor (1884a), 231.

74. Cantor (1884a), 232.

75. Cantor (1884a), 235.

76. Cantor's ternary set was first explained in a note to the *Grundlagen,* Section 10, Note 11. See Cantor (1883c), 207, and Cantor (1884a), 235.

77. Cantor (1883c), 235.

78. Cantor (1883c), 235–236.

79. Cantor (1883c), 236.

80. See Cantor (1883c), 243.

81. Cantor (1883c), 244.

6. Cantor's Philosophy of the Infinite

1. See Cantor's introduction to the *Grundlagen,* which was not reprinted by Zermelo. Cantor (1883c).

2. Gauss wrote to Schumacher from Göttingen on July 12, 1831. See letter #396 (Gauss' letter #177) in Gauss (1860), 269.

3. Cantor (1883c), 173–175, 205; Cantor (1887), 395–396.

4. See, for example, Cantor (1883c), 174.

5. Cantor's analysis of Aristotle's position concerning the infinite was given in Section 4 of the *Grundlagen,* Cantor (1883c), 173–175.

6. Cantor suggested the list in Section 5 of the *Grundlagen.* See Cantor (1883c), 175.

7. See in particular Cantor's discussion in Sections 5 through 8 of the *Grundlagen,* Cantor (1883c), 175–183. The following analysis presents, in its major outline, the view Cantor took of seventeenth-century philosophy and mathematics.

8. Cantor (1883c), 177.

9. Refer to Cantor's letter to Eneström, November 4, 1885, as published in Cantor (1886), 373.

10. Cantor (1883c), 179.

11. Refer to Sections 29 to 33 of Bolzano (1851/1889) and Bolzano (1950); Cantor (1883c), 180.

12. Cantor criticized Bolzano directly in Cantor (1883c), 180–181.

13. Cantor (1886), 371–372. For Cantor's discussion of *categorematic* and *syncategorematic* infinities, see Cantor (1883c), 180. Cantor emphasized his own view of the potentially infinite in a lengthy footnote to his letter to Lasswitz, published in Cantor (1887), 392–393.

14. See Cantor's letter to Eneström: Cantor (1886), 372. For Cantor's treatment of Herbart, refer to Cantor (1886), 375, and Cantor (1887), 392–393.

15. For the French translation, see Cantor (1883d); for English: Cantor (1976). See as well note 8 to Chapter 5 for additional warning about the English translation.

16. Cantor (1883c), 181–183; Dauben (1971b).

17. See Mittag-Leffler's letter to Cantor of November 14, 1884, kept among Mittag-Leffler's papers in the archives of the Institut Mittag-Leffler, Djursholm, Sweden.

18. Cantor (1885a), 275–276.

19. Cantor discussed his dissatisfaction with traditional mechanics and with mathematical physics as it was usually developed, in letters to Mittag-Leffler dated September 22 and October 20, 1884. Both are to be found in the archives of the Institut Mittag-Leffler, Djursholm, Sweden. See as well Cantor's letter to Valson of January 31, 1886: Cantor (I), 43–44.

20. Cantor's metaphor of the road: Cantor (1887), 392–393.

21. Cantor (1887), 393.

22. Refer to Cantor's letter to Vivanti in Cantor (1887), 410–411.

23. Cantor (1887), 387–388.
24. Cantor (1887), 388.
25. Cantor (1887), 395–396.
26. Cantor (1883c), 182.
27. Du Bois Reymond (1875b), (1876), (1887), and (1882). See also Hardy (1954); Stolz (1881), (1885); Vivanti (1891a), (1891b), and (1899). See as well Vivanti's own estimate of infinitesimals as quoted by Cantor in a return letter to Vivanti of December 13, 1893, in Meschkowski (1965), 505.
28. Kerry (1885), 212.
29. Cantor (1883c), 172.
30. Cantor included the argument in two letters, one to Goldscheider dated May 13, 1887; another to Weierstrass three days later, dated May 16, 1887; both in Cantor (1887), 407–409.
31. Cantor (1887), 407.
32. Cantor (1887), 408.
33. Meschkowski (1965), 505.
34. Meschkowski (1965), 505.
35. Meschkowski (1965), 505.
36. Cantor in Meschkowski (1965), 507. This same letter was first published in Cantor (1895b).
37. See Mittag-Leffler's letter to Cantor of February 7, 1883, in Meschkowski (1967), 234.
38. See Cantor's explanation of immanent and transient realities in Section 8 of the *Grundlagen:* Cantor (1883c), 181–183.
39. Cantor (1883c), 181.
40. Cantor (1883c), 181.
41. Cantor (1883c), 182.
42. Cantor (1883c), 182.
43. Cantor (1883c), 182.
44. Cantor (1883c), 183.
45. Cantor (1883c), 183.
46. Cantor in letter #36 to Mittag-Leffler, dated May 5, 1883, archives of the Institut Mittag-Leffler, Djursholm, Sweden.
47. Cantor to Mittag-Leffler, letter #36.
48. Cantor to Mittag-Leffler, letter #45, September 9, 1883, archives of the Institut Mittag-Leffler.
49. Cantor to Mittag-Leffler, letter #65, December 23, 1883. See Schoenflies (1927), 3.
50. Cantor to Mittag-Leffler, letter #68, December 30, 1883.
51. Cantor to Mittag-Leffler, letter #1 in the collection of letters from Cantor to Mittag-Leffler for the year 1884. The letter is dated January 1, 1884. See Schoenflies (1927), 3–4, and Meschkowski (1967), 131–132.
52. Quoted in a letter from Cantor to Mittag-Leffler, letter #10, January 26, 1884, in Schoenflies (1927), 5.
53. Cantor to Mittag-Leffler, Letter #8, January 25, 1884, in Schoenflies (1927), 4–5.

54. Cantor to Mittag-Leffler, letter #9, January 26, 1884, in Schoenflies (1927), 5.

55. Cantor to Mittag-Leffler in a letter dated April 4, 1884. See Schoenflies (1927), 15. This letter, with a number of others from the year 1884, is kept in a separate envelope identified in the archives of the Institut Mittag-Leffler as "wichtigere Briefe und Karten von Cantor, 1884." This group of letters is not identified by any numerical system.

56. Cantor (1883c), 244.

57. See Grattan-Guinness (1971a), 356; Peters (1961), 15 and 27.

58. Cantor to Mittag-Leffler, letter #19, dated June 21, 1884, in Schoenflies (1927), 9; Grattan-Guinness (1971a), 376–377.

59. Kronecker's return letter to Cantor, dated August 21, 1884, is given in Meschkowski (1967), 237–239.

60. Cantor wrote again to Kronecker on August 24, 1884. See Meschkowski (1967), 240–241.

61. Cantor to Mittag-Leffler, letter #32, October 9, 1884, in Schoenflies (1927), 12.

62. Cantor to Mittag-Leffler in a letter of August 26, 1884, in Meschkowski (1967), 243.

63. Schoenflies (1927), 17–18.

64. Cantor to Mittag-Leffler, in a letter of November 14, 1884, in Schoenflies (1927), 17–18.

65. See Cantor (1970). This paper was only recently edited; see in particular the introduction and supporting documents in Grattan-Guinness (1970a).

66. Mittag-Leffler to Cantor in a letter dated March 9, 1885, in Grattan-Guinness (1970a), 102.

67. Cantor to Poincaré, in Grattan-Guinness (1970a), 105.

68. Cantor to Gerbaldi, in Grattan-Guinness (1970a), 104. An indication of Cantor's real feelings, however, is given in a letter written by the French mathematician Painlevé to Mittag-Leffler on January 27, 1896. Painlevé was soliciting signatures from mathematicians in support of *Acta Mathematica,* but Cantor refused to sign. The letter in question is #8 in the collection of Painlevé correspondence now kept in the archives of the Institut Mittag-Leffler.

69. Schoenflies (1927), 23.

70. The best short introduction to Leo XIII's place in modern European history is to be found in C.J H. Hayes (1940), esp. pp. 141–148. Leo XIII's concern for science was indicated by his support of a well-trained staff of physicists and the purchase of the most modern equipment available for the astronomical observatory at the Vatican. As for the problems science posed to religion, Darwin's theory of evolution was a major concern. It was an obvious target of Leo XIII's appeals for Christian philosophers to reassess the foundations and principles of modern science, particularly when it seemed ready to support materialism and atheism. For details, consult Muldoon (1903), chapter 27, 404. Indicative is Thesis 10 from the "Index Thesium" of T. Pesch: "Th. 10: Apparet in rebus naturalibus veri fines sive *causae finales;* unde *systema Darwini,* adeo hisce temporibus celebratum, falsum esse legitima conclusione infertur," in Pesch (1897), volume I. See also Galland (1880), O'Reilly (1887), Dorlodot (1921), and Ong

(1960). Of special interest is Chapter IX: "Leo XIII in a Scientific World," in Wallace (1966), 206–230.

71. The best commentary on *Aeterni Patris* is still Ehrle (1954). See also Alexander (1880). For English translations, consult Wynne (1903); Maritain (1958).

72. For details concerning the revival of Thomism in the nineteenth century, see Bonansea (1954); Collins (1961); Perrier (1909); Wyser (1951), esp. section 40: "Die Thomistische Philosophie des 19. und 20. Jahrhunderts," 46–53; John (1966).

73. Leo XIII (1877).

74. Perrier (1909), chapter 9 in particular.

75. Hugon (1924); and Raeymaeker (1952).

76. Leo XIII (1879), taken from Maritain (1958), 207.

77. Leo XIII (1878), in Wynne (1903), 9–21.

78. For details, see Perrier (1909), 161–164.

79. Cornoldi, as quoted in Perrier (1909), 162.

80. For biographical information concerning Gutberlet, see Perrier (1909), 199.

81. Gutberlet (1886), 179–223.

82. See Cantor's letter to Gutberlet of January 24, 1886, in Cantor (1887), 396–398.

83. Isenkrahe (1885), 73–111, 145–198.

84. For some indication of the context and flavor of the arguments on infinite duration, refer to the first pages of section 2 of Gutberlet (1886), 195–198. The problem of eternity and the infinity of space is also discussed by Désiré (1908); Phillips (1941) in particular 164–172; Greenwood (1944) and (1945).

85. Gutberlet (1886), 180.

86. Gutberlet (1886), 206.

87. The continuum hypothesis is Cantor's celebrated conjecture that the total number of all real numbers is given by the second of his transfinite cardinal numbers $\aleph_1$. Algebraically the continuum hypothesis claims that $2^{\aleph_0} = \aleph_1$. For details, consult Gödel (1947).

88. See the analysis of Cantor's religious sentiments in Meschkowski (1967), 122–129, and Meschkowski (1965), 514–515.

89. See Gutberlet (1886), 206: "But in the absolute mind the entire sequence is always in actual consciousness."

90. Perrier (1909), 199. For specific treatment of neo-Thomism in Germany, consult Wehofer (1879), 1–26; Jansen (1926) in Zybura (1926), 250–275.

91. Pesch (1880); Perrier (1909), 198.

92. Hontheim (1893) and (1895); Perrier (1909), 198; and Ternus (1929).

93. For an assessment of the contribution made by Pesch to the reconciliation of science and philosophy following the promulgation of *Aeterni Patris,* see van Riet (1946), 121.

94. Jeiler (1897); and Galland (1880), 150.

95. Esser (1895); Cantor's letter to Esser is given in Bendiek (1965), 65–73.

96. Gutberlet (1919), 364–370. For the problem of the existence of the actual infinite *in concreto,* refer to Cantor's letter to Gutberlet of January 24, 1886, in Cantor (1887), 396–398.

97. For biographical information concerning Franzelin's life and work, see Walsh (1895), Walsh (1911); and Bonavenia (1887).

98. See Franzelin's letter to Cantor in Cantor (1887), 385.

99. Pius IX (1864).

100. Cantor discussed such matters, for example, in Cantor (1886), 375; (1887), 385–387, 399–400. See also his note to Section 4 of the *Grundlagen,* Cantor (1883c), 205; and Brunnhofer (1892).

101. See Cantor's letter to Franzelin in Cantor (1887), 399–400.

102. Franzelin's response was included as part of the introduction to Cantor (1887), 385–386.

103. See, for example, Cantor's letter to Hontheim, dated December 21, 1893, in Ternus (1929), 570.

104. Cantor to Franzelin, in a letter of January 22, 1886, in Cantor (1887), 400.

105. Cantor (1887), 400.

106. Cantor to Gerbaldi in Grattan-Guinness (1970a), 104, and Grattan-Guinness (1971a), 366–367.

107. See Cantor's letter to Mittag-Leffler of December 23, 1883, as well as the letter of January 31, 1884, in Schoenflies (1927), 15–16.

108. Cantor to Jeiler, in Bendiek (1965), 68.

109. Refer to Cantor's letter to Hermite, dated January 22, 1894, in Meschkowski (1965), 514–515.

110. See Cantor's letter to Mittag-Leffler of January 1, 1884, in Schoenflies (1927), 3–4; and Grattan-Guinness (1971a), 358–359.

111. From part of a letter dated February 15, 1896, from Cantor to Esser, in Meschkowski (1965), 513.

7. From The *Grundlagen* To The *Beiträge:* (1883–1895)

1. Cantor (1884a), 243–244. The closing promise, "Continuation follows" was omitted from Zermelo's edition of the *Gesammelte Abhandlungen* but appears on page 488: "Über unendliche lineare Punktmannigfaltigkeiten" (part 6) *Mathematische Annalen 23* (1884).

2. Cantor (1884b)). His promise to prove the continuum hypothesis in a future paper appears on page 257.

3. Cantor (1885a). The "First Communication" to this "Second Communication" of 1885 was Cantor's earlier letter to the editor of *Acta Mathematica:* Cantor (1883e).

4. Cantor (1884a), 243.

5. Cantor (1970). For details concerning this episode, see Chapter 6.

6. Cantor (1885a), 275.

7. Cantor (1970).

8. Cantor (1970), 86.

9. Cantor (1970), 87.

10. Cantor (1970), 88.

11. Cantor (1970), 89.

12. Cantor devoted section 6 of the *Principien* to an analysis of the familiar arithmetic operations translated into terms of simply-ordered sets. Cantor (1970), 92.

13. Cantor (1970), 92.

14. Cantor (1970), 93.

15. Cantor (1970), 94, 99.

16. Cantor (1970), 94.

17. Cantor (1970), 95.

18. Cantor (1970), 100.

19. The *Theorie der Ordnungstypen* was published as section 8 of Cantor's "Mitteilungen zur Lehre vom Transfiniten." He explained in a note that the ideas were prepared for publication "in another journal" but that a surprising set of circumstances prompted him to withdraw the paper from press. See Cantor (1887), 411.

20. Cantor (1887), 419. Compare with Cantor's notes (numbers 1 and 2) at the end of the *Grundlagen,* Cantor (1883c), 204–205.

21. Cantor (1887), 411–412.

22. For detailed discussion of how the entire process of definition by abstraction came to figure prominently in Cantor's *Beiträge,* refer to Chapter 10.

23. Cantor (1887), 418.

24. Cantor argued that to carry the cardinal number theory with certainty into the domain of the transfinites, it was imperative that the transfinite ordinal numbers be introduced. These were taken as special forms of order types of well-ordered sets, referred to in the 1887–88 paper as ideal numbers, though the terminology did not persist. As for the general theory of order types, Cantor began to sketch its major features in section 8, pp. 417 and following. He also promised that a later paper would go on to outline an application of a special theory of ordinal numbers. Eventually the *Beiträge* would fulfill much more than this promise made in 1887. See Cantor (1887), 420.

25. Cantor (1887), 423–424.

26. Cantor (1887), 428.

27. Cantor (1887), 428–429.

28. See in particular sections 13 and 14 of part 8 of Cantor (1887), 428–439. One of Cantor's students, H.C. Schwarz, published a detailed account of order types in his dissertation done at Halle under Cantor's direction: Schwarz (1888). Vivanti published a detailed account of their work in Italian: Vivanti (1889).

29. Cantor made specific reference to this fact in a letter to Gerbaldi of January 11, 1896, in which he explained that Mittag-Leffler's suggestion to withdraw the *Principien* from press had caused him to turn his back upon the mathematical journals entirely: "Since I was disgusted with the mathematical journals by all this, as you will understand, I therefore began to publish my ideas in the 'Zeitschrift für Philosophie u. philos. Kritik'." See Cantor's letter-book for 1895–96: Cantor (III), 85–87. The draft of the letter to Gerbaldi is transcribed in Grattan-Guinness (1970a), 104.

Similarly, Cantor's major publications in the period 1885–91 stressed that philosophy and theology were directly related to the subject of his transfinite numbers. Though Cantor did include the fundamental principles of what he termed a "Theorie der Ordnungstypen" as section 8 of the "Mitteilungen zur Lehre vom Transfiniten," the material on order types had been drafted in 1884 and intended for the *Acta Mathematica*. When it finally did appear as part of Cantor (1887), it was not in a mathematical journal but in one devoted to philosophy and philosophical criticism. The only item to appear in a mathematical journal was a short note (running only a page in the *Mathematische Annalen*) in reply to criticism of his theory of irrationals: Cantor (1889).

30. For a survey of the most important and representative of Cantor's letters dealing with such topics, consult: Cantor to Gutberlet, January 24, 1886: Cantor (I), 36–38, and February 6, 1887: (I), 94–95; Cantor to Schmidt, March 26, 1887: (I),

98–102; Cantor to Grafen Vitzthum von Eckstadt, August 29, 1887: (I), 104; Cantor to Kiesewetter, September 9, 1891: (II), 79–80, and November 3, 1891: (II), 96–98; Cantor to Hermite, November 30, 1895: (III), 45–50, and February 11, 1896: (III), 144–145.

31. Cantor first indicated a desire to abandon mathematics for philosophy in a letter to Mittag-Leffler of October 20, 1884. Mittag-Leffler wrote back on November 2, 1884, endorsing the idea but adding his hope that Cantor would continue to be the great mathematical author. That Cantor actually began to teach philosophy at Halle is clear from Sophie Kowalevski's letter to Mittag-Leffler of May 21, 1885, #35, Institut Mittag-Leffler, Djursholm, Sweden. Cantor wrote to Hermite on January 22, 1894, explaining that for nearly twenty years mathematics had not been his sole interest, but that "Metaphysics and theology have seized my mind to such a degree that I have relatively little time left over for my *first flame* [*erste Flamme*]," in Cantor (II), 126; transcribed in Meschkowski (1965), 514.

32. See Cantor's letter to Gerbaldi of January 11, 1896, transcribed in Grattan-Guinness (1970a), 104. Note that instead of *Zeilen* the transcription should read *Ideen*. Cantor also cancelled the strident *für immer* in his reference to Mittag-Leffler's journal: "But naturally I don't want to know anything more about *Acta Mathematica*!" The best example of the intensive mathematical research Cantor continued to pursue in this period is the correspondence he maintained with Goldscheider. Twenty-two of these letters, spanning the years 1886–97 and running to 47 pages, are kept in the archives of the Universitätsbibliothek in Göttingen under accession number 1926.8260.

33. Consult Cantor's letter of May 24, 1887, in response to a request from Mittag-Leffler for a paper along number-theoretic lines for *Acta Mathematica*. The letter is #6 among those from the years 1886–1905 of the Cantor to Mittag-Leffler correspondence, in the archives of the Institut Mittag-Leffler, Djursholm, Sweden. Refer as well to notes 34 and 35 below.

34. Cantor to Eneström, September 10, 1886: Cantor (I), 81.

35. Cantor to Eneström, March 6, 1887: Cantor (I), 97.

36. For Oken's role in founding the GDNA, see Ecker (1883) and Pfannenstiel (1953). For a concise history of the organization, including reference to the early omission of any special section for mathematics, see Sudhoff (1922), 23. Gericke also gives a very short account in his recent history of the DMV: Gericke (1966), 48–49.

37. Sudhoff (1922), 25.

38. The first history of the DMV was written by Gutzmer (1904). Refer also to Gutzmer (1909). For a more recent version see Gericke (1966). Additional accounts of Cantor's role in promoting the idea may be found in Fraenkel (1930), 211–212; Meschkowski (1967), 166–167; Grattan-Guinness (1971a), 360.

39. Gutzmer (1904), 1. Gutzmer also appended a document at the end of his article that is of special interest: "Bericht über die Mathematiker-Versammlung zu Göttingen am 16., 17. und 18. April 1873," pp. 19–24. See also Müller (1873). For specific mention of the founding of *Mathematische Annalen* see Lorey (1916), 136–137, and the article celebrating Teubner's 100th anniversary, in Jahresbericht (1911), 81.

40. Gutzmer (1904), 1, 2.

41. Gutzmer (1904), 4; Sudhoff (1922), 40–43.

42. Gutzmer (1904), 3; Sudhoff (1922), 40–41; Gericke (1966), 49.

43. Schoenflies (1928), 561.

44. Cantor (1883c), 182. Cantor emphasized this idea repeatedly in his correspondence. See, for example, Cantor to Valson, January 31, 1886, in Cantor (I), 43.

45. Kronecker, during a visit with Cantor in July 1883, laughingly said he had been writing frequently to Hermite and arguing that Cantor's latest work was humbug. Cantor did not find Kronecker's attitude anything to laugh about and said so to Mittag-Leffler in a letter of September 9, 1883, #45 in the archives of the Institut Mittag-Leffler. See also Cantor's letter to Mittag-Leffler of November 7, 1883, #53, archives of the Institut Mittag-Leffler, and Cantor to Thomae, September 2, 1891: Cantor (II), 73, in which Cantor mentioned that Kronecker had been referring to his set theory as "mathematische Sophistik."

46. Cantor complained of financial difficulties in some detail in a letter to Woker of December 30, 1895: Cantor (III), 88–90. Woker was organizing a congress of Catholic intellectuals in Freiburg, and Hermite had urged Cantor to attend (see Cantor to Hulst, February 25, 1896: (III), 156–157). Cantor told Woker that he would like nothing better, but with six children he could not afford the trip. He then added, bitterly, that professors in Berlin and Göttingen made twice as much yearly; Cantor blamed his own position and financial straits upon independence from Weierstrass and Kronecker.

47. Cantor used the expression explicitly in a letter to Laisant of March 1, 1896: Cantor (III), 160. He complained that his salary was half that of other *ordentliche* professors, because he was a "free mathematician" and now worth nothing. In a letter to Lemoine dated March 17, 1896: (III), 187–190, he was even more specific and wished he had never given up music for mathematics. Instead, his yearly salary as a mathematician was only 4,800 Marks, while Klein and Schwarz, "the chosen," received twice as much. Consult as well Cantor's letter to Valson of January 31, 1886: (I), 43.

48. Consult Cantor's letter to Mittag-Leffler of January 11, 1884 (in the *wichtigere Briefe* envelope, archives of the Institut Mittag-Leffler), and Cantor to St. Hilaire, January 20, 1886: Cantor (I), 39, in which Cantor noted that unjust criticism of his work was particularly injurious to the inexperienced, younger students. Such motives were incorporated into the statutes of the DMV, from the opening paragraph describing the "Purpose of the Union." See Jahresbericht (1891), 12, as well as Schoenflies (1928), 562.

49. See Cantor's letter to Königsberger of June 10, 1891, in the archives of the Staatsbibliothek Preussischer Kulturbesitz, Berlin, Sammlung Darmstaedter.

50. Cantor to Königsberger, June 10, 1891. Cantor made similar comments along these lines in correspondence with Mittag-Leffler and Thomae. See Cantor to Thomae, September 2, 1891: Cantor (II), 73, and Cantor to Mittag-Leffler, September 5, 1891: (I), 74.

51. Cantor read Kronecker's letter addressed to the DMV at its opening session in 1891. It was subsequently printed in the first volume of the union's yearly report: *Jahresbericht* (1891).

52. Cantor wrote to Kronecker immediately after the meetings to tell him that he had been elected to the board of directors of the DMV. Cantor to Kronecker, September 30, 1891: Cantor (II), 88.

53. Cantor wrote to Thomé before the DMV convened and noted with some relief that Kronecker's absence would fortunately preclude any distasteful intrigue. Cantor to Thomé, September 2, 1891: Cantor (II), 73. See also Cantor to Mittag-Leffler, Sep-

tember 5, 1891: (II), 74. The letter Cantor actually sent to Mittag-Leffler is #11 in the envelope of letters from Cantor to Mittag-Leffler for the years 1886–1905, archives of the Institut Mittag-Leffler.

54. Schoenflies (1928), 561.

55. Cantor to Mittag-Leffler, December 30, 1883, #68, archives of the Institut Mittag-Leffler. Among others, see as well Cantor to Laisant, March 1, 1896: Cantor (III), 160; Cantor to Lemoine, March 17, 1896: (III), 188–189. On financial matters, see notes 46 and 47, as well as Cantor's letter to Mittag-Leffler of January 6, 1884, #3 in the archives of the Institut Mittag-Leffler.

56. Refer to Cantor's letter to Poincaré of January 7, 1896: Cantor (III), 102. See also Cantor to Vassilief, February 18, 1896: (III), 149–150.

57. Cantor to Hermite, December 26, 1895: Cantor (III), 82; Cantor to Mittag-Leffler, January 6, 1884, #3 in the archives of the Institut Mittag-Leffler.

58. Cantor wrote to Vassilief on July 4, 1894: Cantor (II), 132, saying he had had the idea for an international congress in mind for five years, at the time that he began promoting the DMV itself. He also told Althoff in 1896 that the idea for an international congress was his, just as the DMV had been his inspiration. But Cantor was appealing for travel funds, and his emphasis on the international organization may not have been entirely straightforward. See Cantor to Althoff, January 30, 1896: Cantor (III), 131–132.

59. Wangerin said as much in his short biographical sketch (1918), 2. Cantor went even further in a letter to Poincaré written on January 7, 1896: Cantor (III), 102.

60. In 1895, at its meeting in Lübeck, the DMV expressed sympathy with the idea of an international congress but expressly refused to take any initiative toward organizing such a meeting. For details, consult Cantor to Poincaré, January 7, 1896: Cantor (III), 124; Cantor to Vassilief, February 1, 1896: (III), 124.

61. The earliest reference I have found to Cantor's efforts in promoting the idea of an international congress is a letter to Vassilief dated July 4, 1894: Cantor (II), 132. See also letters to Laisant of April 25, 1895: (II), 141, September 22, 1895: (II), 181, March 1, 1896: (III), 159–162, March 19, 1896: (III), 193; to Poincaré of December 15, 1895: (III), 70, January 7, 1896: (III), 102, January 22, 1896: (III), 124–125; to Lemoine of December 27, 1895: (III), 84, March 4, 1896: (III), 172–174, March 17, 1896: (III), 187–189. Cantor noted in a letter to Vassilief of February 18, 1896: (III), 150, that Laisant and Lemoine had written to say that the *Société mathématique de France* had agreed to take the initiative in promoting the constituent meeting of the international congress for Zürich in 1897. See also Laisant's article on the idea of the congress in the January issue of *Revue générale des Sciences* for 1896. Geiser was eventually responsible for organizing the Swiss to host the first congress at Zürich in 1897. Cantor first mentioned Geiser in his letter to Lemoine of March 4, 1896. See also the introduction to Congress (1897).

62. Cantor (1891).

63. Cantor (1891), 278.

64. Cantor (1891), 279.

65. Cantor (1891), 278.

66. Cantor (1891), 279.

67. Cantor (1891), 280.

8. The *Beiträge*, Part I: The Study of Simply-Ordered Sets

1. Cantor published Part I of his "Beiträge zur Begründung der transfiniten Mengenlehre" in March 1895. Part II followed in March of 1897. Some measure of the importance attached to the work can be indicated in terms of translations that began to appear almost immediately. The first was Gerbaldi's Italian translation of Part I: Cantor (1895c). Both Part I and Part II were translated into French by Marotte: Cantor (1899). English readers had to wait until 1915 for P.E.B. Jourdain's edition, complete with lengthy introduction: Cantor (1915). The original papers appeared in the *Mathematische Annalen* and are listed in the bibliography as Cantor (1895a) and Cantor (1897).

2. Cantor (1895a), 282; Cantor (1915), 85.

3. Cantor (1883c), 204

4. Cantor (1883c), 204.

5. As early as 1886 Cantor referred to any actually infinite set as constituting a *Ding für sich*. See his letter to Eulenburg of February 28, 1886, reproduced as Section V of Cantor (1887), 400–407, esp. 401. See also a reference to this same terminology in a lengthy note to Section VIII, in Cantor (1887), 419.

6. Cantor (1895a), 282; Cantor (1915), 86.

7. Cantor (1895a), 283; Cantor (1915), 86.

8. Cantor (1887), 411.

9. All of Section 2 of Part I of the *Beiträge* was devoted to the comparison of powers. For details, consult Cantor (1895a), 284–285; and Cantor (1915), 89–91.

10. For the earlier formulation, see Cantor (1887), 411–412, and the discussion in Chapter 7 above.

11. Cantor (1895a), 285; Cantor (1915), 90–91. For the papers by Bernstein and Schroeder dealing with the equivalency of powers, see Schroeder (1896) and Bernstein (1898), Note 1 of Borel (1898), 102 ff; refer also to Zermelo (1901), 34–38.

12. Cantor professed to have an indirect proof of the comparability theorem, of which he sent a copy to Hilbert in 1896 and one to Dedekind in 1899. The proof was developed in the context of his discovery of the antinomies of set theory, and was based on his claim that every cardinal number was an *aleph*, i.e., the power of a well-ordered set. From this conclusion it was possible to establish his conjecture that every set could be well-ordered, and in turn the necessary truth of the comparability theorem followed. The fact that Cantor never published his proof but confined discussion of its merits to his private correspondence suggests that he was not altogether pleased with the rigor of its argument. For an analysis of all these matters in terms of Cantor's letter to Dedekind on the subject in 1899, see Chapter 10.

13. Arithmetic operations for the addition and multiplication of powers were defined in Section 3 of the *Beiträge*. See Cantor (1895a), 285–287; Cantor (1915), 91–94.

14. For Cantor's earlier definition of the multiplication of powers given in the *Mittheilungen*, see Cantor (1887), 414. For the introduction of the *Verbindungsmenge* (M.N) in the *Beiträge*, turn to Cantor (1895a), 286; Cantor (1915), 92–93.

15. See Cantor's use of the idea of covering in his proof of 1891 that the power of the set of all subsets of any given set was always necessarily *greater* than the parent set itself: Cantor (1891), 278–280.

16. Cantor (1895a), 287; Cantor (1915), 94. Jourdain's opinion concerning the importance of Cantor's idea of covering was offered at the very end of his introduction to the English translation of the *Beiträge:* "The introduction of the concept of 'covering' is the most striking advance in the principles of the theory of transfinite numbers from 1885 to 1895," in Cantor (1915), 82.

17. Cantor (1895a), 288; Cantor (1915), 95.

18. Cantor to Klein, letter #451 of July 19, 1895, in the archives of the Niedersächsige Staats- und Universitätsbibliothek, Göttingen.

19. Cantor (1895a), 289; Cantor (1915), 97. These same words appeared in Cantor's letter of July 19, 1895, to Klein.

20. Cantor devoted all of Section 5 of the *Beiträge* to his discussion of the finite cardinal numbers. He was careful to express in explicit terms how his set theory made possible "the most natural, shortest, and most rigorous foundation of the theory of finite numbers," Cantor (1895a), 289; Cantor (1915), 98.

21. Nowhere in his writings about set theory did Cantor ever discuss the empty set, nor did he ever define the number zero. He doubtless regarded both as straighforward concepts posing no special difficulties and requiring no separate definition.

22. Von Neumann (1925) and (1928).

23. Cantor to Peano, September 21, 1895: Cantor (II), 185–186.

24. Cantor to Peano, September 14, 1895: Cantor (II), 178–179; and September 21, 1895: (II), 185–186.

25. Dedekind (1888).

26. See Cantor's Theorems A through G, in Cantor (1895a), 290–292; Cantor (1915), 99–103.

27. Cantor to Peano, September 14, 1895: Cantor (II), 178–179.

28. See Cantor's remarks concerning a theory of finite numbers in Section VIII of the "Mittheilungen zur Lehre vom Transfiniten": Cantor (1887), 414–415, 418–419; and see the discussion of this aspect of Cantor's research in Chapter 10.

29. Cantor's words were: "By a *finite* set we understand a set *M* which arises from *one* original element through the successive addition of new elements such that the original element can be retrieved *backwards* from *M* by the *successive* removal of elements *in the reverse order,*" Cantor (1887), 415. This point of view was sharply criticized by Frege as being entirely too psychological. For details of his review of Cantor's "Mittheilungen," see Frege (1892) and the evaluation in Chapter 10.

30. Cantor to Peano, September 14, 1895: Cantor (III), 178–179.

31. Cantor (1895a), 292; Cantor (1915), 103.

32. Cantor did so in a letter to Felix Klein: April 30, 1895: (II) 142–143; also #499 in the collection held by the archives of the Universitätsbibliothek, Göttingen.

33. Only when he began to approach his set theory more abstractly (by regarding the transfinite numbers as *abstractions* from the specific properties of actual sets) was this probably even possible. This must have happened sometime after October 1882, when Cantor completed and dated the *Grundlagen,* but before September 1883, when he first publicized the new process of defining the transfinite numbers by abstraction at the GDNA meeting in Freiburg. See Cantor (1887), 387.

34. Cantor (1883c), 168–169. The emphasis upon the word *Attribut* is mine.

35. Cantor (1887), 387.

36. Cantor to Goldscheider, June 18, 1886: Cantor (I), 69–75, esp. p. 75; also #3 in the collection of Cantor's letters to Goldscheider in the archives of the Universitätsbibliothek, Göttingen.

37. Cantor to Vivanti, November 6, 1886: Cantor (I), 87–88. Cantor made use of the superscripted bar to denote cardinal numbers in correspondence with Vivanti dated June 26, 1888: (I), 180–181, and with Veronese, September 7, 1890: (II), 3–6. His first known mention of the new notation of *alephs* occurred in a letter to Vivanti of December 13, 1893: (II), 108–114, reproduced in Meschkowski (1965), 508.

38. Cantor (1895a), 293; Cantor (1915), 104.

39. For Cantor's proofs of Theorems A and B of Section 6, turn to Cantor (1895a), 293–294; Cantor (1915), 105–106. For additional discussion of Cantor's work and well-ordering, including the importance of the Axiom of Choice, refer to the material presented in Chapter 10.

40. Cantor (1895a), 294; Cantor (1915), 106.

41. Cantor (1895a), 295–296; Cantor (1915), 109–110.

42. Cantor to Janssen van Ray, July 10, 1894: Cantor (II), 135: "As for your comment, 'It is extremely difficult to define the number ω,' I take the liberty of pointing out that I have long ago overcome the difficulties in question. I ask that [you] simply examine with care the first essay in the collection 'Zur Lehre vom Transfiniten' (Halle, 1890)." Cantor also made the same point at the end of Section 6 of the *Beiträge:* Cantor (1895a), 296; Cantor (1915), 109–110.

43. It should be noted, however, that Cantor's *Beiträge* was not the first work to be published concerning Cantor's theory of order types. Even before Cantor's own research had been published in the *Zeitschrift für Philosophie und philosophische Kritik:* Cantor (1887), esp. Section VIII, 411–439, various aspects of the work had been published by Gutberlet (1886), at Cantor's specific request. See Cantor's remarks warning Benno Kerry about certain aspects of Gutberlet's article: Cantor to Kerry, March 21, 1887: Cantor (I), 88–89. Similarly, see Cantor to Bendixson, November 11, 1886: (I), 88–89. Cantor himself mentioned a reference to Gutberlet's publication in Cantor (1887), 388. Moreover, his students had begun to bring attention to the theory of order types in the mathematical literature. See in particular the work by Schwarz (1888) and Vivanti (1889).

44. Though Cantor had already explored the diversity of multiply-ordered sets in the *Mittheilungen,* the *Beiträge* limited its consideration to only simply-ordered sets. All references to ordered sets in the duration of Part I of the *Beiträge* were taken to mean simply-ordered sets, and the same convention is followed here. The terminology "simple order" and "simply-ordered" set has become standard for Cantor's *einfach geordnete Menge.* Thus I have resisted the temptation to revise standard practice and to introduce *singly*-ordered set, which in the context of Cantor's work seems a more appropriate and exact translation.

45. Cantor (1895a), 297; Cantor (1915), 111.

46. Cantor (1895a), 297; Cantor (1915), 111–112.

47. Cantor (1895a), 298; Cantor (1915), 114.

48. Cantor (1895a), 298–299; Cantor (1915), 114. This fact became extremely important later, in Part II of the *Beiträge,* when Cantor came to discuss the character of the class of order types of well-ordered sets of power $\aleph_0$, i.e., the class $Z(\aleph_0)$. Although the class itself was determined by well-ordered sets of power $\aleph_0$, the *power* of the

set $Z(\aleph_0)$ was greater than $\aleph_0$, and as Cantor demonstrated, its power represented the next larger transfinite cardinal number $\aleph_1$. But all this waited upon specific treatment of the concept of well-ordered sets.

49. For Cantor's earlier discussion of various conjugate transformations in the *Principien* and *Mittheilungen,* see Cantor (1970), 88–90, and Cantor (1887), 428. For Cantor's treatment of inverse order types in the *Beiträge,* see Cantor (1895a), 299; Cantor (1915), 114–115.

50. Cantor (1895a), 299–300; Cantor (1915), 115–116.

51. Cantor (1895a), 300; Cantor (1915), 116–117.

52. Cantor defined addition and multiplication of order types in Section 8 of the *Beiträge:* Cantor (1895a), 301–303; Cantor (1915), 119–122.

53. Cantor (1895a), 303; Cantor (1915), 122.

54. Cantor's analysis of the type η occupied all of Section 9: Cantor (1895a), 303–307; Cantor (1915), 122–128. Cantor had first discussed properties of the set of rational numbers specifically in the *Punktmannigfaltigkeitslehre* series, Cantor (1879b), 139–145, 140 in particular. Compare Cantor's discussion in Number 6 of the *Punktmannigfaltigkeitslehre* series, Section 19: Cantor (1884a), 238–240, with the later analysis of η in the *Beiträge*.

55. Cantor (1895a), 304; Cantor (1915), 124.

56. Cantor (1895a), 305; Cantor (1915), 126.

57. Cantor (1970), 87–88.

58. Cantor (1895a), 306–307; Cantor (1915), 128. Related conclusions were drawn in the *Principien:* Cantor (1970), 88.

59. See Cantor (1970), 88. Cantor first discussed order of the type θ of the linear continuum in a letter to Goldscheider of April 9, 1888: Cantor (I), 162–164.

60. Cantor (1970), 92–93.

61. Cantor's analysis of fundamental sequences and his definition of *Hauptelements* were detailed in Section 10 of the *Beiträge:* Cantor (1895a), 307; Cantor (1915), 129.

62. Cantor (1895a), 308; Cantor (1915), 130–131.

63. Cantor (1895a), 310; Cantor (1915), 134.

64. Cantor (1895a), 310; Cantor (1915), 134. Cantor's proof followed in major outline similar arguments devised as early as 1888 and communicated by letter to Goldscheider. Compare the published version with Cantor to Goldscheider, April 9, 1888: Cantor (I), 162–164.

65. Cantor pointed to this advance himself, explicitly, at the very end of the *Principien:* Cantor (1970), 100.

66. Full discussion of Cantor's discovery of the paradoxes of set theory may be found in Chapter 11.

9. The *Beiträge,* Part II: The Study of Well-Ordered Sets

1. Cantor referred specifically to Part II of the *Beiträge,* the major part of which he said was ready for press, in a letter to Stäckel dated October 12, 1895: Cantor (II), 192–193.

2. Cantor wrote to Valerian von Derwies, who was then in St. Petersburg, as

follows: "You are correct to emphasize the importance of a rigorous proof of the theorem $2^{\aleph_0} = \aleph_1$, which I stated in another form in 1877." Cantor added that he hoped to be able to prove the continuum hypothesis in Part II of the *Beiträge*. See Cantor to Valerian von Derwies, October 17, 1895: Cantor (III), 3.

3. Cantor (1897). Translations appeared, together with Part I, in French: Cantor (1899); and in English: Cantor (1915).

4. Cantor (1895a), 296; Cantor (1915), 109.

5. Cantor (1897), 312; Cantor (1915), 137–138.

6. Cantor (1897), 312–313; Cantor (1915), 138–139.

7. Cantor (1897), 314–315; Cantor (1915), 141–142.

8. Cantor proved this in Theorem A of Section 13: Cantor (1897), 317; Cantor (1915), 143.

9. Cantor (1897), 316; Cantor (1915), 144.

10. This was given in the *Beiträge* as Theorem E of Section 13: Cantor (1897), 317; Cantor (1915), 146.

11. Cantor (1897), 318; Cantor (1915), 147.

12. Cantor (1897), 318; Cantor (1915), 148.

13. Cantor (1897), 319; Cantor (1915), 149.

14. Cantor (1897), 319; Cantor (1915), 150.

15. Cantor devoted all of Section 14 to the definition and analysis of the ordinal numbers of well-ordered sets: Cantor (1897), 320–325; Cantor (1915), 151–159. The application of Theorem N from Section 13 to establish the comparability of all transfinite ordinal numbers appeared on pages 321 and 152, respectively.

16. For Cantor's discussion of these inequalities, see Cantor (1897), 322; Cantor (1915), 153–154.

17. Cantor had first considered the inverse operations of subtraction and division of order types in the *Grundlagen,* Section 14. For his earlier definition of subtraction, see Cantor (1883c), 201–202. For comparison, consult his later formulation in the *Beiträge:* Cantor (1897), 323; Cantor (1915), 155–156.

18. Cantor's analysis of infinite sums of ordinal numbers was designed to introduce his definition of *Fundamentalreihen* and their limits. Cantor (1897), 323–324; Cantor (1915), 156–158.

19. Cantor's *Fundamentalreihen* were introduced in Section 10 of the *Beiträge,* Part I: Cantor (1895a), 307–310; Cantor (1915), 128–133. For fuller discussion of their significance, see the discussion in Chapter 8. For the details of Cantor's use of fundamental sequences in Section 14 of Part II, see Cantor (1897), 324; Cantor (1915), 157–158.

20. Cantor (1897), 324; Cantor (1915), 158.

21. Cantor (1897), 324; Cantor (1915), 158.

22. Cantor had shown in Section 7 of the *Beiträge* that "for one and the same finite cardinal number ν, all simply-ordered sets are similar to one another, thus they have one and the same type," Cantor (1895a), 298; Cantor (1915), 113. For Cantor's discussion of finite ordinal numbers in Part II, see Cantor (1897), 324–325; Cantor (1915), 158–159.

23. Cantor (1897), 325; Cantor (1915), 159.

24. Cantor to Janssen van Ray, July 10, 1894: Cantor (II), 135. For further details consult note 42 in Chapter 8.

25. Cantor realized the limitations of the presentation of his theory of order types in

the *Mittheilungen*, and promised to elaborate the special theory of ordinal numbers in a later article. See Cantor (1887), 420.

26. Cantor (1897), 325; Cantor (1915), 160.

27. Cantor (1897), 325; Cantor (1915), 160.

28. This Cantor proved as Theorem B of Section 15 of the *Beiträge:* Cantor (1897), 326; Cantor (1915), 161.

29. See Theorem C: Cantor (1897), 326; Cantor (1915), 161.

30. For Cantor's definition of the conditions under which two fundamental sequences were *zusammengehörig*, see Cantor (1897), 327; Cantor (1915), 162–163.

31. Cantor (1897), 327; Cantor (1915), 163.

32. These were drawn in Theorems E, F, and G of Section 15: Cantor (1897), 327–329; Cantor (1915), 163–165.

33. Cantor (1897), 329; Cantor (1915), 165.

34. Cantor (1897), 329; Cantor (1915), 167.

35. Cantor (1897), 331; Cantor (1915), 169.

36. Cantor (1883c), 195.

37. Cantor (1897), 331; Cantor (1915), 170.

38. Cantor (1897), 332; Cantor (1915), 171.

39. Cantor (1897), 333; Cantor (1915), 172.

40. Theorem J of Section 15, Cantor (1897), 329; labeled Theorem I in Cantor (1915), 166.

41. Cantor (1897), 333; Cantor (1915), 173.

42. Cantor (1897), 333; Cantor (1915), 173.

43. The single reference to any transfinite cardinal number after Section 16 was made in Theorem F of Section 20, in which Cantor argued that the set of all ϵ-numbers of $Z(\aleph_0)$ was a well-ordered set of type Ω and cardinality $\aleph_1$. Cantor (1897), 349; Cantor (1915), 199.

44. The more Cantor analyzed the properties of the second number class in hopes of reaching a suitable characterization of the cardinal number $\aleph_1$, and perhaps in hopes of producing thereby a positive solution to his continuum conjecture, the more he was fascinated by the very intricate nature of the transfinite ordinal numbers. Much of his correspondence in the decade between 1885 and 1895 reflected the refined interest he had developed in the abstract number-theoretic problems he could pose concerning the class of numbers $Z(\aleph_0)$. See in particular his correspondence with Goldscheider and Vivanti, and the thesis by Schwarz (1888). Cantor to Goldscheider, September 2, 1887; January 29, 1888; March 9, 1888; all in the collection of Cantor's letters to Goldscheider in the archives of the Universitätsbibliothek, Göttingen. Cantor to Vivanti, January 30, 1888: Cantor (I), 148–150; April 2, 1888: (I), 159–161. In addition to Schwarz's *Dissertation* of 1888, see Vivanti (1889) and Cantor's own presentation: Cantor (1887), 428–438.

45. All of Section 17 of the *Beiträge* was devoted to the numbers in $Z(\aleph_0)$ of the form $\omega^\mu \nu_0 + \omega^{\mu-1}\nu_1 + \ldots + \nu_\mu$. Cantor (1897), 333–336; Cantor (1915), 174–178.

46. Similar results were expressed somewhat differently in the *Grundlagen* because there the order of multiplication was reversed from the form adopted in the *Beiträge*. Thus in the *Grundlagen* Cantor wrote $\omega_0\omega^\mu + \nu_1\omega^{\mu-1} + \ldots + \nu_{\mu-1}\omega + \nu_\mu$. See Cantor (1883c), 196–197, and all of Section 14, 201–204. Cantor made specific reference to this change at the end of Section 17 of the *Beiträge:* Cantor (1897), 336; Cantor (1915), 177–178.

47. Cantor (1897), 335; Cantor (1915), 177.
48. Cantor (1883c), 204; Cantor (1897), 336; Cantor (1915), 177.
49. For Theorem A, see Cantor (1897), 336; Cantor (1915), 178.
50. Cantor (1897), 337; Cantor (1915), 180.
51. Cantor (1897), 340–347; Cantor (1915), 183–195.
52. Cantor (1897), 340; Cantor (1915), 184.
53. Theorem A appears in Cantor (1897), 341; Cantor (1915), 185–186.
54. Cantor (1897), 341; Cantor (1915), 187.
55. Cantor (1897), 343; Cantor (1915), 189.
56. Cantor (1897), 347–351; Cantor (1915), 195–201.
57. Cantor to Goldscheider, October 11, 1886, in the archives of the Universitätsbibliothek, Göttingen. Note that Cantor, instead of using the notation Lim_{ν} to indicate the limits of infinite sequences, used the abbreviated notation *Li* in his letters on the subject of *Giganten* in 1886. See also his letters to Goldscheider of September 3, 1886: Cantor (I), 80; and to Eneström of September 26, 1886: (I), 82–83.
58. Cantor (1897), 347; Cantor (1915), 195.
59. Theorem H, Cantor (1897), 350; Cantor (1915), 200.
60. The analysis which follows draws largely upon the work of Jacobsthal (1908).
61. Jacobsthal (1908), 163, 168.
62. Jacobsthal (1908), 181.
63. Jacobsthal (1908), 183.
64. Cantor noted in Section 7 of his letter to Goldscheider (October 11, 1886) that $\omega^{\xi} = \xi$ was equivalently formulated by any of the equations $2^{\epsilon} = 3^{\epsilon} = \ldots = \nu^{\epsilon} = \ldots = \omega^{\epsilon} = \epsilon$. For details, see note 57 above and Cantor's letter to Goldscheider transcribed in Appendix A.
65. Cantor (1897), 350; Cantor (1915), 200. *Hauptzahlen* were thus especially significant in the study of prime numbers, division, and exponentiation. Above all, the nature of the additive *Hauptzahlen* suggested that the number theory of the transfinite numbers was essentially *additive,* while for finite numbers the character of their corresponding number theory was clearly *multiplicative*. For example, any finite number was uniquely representable as a product of prime factors, i.e., $12 = 3 \cdot 2^2$. Prime decompositions were of course possible in additive terms, but these were not necessarily *unique* factorizations, i.e.: $12 = 1 + 11 = 5 + 7$. However, the transfinite numbers, by contrast, were always uniquely representable in terms of additive primes, as Cantor's normal form made clear. This was true because the additive *Hauptzahlen* were all prime with respect to addition.
66. Cantor (1884a), 244.

10. The Foundations and Philosophy of Cantorian Set Theory

1. Cantor (1895a), 282; Cantor (1915), 85.
2. Cantor once wrote to Vivanti that despite his admiration for attempts to provide a purely logical foundation for arithmetic, such an approach was "not to my taste," Cantor to Vivanti, April 2, 1888: Cantor (I), 159. For Frege's major works, see Frege (1879), (1884), (1893) and (1903).
3. Frege (1885a), 94; Frege (1967), 103. Cantor noted his strong agreement with Frege's view expressed in the *Grundlagen der Arithmetic* that all psychological aspects

had to be removed from mathematics if logical purity was to be achieved. See Cantor (1885d), 440.

4. Hilbert's famous remark was prompted by a paper which Hermann Wiener had delivered at the first meeting of the Deutsche Mathematiker-Vereinigung in Halle, 1891. Hilbert was waiting for a train in Berlin and happened to be discussing Wiener's paper on the "Grundlagen und Aufbau der Geometrie" with Arthur Schoenflies and Ernst Kötter when he remarked that "Man muss jederzeit an Stelle von 'Punkte, Geraden, Ebenen' 'Tische, Stühle, Bierseidel' sagen können." For details, refer to Otto Blumenthal (1935), 402–403. See also Weyl (1944b), 635.

5. Frege (1884). For Cantor's review, see Cantor (1885d).

6. The following extracts from Montgomery Furth's translation of Frege's *The Basic Laws of Arithmetic* specify what Frege meant by "the extension of a concept": "I wish here to substantiate in actual practice the view of Number that I expounded in the latter book [*Grundlagen der Arithmetik,* 1884]. The most fundamental of my results I expressed there, in §46, by saying that a statement of number expresses an assertion about a concept; and the present account rests upon this . . . In a statement of number there is invariably mention of a concept—not a group, or an aggregate, or the like—or that, if a group or aggregate is mentioned, it is invariably determined by a concept, that is, by the properties an object must have in order to belong to the group; while what makes the group into a group, or a system into a system—the relations of the members to one another—is for the Number wholly irrelevant," Frege (1893/1964), 5. "At this point we may also use an expression from logic: 'the concept *square root of 4* has the same extension as the concept *something whose square trebled is 12*'. With such functions, whose value is always a truth-value, one may accordingly say, instead of 'course-of-values of the function,' rather 'extension of the concept'; and it seems appropriate to call directly a *concept* a function whose value is always a truth-value," Frege (1893/1964), 36.

7. Cantor (1885d), 440.

8. Cantor (1887). Frege reviewed the version issued as a separate monograph three years later: Cantor (1890). See Frege (1892).

9. Cantor (1887), 387. Cantor also gave a similar definition for powers (or cardinal numbers) in Section VIII of the *Mittheilungen:* Cantor (1887), 411–412.

10. Frege (1890/1969), 76–80. The exact date of this draft is uncertain but must have fallen sometime between the appearance of Cantor's publication and the published version of Frege's review in 1892. See Frege (1892), 163–166.

11. Frege (1890/1969), 77.

12. Frege (1890/1969), 78.

13. For Frege's version of the imagined dialogue, see Frege (1890/1969), 79–80.

14. Frege (1890/1969), 80.

15. Frege (1890/1969), 80.

16. Frege (1892).

17. This phrase echoed a line in Cantor's largely critical review of Frege's *Die Grundlagen der Arithmetik.* Cantor had written, after a short paragraph in praise of Frege's approach in general: "Weniger erfolgreich dagegen scheint mir sein eigener Versuch zu sein, den Zahlbegriff streng zu begründen," Cantor (1885d), 440. Frege's words, for the sake of comparison, were: "Weniger glücklich ist Herr Cantor, wo er definiert," Frege (1892), 269.

18. Frege (1892), 270.

19. Frege (1884), section 46, pp. 59, 59e in Frege (1884/1950).

20. Frege (1892), 270–271.

21. Cantor (1887), 415; cited in Frege (1892), 271.

22. Frege (1892), 272.

23. Frege (1892), 272. Cantor claimed that the reaction in Germany to the Idealism of Kant-Fichte-Hegel-Schelling had resulted in the "popular and thriving academic-positivistic skepticism," which had adversely affected even arithmetic. See Cantor (1887), 383–384.

24. Frege (1892), 272.

25. Cantor (1895a), 282; Cantor (1915), 85. Refer as well to the discussion in Chapter 11 and to Bernstein (1905a), 187.

26. Cantor devoted all of Section 5 of the *Beiträge* to an analysis of finite cardinal numbers. See Cantor (1895a), 289–292; Cantor (1915), 97–103. Strictly mathematical aspects of Cantor's development of the finite numbers is given in Chapter 8.

27. Cantor (1887), 415. Frege's criticism of this approach has already been noted in the previous section of this chapter. For Frege's own words on the matter, consult Frege (1892), 271.

28. See the development given in *Die Grundlagen der Arithmetik:* Frege (1884), and Sections 41–46 of Frege (1893/1964), 57–60.

29. Cantor to Peano, September 14, 1895: Cantor (II), 178–179.

30. Cantor to Peano, September 14, 1895: Cantor (II), 179.

31. Peano (1894–1908). Compare Dedekind (1888) with Peano (1889), especially Peano's Axiom 4 of the *Arithmetices Principia*.

32. Cantor (1895a), 283; Cantor (1915), 86.

33. Cantor to Hermite, November 30, 1895: Cantor (III), 45–50. Above all, the letter reflects the many and various interests mulling together in Cantor's mind at precisely the same period in which the *Beiträge* appeared. The early pages of the letter explained that his real interest in Francis Bacon had less to do with the Shakespeare problem than with the fact that Bacon after age forty approached more and more a strictly Roman Catholic point of view, one which the Enlightenment tried (successfully, he added) to cover up. This in itself may seem rather inconsequential, but it points to the fact that Cantor could feel a religious dimension in everything he did. Mathematics was by no means an exception, as the rest of the letter made clear. For further details, see Chapter 12.

34. Cantor to Hermite, November 30, 1895: Cantor (III), 47. Hermite's original is unknown. It does not appear among the collection of papers and manuscripts of Cantor's surviving *Nachlass*.

35. Cantor to Hermite, November 30, 1895: Cantor (III), 48.

36. Cantor (1869), 62. The underlining is based upon Cantor's emphasis given in his letter to Hermite, November 30, 1885: (III), 48.

37. Cantor to Hermite, November 30, 1895: Cantor (III), 48. Cantor reproduced the passage from St. Augustine in a lengthy note to Section V of his *Mittheilungen:* Cantor (1887), 401.

38. Cantor (1887), 404.

39. See for example, Cantor (1887), 405–406. For a more complete discussion of the important connections between Cantor's religious views and his mathematics, refer to Chapter 12. One is tempted to wonder if this view of God's role in ensuring the reality

and existence of Cantor's *Transfinitum* was responsible for his discovery of the contradictory nature of that very concept. For if Cantor believed God was capable of comprehending the *Transfinitum* in its entirety, it was natural to ask whether there was a greatest cardinal number? Surely the cardinal number of the set of *all* cardinal numbers had to be the largest, but Cantor had shown earlier that no largest cardinal could exist. In 1895 Cantor was only beginning to come to grips with the agonizing dilemma of inconsistent sets. Still, it is suggestive in light of his reliance upon the Divine Intellect as the "certain repository of the Transfinitum," that he first called such inconsistent sets *absolutely infinite* sets. In fact, Cantor had always thought such an absolute actual infinity did exist, one which the finite mind of man could never hope to understand fully but was nevertheless realized as an *Absolutum* in God's understanding. But the full implication of Cantor's view of the antinomies is discussed in further detail in Chapter 11.

40. In additon to the transcriptions of Cantor's two letters to Cardinal Franzelin in Cantor (1887), 399–400, see Cantor to Illigens, May 21, 1886: Cantor (I), 52–53.

41. Harnack (1886).

42. Cantor to Harnack, November 3, 1886: Cantor (I), 86.

43. Cantor to Kerry, March 21, 1887: Cantor (I), 106.

44. Cantor to Harnack, November 3, 1886: Cantor (I), 86.

45. Cantor to Veronese, September 7, 1890: Cantor (II), 3.

46. Consult the letters Cantor sent to Veronese: September 7, 1890: Cantor (II), 3, and October 6, 1890: (II), 16. Refer as well to the discussion in Chapter 8.

47. Cantor to Veronese, October 6, 1890: Cantor (II), 17.

48. Cantor to Jeiler, Pfingsten, 1888: Cantor (I), 167–174.

49. Cantor to Jeiler, Pfingsten, 1888: Cantor (I), 169.

50. See Cantor's letter to Mittag-Leffler of December 23, 1883, #65, archives of the Institut Mittag-Leffler, Djursholm, Sweden, as well as the letter to Mittag-Leffler of January 31, 1884, #11. For further discussion of Cantor's religious views connected with his view of himself as messenger, see the following sections of this chapter and the discussion in Chapter 12.

51. Cantor to Hermite, January 22, 1894: Cantor (II), 126–127.

52. St. Thomas Aquinas: *Summa contra gentiles*. Cantor cited in particular the words "Error circa creaturas redundat in falsam de Deo scientiam," in Cantor to Jeiler, Pfingsten, 1888: Cantor (I), 169.

53. On Cantor and his religious views see in particular Meschkowski (1965), 514–518, and Meschkowski (1967), 122–129. See as well the discussion in Chapter 12.

54. Cantor to Veronese, October 6, 1890: Cantor (II), 16. See as well Cantor to Peano, July 27, 1895: Cantor (II), 165. Cantor consequently took his characterization of the continuum to be *absolute:* see Cantor to Kerry, December 20, 1886: (I), 89.

55. Cantor to Vivanti, December 13, 1893: Cantor (II), 108; transcribed in Meschkowski (1965), 505.

56. Veronese (1891); the German translation by A. Schepp appeared three years later: Veronese (1894). Cantor in a letter of June 3, 1895, told Killing it was appropriate to warn everyone of Veronese's errors: Cantor (II), 156. Cantor's was not the only attack upon Veronese's ideas. For others, see Schoenflies (1896/1897) and Schoenflies (1906).

57. Mittag-Leffler raised the possibility in a letter to Cantor of February 7, 1883, in the archives of the Institut Mittag-Leffler, Djursholm, Sweden. Cantor responded to the matter of infinitesimals in Section 4 of the *Grundlagen:* Cantor (1883c), 171–175, and

later he devised a proof that infinitesimals could not exist. He also communicated his argument in letters to Goldscheider of May 13, 1887: Cantor (I), 118–120, and to Weierstrass of May 16, 1887. The proof was later published in Cantor (1887), 407–409. The original copy of Cantor's letter to Goldscheider is in the archives of the Universitätsbibliothek, Göttingen.

58. Cantor referred to Vivanti's letter and offered a lengthy rebuttal to the challenge that in light of Veronese's work his own was deficient for its lact of infinitesimals, in a letter of December 13, 1893, to Vivanti, in Cantor (II), 108–111; in Meschkowski (1905), 505. See also Cantor to Veronese, September 7, 1890: Cantor (II), 5.

59. Cantor (1895a), 300–301; Cantor (1915), 117. Again, this echoes Frege's comment that one could count anything that was an object of thought. Frege (1885), 94.

60. Veronese (1894), 31. Cantor cited this passage from Veronese's *Grundzüge der Geometrie* in Cantor (1895a), 301; Cantor (1915), 117–118. He actually accused Veronese of plagiarism in a letter to Killing of June 28, 1895: Cantor (II), 168. See also Cantor to Killing, July 26, 1895: Cantor (II), 162–163.

61. The italics are based upon Cantor's underlining of a similar passage quoted in a letter to Peano, July 28, 1895: Cantor (II), 169–172.

62. Cantor (1895a), 301; Cantor (1915), 118. See also Cantor's discussion of Veronese's research in letters to Veronese of September 7, 1890: Cantor (II), 3–6; October 6, 1890: (II), 14–17; November 13, 1890: (II), 24; November 17, 1890: (II), 29–30; November 26, 1890: (II), 30–31; to Killing of April 5, 1895: (II), 138; June 3, 1895: (II), 156; July 26, 1895: (II), 162–163; June 28, 1895: (II), 168–169; to Vivanti of July 21, 1895: (II), 163–164; to Peano of July 27, 1895: (II), 165–167; July 28, 1895: (II), 169–172; November 9, 1895: (III), 21–22; December 6, 1895: (III), 58–59; and to Gerbaldi, January 11, 1896: (III), 85–86.

63. Cantor to Peano, July 28, 1895: (II), 171, from a portion of the draft version of the letter, later cancelled.

64. Cantor (1895a), 301; Cantor (1915), 118. Cantor to Peano, December 6, 1895: Cantor (III), 59.

65. Cantor to Peano, July 27, 1895: Cantor (II), 165.

66. Cantor to Veronese, November 26, 1890: Cantor (II), 30; Cantor to Killing, April 5, 1895: (II), 138.

67. Cantor to Kerry, March 3, 1887: Cantor (I), 96; Cantor to Veronese, October 6, 1890: (II), 16; Cantor to Weber, March 21, 1895: (II), 137; and Cantor (1887), 409.

68. Hilbert (1899).

69. Peano (1892).

70. Russell (1903), 334–337, in particular 335. Compare with Cantor's statement in the *Grundlagen:* "Wenn anders überhaupt eigentlich-unendlichkleine Grössen existiren, d.h. definierbar sind, so stehen sie sicherlich in keinem unmittelbaren Zusammenhange mit den gewöhnlichen, unendlich klein *werdenden* Grössen," Cantor (1883c), 172.

71. Cantor to Veronese, October 6, 1890: Cantor (II), 15.

72. Veronese was admonished to admit his error: "It would certainly be best if Veronese heroically decided to say 'I have erred,' which is only human: 'Errare humanum est.' " In Cantor to Peano, November 9, 1895: Cantor (III), 22.

73. Newton, as quoted in Cantor (1895a), 282; Cantor (1915), 85.

74. Cantor (1895a), 301; Cantor (1915), 118.

75. Cantor to Veronese, November 17, 1890: Cantor (II), 29.

76. Cantor to Veronese, November 17, 1890: (II), 29.

77. Cantor to Veronese, November 17, 1890: (II), 29.

78. Cantor to Veronese, November 17, 1890: (II), 29.

79. Cantor to Killing, April 5, 1895: Cantor (II), 138: "I take what Veronese presents in his book as fantasy [*Phantasterei*], and what he raises there against me is unfounded. Of his *infinitely large numbers* he says that they are introduced *on other hypotheses* than mine. But mine depend upon *absolutely no hypotheses,* but are immediately derived from the natural concept of set. They are just as necessary and free from arbitrariness as the finite whole numbers." See also Cantor to Veronese, November 17, 1890: Cantor (II), 29.

80. Cantor to Veronese, November 26, 1890: Cantor (II), 30–31.

81. Cantor to Vivanti, December 13, 1893: Cantor (II), 108–114; transcribed in Meschkowski (1965), 507.

82. Bacon, as quoted in Cantor (1895a), 282; Cantor (1915), 85.

83. See Cantor's letter to Vivanti, December 13, 1893: Cantor (II), 108; in Meschkowski (1965), 504.

84. Cantor equated Nature with "the possible" in a letter to Vivanti of December 13, 1893, cited in Meschkowski (1965), 506.

85. Very early Cantor had commented to Mittag-Leffler that he was only a messenger and not the true discoverer of his transfinite set theory: Cantor to Mittag-Leffler, December 23, 1883, letter #65, and Cantor to Mittag-Leffler, January 31, 1884, letter #11, both in the archives of the Institut Mittag-Leffler, Djursholm, Sweden.

86. The *Bible,* as quoted in Cantor (1895a), 282; Cantor (1915), 85.

87. Georg Cantor to Georg Woldemar Cantor, May 25, 1862, cited in Fraenkel (1930), 193; Cantor (1932), 453; Meschkowski (1967), 5.

11. The Parodoxes and Problems of Post-Cantorian Set Theory

1. Frege (1903), 127.

2. Bernstein (1905c), 119.

3. Frege (1892), 272.

4. Frege (1893), 25.

5. See Frege's discussion of the impact Russell's letter had upon his *Grundgesetze* and Frege's attempt to repair the damage, in Appendix II to Frege (1903), 127–143. See also Russell's letter to Frege, in van Heijenoort (1967), 124–125.

6. Cantor (1895a), 285; Cantor (1915), 90.

7. Cantor argued the point in his letter to Dedekind of August 31, 1899: Cantor (1932), 448.

8. Cantor (1891), 280. Cantor drew a similar conclusion in his letter to Dedekind of August 3, 1899; Cantor (1932), 446.

9. Cantor (1883c), 167, 176, 196. See in particular Cantor's comments in Note 2 to Section 4 of the *Grundlagen:* p. 205.

10. Thomae (1906), 438: "Mathematics is the unclearest of all sciences."

11. Cantor to Dedekind, August 3, 1899: "One can see that this process of generating the alephs and the corresponding number classes of the system Ω is abso-

lutely limitless," Cantor (1932), 446. Note that Zermelo has merged two of Cantor's letters under the date of the earlier: July 28, 1899. Actually, only the first four paragraphs of page 443 were written on July 28th! Beginning with the fifth paragraph, "Gehen wir von dem Begriff . . . ," Zermelo should have indicated that he was transcribing Cantor's letter to Dedekind of August 3, 1899, the draft of which survives in Cantor's *Nachlass*.

The letters were first edited rather poorly by Zermelo and included in Cantor (1932). They were not included as part of Cantor/Dedekind (1937). Only recently has the original correspondence been rediscovered. Apparently Emmy Noether took the letters with her to Bryn Mawr when she left Göttingen in the late 1930s. Clark Kimberling discovered them: Kimberling (1972), and Ivor Grattan-Guinness recently published an account of the letters, with partial transcriptions, adding that Zermelo's edition of the correspondence from 1899 was remarkably inexact. See Grattan-Guinness (1974), 126–131; and Dugac (1976a), 259–262.

12. Cantor to Dedekind, August 3, 1899: Cantor (1932), 445.

13. Cantor to Dedekind, August 3, 1899: Cantor (1932), 446.

14. Cantor to Dedekind, August 3, 1899: Cantor (1932), 447. The difficulty in Cantor's proof involved his assumption, never verified, that under the conditions he set forth the entire system Ω could be mapped into the system V, so that a subset $V' \subset V$ had to exist such that $V' \simeq \Omega$. See Zermelo (1908a), 121.

15. Cantor to Dedekind, August 3, 1899: Cantor (1932), 447.

16. Dedekind to Cantor, August 29, 1899: Cantor (1932), 449. For Dedekind's theory of chains used to provide the finite numbers with a rigorous foundation, see Dedekind (1888).

17. Cantor to Dedekind, August 30, 1899: Cantor (1932), 449–450. Theorem A in Section 2 of the *Beiträge* read as follows: "Sind $\mathfrak{a}$ und $\mathfrak{b}$ zwei beliebige Kardinalzahlen, so ist entweder $\mathfrak{a} = \mathfrak{b}$, $\mathfrak{a} < \mathfrak{b}$ oder $\mathfrak{a} > \mathfrak{b}$," Cantor (1895a), 285; Cantor (1915), 90.

18. Cantor to Dedekind, August 30, 1899: Cantor (1932), 450.

19. Cantor to Dedekind, August 3, 1899: Cantor (1932), 443.

20. Cantor (1883c), 204.

21. See Cantor's letter to Eulenburg of February 28, 1886, in Part V of Cantor's "Mittheilungen zur Lehre vom Transfiniten," Cantor (1887), 401.

22. Albrecht von Haller, cited in Cantor (1883c), 205.

23. Cantor (1886), 372 and 376.

24. Cantor to Franzelin, January 22, 1886; Part III in Cantor (1887), 399.

25. Cantor to Franzelin, January 29, 1886; Part IV in Cantor (1887), 400.

26. Cantor to Eulenberg, February 28, 1886; Part V in Cantor (1887), 407. At the very end of his letter Cantor compared his theory of the infinite with the erroneous theory of Fontenell.

27. Hurwitz (1897). See as well Hadamard's use of set theory in his paper to the congress. It is clear from Hadamard's remarks that he assumed that Cantor's work was already known and of indispensable use. See Hadamard (1897) and Fraenkel's comments about the paper: Fraenkel (1930), 213.

28. Hilbert (1900).

29. Cantor to Klein, July 19, 1895: Cantor (II), 160. This letter is also #451 in the collection held by the Universitätsbibliothek, Göttingen.

30. Schoenflies (1899).

31. Vivanti (1892).

32. Russell (1903) and W.H. and G.C. Young (1906).

33. See Cantor's letter of acceptance, dated December 13, 1904, accession number 04175, archives of the Royal Society, London.

34. Burali-Forti (1897).

35. For a reference to Cantor's remarks in Kassel, see the note in Schoenflies (1911), 251, and *Jahresbericht* (1903).

36. König (1904).

37. Schoenflies (1928), 560.

38. Kowalewski (1950), 202.

39. König (1904); König thought the proof might be salvaged, despite the error in his original application of Bernstein's theorem; see König (1905), 180.

40. Bernstein's *Inaugural-Dissertation:* "Untersuchungen aus der Mengenlehre" (1901), 49.

41. Kowalewski (1950), 202.

42. Schoenflies (1928), 560.

43. Pittard-Bullock (1905), 14, suggested that Cantor had found the proof himself, but more reliable sources agree that it was Zermelo who made the discovery: see in particular Kowalewski (1950), 202.

44. Schoenflies (1922), 100–101. Meschkowski added Hausdorff and Hensel to the group in (1967), 166.

45. Schoenflies (1928), 560. Fraenkel noted Cantor's habit of working through all the immediate ramifications of a given problem himself, leaving little in the way of material for students to work on. This, Fraenkel suggested, explained why so few students wrote their dissertations under Cantor and why he left no significant unpublished papers at the time of his death. Fraenkel (1930), 217.

46. Zermelo (1904).

47. Zermelo noted this at the end of the unpublished version of the proof. See Zermelo (1904), 516.

48. Zermelo referred to these as *ausgezeichnete* elements. Zermelo (1904), 514.

49. Zermelo (1908a), 128.

50. Cantor to Dedekind, August 3, 1899; Cantor (1932), 447. See Jourdain's analysis of Cantor's method in Jourdain (1904), 70, as well as a letter from Cantor to Jourdain dated November 4, 1903, in Grattan-Guinness (1971b), 116–117. Cantor believed any set could be well-ordered in terms of successive choices of elements; see Zermelo (1908a), 121.

51. Zermelo (1904), 516:

52. König (1905b), 156–160.

53. Jourdain (1905).

54. Bernstein (1905a).

55. Schoenflies (1905).

56. Borel (1905).

57. Borel (1898).

58. Borel (1905), 194.

59. Borel (1905), 194.

60. Borel (1905), 195.

61. Note IV was devoted to "Cinq lettres sur la théorie des ensembles," Borel

(1928; 3*rd* edition). The letters were first published in the *Bulletin de la Société mathématique de France 33* (1905), 261–273.

62. Hadamard to Borel, Borel (1928), 150–152.

63. Hadamard to Borel, Borel (1928), 151.

64. Tannery (1897), cited by Hadamard in Borel (1928), 151.

65. Hadamard to Borel, in Borel (1928), 151.

66. Hadamard to Borel, in Borel (1928), 151.

67. Hadamard to Borel, in Borel (1928), 152.

68. Baire to Hadamard, in Borel (1928), 152.

69. Baire to Hadamard, in Borel (1928), 152.

70. Lebesgue to Borel, in Borel (1928), 153–156.

71. Lebesgue to Borel, in Borel (1928), 153.

72. Lebesgue to Borel, in Borel (1928), 154.

73. Lebesgue to Borel, in Borel (1928), 154.

74. Lebesgue to Borel, in Borel (1928), 154.

75. Lebesgue to Borel, in Borel (1928), 154–155.

76. Lebesgue to Borel, in Borel (1928), 155.

77. Lebesgue to Borel, in Borel (1928), 155.

78. Lebesgue to Borel, in Borel (1928), 156.

79. Hadamard to Borel, in Borel (1928), 156–159; esp. 157.

80. Hadamard (1905); cited in Hadamard to Borel, in Borel (1928), 157.

81. Borel to Hadamard, in Borel (1928), 159–160.

82. Hadamard to Borel, in Borel (1928), 158.

83. Burali-Forti (1897), 154–164.

84. Burali-Forti (1897).

85. Russell (1903), paragraph 301 of chapter 38 on ''Transfinite Ordinals,'' 323.

86. Russell (1903), 323.

87. Jourdain (1905), 465–470; Jourdain (1904), 65.

88. Jourdain (1904), 67. Note that Jourdain's use of "consistent manifold" was independent of Cantor's designation of consistent and inconsistent collections. For details, see Grattan-Guinness (1971d), 116.

89. Bernstein (1905a).

90. Bernstein (1905a), 188.

91. Bernstein (1905a), 193. Moreover, Bernstein, in taking W to be the system of all order types of all continuable well-ordered sets, added that W was not itself continuable; at the same time he produced from W the system Z comprising all subsets of W. Bernstein claimed that Z was the simplest example of a set that was not capable of being well-ordered. This meant that Zermelo's proof that *every* set could be well-ordered was definitely incorrect. The error in Zermelo's reasoning, as Bernstein interpreted it, concerned the noncontinuable character of W. Specifically, the fact that Zermelo might have unwittingly introduced the impossible continuation (W, e) rendered the proof invalid.

92. Zermelo (1908a), 122.

93. Zermelo (1908a), 122–123.

94. Zermelo (1908a), 123.

95. Russell (1903). For Russell's appraisal of Frege, see his Appendix A: "The Logical and Arithmetical Doctrines of Frege," 501–522.

96. Russell (1907), 31–32. See as well the recent paper by Crossley (1973).

97. Russell (1907), 32.

98. Russell (1907), 33.

99. Russell introduced the distinction between predicative and nonpredicative norms in Russell (1907), 34. For an early and general discussion of Russell's doctrine of types, see Appendix B of Russell (1903), 523–528. The mature theory was developed in *Principia Mathematica*. The basic goal of Russell's *Principia Mathematica,* written with A.N. Whitehead, was directed toward showing how it was necessary to distinguish between various *types* of objects, for example classes as opposed to classes of classes. By restricting propositional functions to domains of one and only one type, the theory of types effectively excluded such paradoxical concepts as the class of all classes which contained themselves, since $x \in x$ could not be permitted between elements of different type.

100. Russell/Whitehead (1925–1928; 2nd edition). Russell and Whitehead were themselves painfully aware of this fact and in their introduction to the second edition of the *Principia* lamented the unsatisfactory point they had reached. In order to avoid the contradictions, Russell advocated his theory of types, as has been noted. In the *Principia Mathematica* this eventually took the form of a "ramified" type theory designed to handle different sorts of contradictions, both mathematical and semantical. This in turn required introduction of an axiom by which Whitehead and Russell identified propositional and predicative functions as being formally equivalent, but there was no attempt to hide their own dissatisfaction with the artificial and arbitrary nature of the Reducibility Axiom. As they noted specifically, "This axiom has a purely pragmatic justification: it leads to the desired results, and to no others. But clearly it is not the sort of axiom with which we can rest content," Russell/Whitehead (1925), vol. 1, *xiv*.

101. Hilbert (1899).

102. Hilbert (1900a), 181; see also Hilbert (1905).

103. Hilbert (1900a), 180–181.

104. For details of Hilbert's formalism, see Carathéodory (1943), Weyl (1944), and Beth (1959).

105. Hilbert (1900a), 184.

106. Hilbert (1900a), 184.

107. Hilbert (1900a), 184.

108. Zermelo (1908b), 261–281.

109. Zermelo (1908b), 261.

110. Zermelo (1908b), 262.

111. Poincaré (1908).

112. See the exchange of articles between Poincaré and Russell in the *Revue de métaphysique et de morale*.

113. Poincaré (1908), 182.

114. Borel to Hadamard, in Borel (1928), 158.

115. Refer to the exchange between Poincaré and Russell listed in note 112.

116. Thomae (1906), 434.

117. Hessenberg (1908), 145.

118. Hessenberg (1908), 145.

119. Zermelo (1908a), 128.

120. Gödel (1931).

121. Gödel (1936).

122. Cohen (1963).

123. Cohen (1966).

124. Cohen (1966), 151.

125. Cohen (1966), 151.

126. Baire to Borel, as published in Borel (1905), 195. König noted incomparability on other grounds. $Z(\aleph_0)$ was self-generating in terms of Cantor's first and second principles of generation and therefore could never be regarded, said König, as completed. This he took to be the origin of the paradoxes, in particular of Burali-Forti's paradox of the greatest ordinal number. The continuum, by contrast, was a completed set. Thus the two concepts, $Z(\aleph_0)$ and the continuum, could not be equated. They were entirely noncomparable concepts. König was careful to add, however, that none of his conclusions were meant to oppose Cantor's "ingenious" work but were meant only to question a few conjectures, the continuum hypothesis being the most prominent. The content of Cantor's proven theorems remained unaffected. As König explained, the distinction Cantor had introduced between consistent and inconsistent systems, between sets and classes, completely explained the paradoxes.

127. Borel (1905), 195. For König's position, consult König (1905b), 159–160.

12. Epilogue: The Significance of Cantor's Personality

1. For biographical details of Cantor's family history, refer to Fraenkel (1930), 189–190; Peters (1961); Meschkowski (1967), 1–4; Grattan-Guinness (1971a), 351–352; and Kertész (1977).

2. From the report of the Danish Genealogical Institute, prepared in 1937 by Th. Hauch-Tausböll. In *Nachlass Cantor I,* as listed in Grattan-Guinness (1971a), 348.

3. Cantor related much of his family history in the course of his correspondence. The information recounted below is gathered from numerous letters to be found throughout Cantor (I), (II), and (III). See in particular the biographical details Cantor wrote out on visiting cards in 1899, transcribed in Grattan-Guinness (1971a), 379–380.

4. Reproduced in Fraenkel (1930), 191–192. In the next to last paragraph, however, Fraenkel transcribes "ein leuchtendes *Gestirn* am Horizonte der *Ingenieure,*" whereas the typescript version of the letter in *Nachlass Cantor I* reads *Wissenschaft* instead of *Ingenieure*. Part of the confirmation letter is reproduced in Meschkowski (1967), 3.

5. Hauch-Tausböll (1937), from the German typescript.

6. Wilmanstrand was then part of Finland. The citation from Georg Woldemar's will comes from Hauch-Tausböll (1937).

7. Fraenkel (1930), 192.

8. Fraenkel (1930), 192.

9. Fraenkel (1930), 193; quoted in Meschkowski (1967), 5.

10. Georg Woldemar Cantor to Georg Cantor, in a letter of August 18, 1862, *Nachlass Cantor I*.

11. Fraenkel (1930), 193.

12. See Georg Woldemar Cantor to Georg Cantor, August 26, 1862, and May 9, 1861, in *Nachlass Cantor I*.

13. Georg Woldemar Cantor to Georg Cantor, November 3, 1862, in *Nachlass Cantor I*.

14. Bell (1937), chapter 29.

15. Bell (1937), chapter 29.

16. Schoenflies (1927).

17. These events are outlined above, in Chapter 6. For literature dealing with Cantor's nervous breakdowns, see Schoenflies (1927); Fraenkel (1930), 198, 207–210; Peters (1961); and Grattan-Guinness (1971a), 355–358, 368–369. Cantor mentioned his trip to Paris, shortly before his breakdown in 1884, in two letters: Cantor to Mittag-Leffler, May 4, 1884, letter #18, in the archives of the Institut Mittag-Leffler, Djursholm, Sweden. Cantor to Klein, May 10, 1884, letter #437, in the archives of the Universitätsbibliothek, Göttingen.

18. Schoenflies (1927).

19. Schoenflies (1927).

20. Cantor to Mittag-Leffler, August 18, 1884, letter #22, archives of the Institut Mittag-Leffler, Djursholm, Sweden.

21. See Cantor's letter to Mittag-Leffler of August 18, 1884.

22. Kronecker to Cantor, August 21, 1884, transcribed in Meschkowski (1967), 237–239.

23. Cantor to Mittag-Leffler, August 26, 1884, printed in Meschkowski (1967), 242–243. (Letter #25, Institut Mittag-Leffler.)

24. Cantor to Klein, May 10, 1884, letter #437 in the archives of the Universitätsbibliothek, Göttingen; Cantor to Mittag-Leffler, June 21, 1884, letter #19 in the archives of the Institut Mittag-Leffler. See Schoenflies (1927), 9; Grattan-Guinness (1971a), 376–377.

25. Peters (1961), 15, 27.

26. For details concerning the amount of correspondence Cantor devoted to such matters, consult the listing in note 30 to Chapter 7. Note also Cantor's work on the Goldbach Theorem, discussed in Meschkowski (1967), 168–172; Grattan-Guinness (1971a), 360–361; and Cantor's letter to Hermite, November 30, 1895, transcribed in Meschkowski (1967), 262–263.

27. Cantor to Mittag-Leffler, October 20, 1884. For details consult note 31 to Chapter 7.

28. Grattan-Guinness (1971a), 368–369.

29. Consult the analysis of the Cantor-Dedekind correspondence in Chapter 11.

30. Cantor to Graf von Posadowsky-Wehner, November 10, 1899; transcribed in Grattan-Guinness (1971a), 378–379.

31. Cantor to Graf von Posadowsky-Wehner, November 10, 1899.

32. See Document VI of Grattan-Guinness (1971a), 380–381.

33. Grattan-Guinness (1971a), 380.

34. Cantor to Klein, December 31, 1899, letter # 455, in the archives of the Universitätsbibliothek, Göttingen; transcribed in Grattan-Guinness (1971a), 381.

35. Cantor to Klein, December 31, 1899, letter # 455.

36. Grattan-Guinness (1971a), 368–370.

37. Turn to the discussion in Chapter 11; and to Kowalewski (1950), 202.

38. For full details, refer to the analysis in Chapter 11; and to Schoenflies (1928), 560.

39. Grattan-Guinness (1971a), 368. From time to time, Cantor apparently visited other sanitaria, though no official records are known by which exact dates of admission or duration of stay may be documented for periods other than those listed here.

40. Cantor to Vally Guttmann Cantor, May 3, 1917; *Nachlass Cantor II.*

41. Landau to Vally Guttmann Cantor, January 8, 1918, *Nachlass Cantor XVII;* transcribed in Meschkowski (1967), 270.

42. Schoenflies (1927).

43. Bell (1937), chapter 29.

44. Russell (1967), 217.

45. Grattan-Guinness (1971a), 368–369.

46. Kowalewski (1950), 106.

47. Kowalewski (1950), 124.

48. Cantor to Graf von Posadowsky-Wehner, November 10, 1899; in Grattan-Guinness (1971a), 378–379.

49. Kowalewski (1950), 124.

50. Wangerin (1918), 2.

51. Schoenflies (1928), 561.

52. Schwarz to Noevius, October 13, 1888; Berlin, copybook 4, 67–70, archives of the Akademie der Wissenschaften der DDR, Berlin; transcribed in Grattan-Guinness (1971a), 377–378.

53. Georg Woldemar Cantor to Georg Cantor, January 21, 1863, *Nachlass Cantor I;* cited in Meschkowski (1967), 2: "Grillen sind mir böse Gäste/ immer mit leichtem Sinn/ tanzen durch' Leben hin,/ das nur ist Hochgewinn!" Note that *Grillen* may refer both to crickets and to a bout of depression.

54. Georg Woldemar Cantor to Georg Cantor, January 21, 1863, in *Nachlass Cantor I.*

55. Georg Cantor to Georg Woldemar Cantor, May 25, 1862; cited in Fraenkel (1930), 193; in Meschkowski (1967), 5.

56. Kowalewski (1950), 106.

57. Ivor Grattan-Guinness has pointed to this dichotomy of Cantor's personality in his (1971a), 358–359.

58. Peters (1961), 65. Compare such sentiments with the poem Cantor sent his wife shortly before his death: cited in note 40 above.

59. Cantor to P.E.B. Jourdain, March 29, 1905; notebook 1 of the two Jourdain notebooks in the archives of the Institut Mittag-Leffler, Djursholm, Sweden. Cantor's letter appears at folio 82. See transcriptions of the document in Grattan-Guinness (1971a), 384–385; Grattan-Guinness (1971b), 123–124.

60. Cantor (1905a). On April 5, 1905, Cantor wrote in English to Grace Chisholm Young, describing as he had to Jourdain the results of his being "hermetically secluded" in the Halle Nervenklinik: "As you know, I had been hermeticly secluded 5½ months (from 17. Sept. to 1. March) from the world, except few visites from my family. But I can not say, that I am by this long fire-baptism embittered because I do know the great pressure, that has been practiced by the 'Ministerium' and the 'amiable' german colleagues upon my wife and my children! Farther I had a great interest to study the quite unreasonable puerile treatment and soitdisant cure of the lamentable patients. The Muse afforded to me I employed to a renewed study of our Bible with opened eyes and postponing all prejudices. The result has been highly remarkable, as you will see by a little pamphlet (anonymous) of half a sheet, that I will send you perhaps in a week; it is now in the printing office. The title is: 'Ex Oriente Lux,' " in Grattan-Guinness (1971a), 385.

61. The full title read: "Ex Oriente Lux, Gespräche eines Meisters mit seinem Schüler über wesentliche Puncte des urkundlichen Christenthums. Berichtet vom Schüler selbst." Cantor (1895d).

62. Cantor to Grace Chisholm Young, June 20, 1908, transcribed in Meschkowski (1971), 30–34.

63. Cantor to Grace Chisholm Young, June 20, 1908; in Meschkowski (1971), 30–34. See as well Poincaré (1908), and the Youngs' review of Hobson in the *Mathematical Gazette* for 1928: Young (1928).

64. From a letter to Kurd Lasswitz, February 15, 1884; constituting Part I of Cantor (1887), 387.

65. See Cantor to Mittag-Leffler, November 6, 1884 (letter #33, archives of the Institut Mittag-Leffler, Djursholm, Sweden); transcribed in Grattan-Guinness (1970a), 78–80. Cantor also alluded to the applications of his theoretical advances of set theory, in particular his concept of *content,* in Part VI of the *Punktmannigfaltigkeitslehre:* Cantor (1884a), 235.

66. Cantor (1885a), 275.

67. Cantor (1885a), 275; Cantor to Eneström, November 4, 1885, in Cantor (1886), 373–374.

68. Cantor (1885a), 275–276. See as well Cantor's correspondence on the subject, in particular, Cantor to Mittag-Leffler, October 20, 1884, in the archives of the Institut Mittag-Leffler, Djursholm, Sweden.

69. Cantor to Mittag-Leffler, November 16, 1884; November 22, 1884, letter #37, in the archives of the Institut Mittag-Leffler.

70. Cantor (1885a), 276.

71. Cantor (1885a), 275.

72. Cantor (1883c), 205.

73. Cantor (1883c), 177.

74. See in particular Cantor's letter to Valson of January 31, 1886: Cantor (I), 43–44. Cantor also discussed his dissatisfaction with traditional mechanics and with mathematical physics as it was usually developed, in letters to Mittag-Leffler: October 20, 1884; September 22, 1884, letter #29, in the archives of the Institut Mittag-Leffler.

75. Cantor (1887), 380.

76. Cantor (1887), 421–422.

77. Cantor (1887), 422–423. The belief that set theory made it possible to establish higher unities among the most diverse entities was suggested early in the course of Cantor's research: Cantor (1882b), 152.

78. Cantor to Mittag-Leffler, September 22, 1884, letter #29, in the archives of the Institut Mittag-Leffler. Refer as well to Cantor (1884/1970), 85–86. Three other letters Cantor wrote to Mittag-Leffler are also relevant: October 20, 1884; November 16, 1884; November 22, 1884, letter #37, Institut Mittag-Leffler.

79. Cantor (1884/1970), 86.

80. Cantor wrote to Cardinal Franzelin on December, 17, 1885: Cantor (I), 39, and requested a thorough critique of his theories. Franzelin replied on December 25, 1885, and Cantor copied out the entire letter in his *Briefbuch:* Cantor (I), 39. Cantor wrote again to Franzelin on January 22, 1886: Cantor (I), 29–41, and on January 29, 1886: Cantor (I), 42. These last two letters were included as Parts III and IV of Cantor (1887), 399–400. Cardinal Franzelin died on December 11, 1886.

81. Cantor to Franzelin, January 22, 1886: Cantor (I), 39–41; in Cantor (1887), 339–400.

82. Cantor to Valson, January 31, 1886: Cantor (I), 43–44.

83. From the manuscript version of Cantor's "Zweite Mittheilung," kept in a bound volume with the other originals comprising volume 7 of *Acta Mathematica* for the year 1885. The passage appears on page 15 of the original manuscript and is preserved in the archives of the Institut Mittag-Leffler, Djursholm, Sweden.

84. Eneström to Cantor, April 16, 1884; the original is in the Kungl. Vetenskaps-akademiens Bibliothek, Stockholm, Sweden.

85. Cantor to Valson, January 31, 1886: Cantor (I), 43–44.

86. Cantor to the Abbé E. Blanc, May 22, 1887: Cantor (I), 123–124.

87. Cantor to Schmid, August 5, 1887: Cantor (I), 130.

88. Cantor to Schmid, August 5, 1887: Cantor (I), 130–131.

89. Cantor to Heman, July 28, 1887: Cantor (I), 128–129; Cantor to Schmid, August 5, 1887: Cantor (I), 131; Cantor to Jeiler, October 13, 1895: Cantor (II), 194–197, esp. pp. 196–197. Unfortunately, Cantor never seems to have prepared an exact account of his proof that the existence of eternal time, space, or matter was impossible, despite promises that he would. In one letter to Heman in which he discussed such matters, he led up to the critical proof only to break off his letter with the excuse that Heman, being on vacation at the time, should not be bothered with such details.

90. Cantor to Jeiler, October 27, 1895: Cantor (III), 8–9. Compare with similar sentiments in Cantor's letter to Killing, October 1, 1895: Cantor (II), 190–191.

91. Cantor to Esser, December 19, 1895: Cantor (III), 75–76; Cantor to Woker, December 30, 1895: Cantor (III), 88–90.

92. From Leo XIII: *Aeterni Patris,* cited in Cantor's letter to Esser, December 25, 1895: Cantor (III), 79.

93. Cantor (1883c), 175.

94. Cantor to Klein, February 7, 1883, letter #432, Universitätsbibliothek, Göttingen; Mittag-Leffler to Cantor, March 11, 1883, in the archives of the Institut Mittag-Leffler, Djursholm, Sweden.

95. When the *Jahrbuch über die Fortschritte der Mathematik* first began to include the literature of set theory (at Vivanti's behest) in 1894, it was reviewed under the heading "Philosophy"; in 1904 it was given a subsection between "Philosophy" and "Pädagogik"; only after the First World War was set theory given a separate, independent position in the *Jahrbuch*. For further details, see Fraenkel (1930), 215.

96. Cantor to Heman, June 21, 1888: Cantor (I), 179.

Bibliography

Alexander, A. 1880. "Thomas Aquinas and the Encyclical Letter," *The Princeton Review* (March), 245–261.

Angelelli, I. 1967. *Gottlob Frege: Kleine Schriften*. Hildesheim: Olms.

Aristotle. 1956. *The Metaphysics*. Trans. and ed. H. Tredennick. Cambridge, Mass.: Harvard University Press.

——— 1963. *The Physics*. Trans. and ed. P. H. Wicksteed and F. M. Cornford. Cambridge, Mass.: Harvard University Press.

Augustine, A. 1957. *The City of God against the Pagans*. Trans. and ed. G. E. McCracken. Cambridge, Mass.: Harvard University Press.

Ascoli, G. 1873. "Ueber trigonometrische Reihen," *Mathematische Annalen 6*, 231–240.

——— 1877. "Nuove Ricerche sulla Serie di Fourier," *Atti della R. Accademia dei Lincei: Memoire della Classe di Scienze Fisiche, Matematiche e Naturali 2*. 2, 584–651.

Bachmann, F. 1933, and H. Scholz: *Die logischen Grundlagen der Arithmetik*. Münster.

Bachmann, F. 1934. *Untersuchungen zur Grundlegung der Arithmetik*. Leipzig: Forschungen zur Logistik No. 1.

Bachmann, H. 1955. *Transfinite Zahlen*. Berlin: Springer.

Ballauff, L. 1883. "Recensionen: Georg Cantor. Grundlagen einer allgemeinen Mannigfaltigkeitslehre," *Zeitschrift für exakte Philosophie im Sinne des neuem philosophischen Realismus 12*, 375–395.

Bary, N. K. 1964. *A Treatise on Trigonometric Series*. Trans. M. Mullins. New York: Macmillan. 2 vol.

Becker, O. 1954. *Grundlagen der Mathematik in geschichtlicher Entwicklung*. München: K. A. Freiburg Verlag.

Bell, E. T. 1937. *Men of Mathematics*. New York: Simon and Schuster.

Benacerraf, P. 1964, and H. Putnam, eds. *Philosophy of Mathematics, Selected Readings*. New Jersey: Prentice-Hall.

Benardete, J. A. 1964. *Infinity: An Essay in Metaphysics*. Oxford: Clarendon Press.

Bendiek, J. 1965. "Ein Brief Georg Cantors an P. Ignatius Jeiler O.F.M.," *Franziskanische Studien 47,* 65–73.

Bendixson, I. 1883. "Quelques théorèmes de la théorie des ensembles de points," *Acta Mathematica 2,* 415–429.

Berg, J. 1962. *Bolzano's Logic.* Stockholm: Almqvist and Wiksell.

Bernstein, F. 1898. see Borel (1928), 102–107.

——— 1901. "Untersuchungen aus der Mengenlehre." *Dissertation,* Göttingen; *Mathematische Annalen 61,* 111–155.

——— 1905a. "Über die Reihe der transfiniten Ordnungszahlen," *Mathematische Annalen 60,* 187–193.

——— 1905b. "Zum Kontinuumproblem," *Mathematische Annalen 60,* 463–464.

——— 1905c. "Untersuchungen aus der Mengenlehre," *Mathematische Annalen 61,* 117–155.

Beth, E. W. 1959. *The Foundations of Mathematics.* Amsterdam: North-Holland Publishing Co.

Bettazzi, R. 1891. "Sull'infinitesimo attuale," *Rivista di Matematica 1,* 174–182.

——— 1892. "Sull' infinitesimo attuale," *Rivista di Matematica 2,* 38–41.

Biermann, K.-R. 1959a. "Über die Förderung deutscher Mathematiker durch Alexander von Humboldt." In *Alexander von Humboldt. Gedenkenschrift zur 100. Wiederkehr seines Todestages.* Berlin: Akademie-Verlag.

——— 1959b. "Johann Peter Gustav Lejeune Dirichlet. Dokumente für sein Leben und Wirken," *Abhandlungen der deutschen Akademie der Wissenschaften zu Berlin, Klasse für Mathematik, Physik und Technik* No. 2. Berlin.

——— 1968. *Die Mathematik und ihre Dozenten an der Berliner Universität. 1810–1920. Habilitationsschrift,* Berlin; published Berlin: Akademie-Verlag, 1973.

Biermann, O. 1887. *Theorie der analytischen Functionen.* Leipzig: B. G. Teubner.

Blumenthal, O. 1935. "Lebensgeschichte" in D. Hilbert: *Gesammelte Abhandlungen.* Berlin: J. Springer. *3,* 388–429.

Bolzano, B. 1851/1889. *Paradoxien des Unendlichen,* ed. F. Prihonský. Photoreproduction of the first edition of 1851. Berlin: Mayer and Müller, 1889.

——— 1905. "Rein analytischer Beweis des Lehrsatzes, dass zwischen je zwey Werthen, die ein entgegengesetztes Resultat gewähren, wenigstens eine reelle Wurzel der Gleichung liege." Ed. P. E. B. Jourdain for *Ostwald's Klassiker der exakten Wissenschaften* No. 153. Leipzig: W. Engelmann.

——— 1950. *Paradoxes of the Infinite.* Trans. D. A. Steele. New Haven: Yale University Press.

——— 1962. *Theorie der reellen Zahlen im Bolzano's handschriftlichen Nachlasse.* Ed. K. Rychlik. Prague: Verlag der Tschechoslowakischen Akademie der Wissenschaften.

Bonansea, B. M. 1954. "Pioneers of the Nineteenth Century Scholastic Revival in Italy," *The New Scholasticism 27,* 1–37.

Bonavenia, G. 1887. *Raccolta di memorie intorno alla vita dell'emo Cardinale Giovanni Battista Franzelin.* Roma: Tipografia Poliglott.

Borchardt, C. W. 1860. "Gustav Lejeune-Dirichlet," *Journal für die reine und angewandte Mathematik* (Crelle's *Journal) 57,* 91–92.

Borel, É. 1898. *Leçons sur la théorie des fonctions.* Paris: Gauthier-Villars.

——— 1904. *Bulletin de la société mathématique de France 33* (1905), 261–273.

——— 1905. Quelques remarques sur les principes de la théorie des ensembles," *Mathematische Annalen 60,* 194–195.

——— 1908. "Sur les principes de la théorie des ensembles." See Congress (1908). Volume II, 15–17.

——— 1928. *Leçons sur la théorie des fonctions*. 3rd ed. Paris: Gauthier-Villars.

——— 1949. *Eléments de la théorie des ensembles*. Paris: A. Michel.

Boutroux, P. 1905. "Correspondance mathématique et relation logique," *Revue de métaphysique et de morale 13,* 620–637. See also Poincaré (1905) and Russell (1905).

Boyer, C. B. 1939. *The Concepts of the Calculus*. New York: Columbia University Press. Reprinted under the title *The History of the Calculus and Its Conceptual Development*. New York: Dover, 1959.

——— 1968. *A History of Mathematics*. New York: Wiley and Sons.

Brouwer, L. E. J. 1911. "Beweis der Invarianz der Dimensionenzahl," *Mathematische Annalen 70,* 161–165.

——— 1913. "Über den natürlichen Dimensionsbegriff," *Journal für die reine und angewandte Mathematik* (Crelle's *Journal*) *142,* 146–152.

——— 1919. *Begründung der Mengenlehre unabhängig von logischen Satz vom Ausgeschlossenen Dritten*. Amsterdam: J. Muller.

Brunnhofer, H. 1882. *Giordano Bruno's Weltanschauung und Verhängnis*. Leipzig: Fues's Verlag.

Bunn, R. 1977. "Quantitative Relations between Infinite Sets," *Annals of Science 34,* 177–191.

Burali-Forti, C. 1897. "Una questione sui numeri transfiniti," *Rendiconti del circolo matematico di Palermo 11,* 154–164.

Burkhardt, H. 1901. "Entwicklung nach oscillirenden Functionen und Integration der Differentialgleichungen der mathematischen Physik," *Jahresbericht der Deutschen Mathematiker-Vereinigung 10* (1901–1908), 1–1804.

——— 1914. "Trigonometrische Reihen und Integrale (bis etwa 1850)," *Encyklopädie der mathematischen Wissenschaften* 2.7. Leipzig: B. G. Teubner, 819–1354.

Cahill, M. C. 1939. *The Absolute and the Relative in St. Thomas and in Modern Philosophy*. Washington: Catholic University of America Press.

Cantor, G. 1867. *De aequationibus secundi gradus indeterminatis. Dissertation,* Berlin: Schultz. In Cantor (1932), 1–31.

——— 1869. *De transformatione formarum ternariarum quadraticarum. Habilitationsschrift,* Halle: Hendel. In Cantor (1932), 51–62.

——— 1870a. "Über einen die trigonometrischen Reihen betreffenden Lehrsatz," *Journal für die reine und angewandte Mathematik* (Crelle's *Journal*) *72,* 130–138. In Cantor (1932), 71–79.

——— 1870b. "Beweis, dass eine für jeden reellen Wert von x durch eine trigonometrische Reihe gegebene Funktion $f(x)$ sich nur auf eine einzige Weise in dieser Form darstellen lässt," *Journal für die reine und angewandte Mathematik 72,* 139–142. In Cantor (1932), 80–83.

——— 1871a. "Notiz zu dem Aufsatze: Beweis, dass eine für jeden reellen Wert von x durch eine trigonometrische Reihe gegebene Funktion $f(x)$ sich nur auf eine einzige Weise in dieser Form darstellen lässt," *Journal für die reine und angewandte Mathematik 73,* 294–296. In Cantor (1932), 84–86.

——— 1871b. "Über trigonometrische Reihen," *Mathematische Annalen 4,* 139–143. In Cantor (1932), 87–91. Translated as Cantor (1871b/1883).

——— 1871b/1883. "Sur les séries trigonométriques," *Acta Mathematica 2,* 329–335.

——— 1871c. [A review of Hankel (1870)] *Literarisches Centralblatt* no. 7 (February 18), 150–151.

——— 1872. "Über die Ausdehnung eines Satzes aus der Theorie der trigonometrischen Reihen," *Mathematische Annalen 5,* 123–132. In Cantor (1932), 92–102. Translated as Cantor (1872/1883).

——— 1872/1883. "Extension d'un théorème de la théorie des séries trigonométriques," *Acta Mathematica 2,* 336–348.

——— 1874. "Über eine Eigenschaft des Inbegriffes aller reellen algebraischen Zahlen," *Journal für die reine und angewandte Mathematik 77,* 258–262. In Cantor (1932), 115–118. Translated as Cantor (1874/1883).

——— 1874/1883. "Sur une propriété du système de tous les nombres algébriques réels," *Acta Mathematica 2,* 305–310.

——— 1878. "Ein Beitrag zur Mannigfaltigkeitslehre," *Journal für die reine und angewandte Mathematik 84,* 242–258. In Cantor (1932), 119–133. Translated as Cantor (1878/1883).

——— 1878/1883. "Une contribution à la théorie des ensembles," *Acta Mathematica 2,* 311–328.

——— 1879a. "Über einen Satz aus der Theorie der stetigen Mannigfaltigkeiten," *Nachrichten von der Königlichen Gesellschaft der Wissenschaften und der Georg-Augusts-Universität zu Göttingen,* 127–135. In Cantor (1932), 134–138.

——— 1879b. "Ueber unendliche, lineare Punktmannigfaltigkeiten," Part 1, *Mathematische Annalen 15,* 1–7. In Cantor (1932), 139–145. Translated as Cantor (1879b/1883).

——— 1879b/1883. "Sur les ensembles infinis et linéares de points," *Acta Mathematica 2,* 349–356.

——— 1880a. "Bemerkung über trigonometrische Reihen," *Mathematische Annalen 16,* 113–114. In Cantor (1932), 102.

——— 1880b. "Fernere Bemerkung über trigonometrische Reihen," *Mathematische Annalen 16,* 267–269. In Cantor (1932), 104–106.

——— 1880c. "Zur Theorie der zahlentheoretischen Funktionen," *Mathematische Annalen 16,* 583–588. In Cantor (1932), 67–70.

——— 1880d. "Ueber unendliche, lineare Punktmannigfaltigkeiten," Part 2, *Mathematische Annalen 17,* 355–358. In Cantor (1932), 145–148. Translated as Cantor (1880d/1883).

——— 1880d/1883. "Sur les ensembles infinis et linéares de points," *Acta Mathematica 2,* 357–360.

——— 1882a. "Über ein neues und allgemeines Kondensationsprinzip der Singularitäten von Funktionen," *Mathematische Annalen 19,* 588–594. In Cantor (1932), 107–113.

——— 1882b. "Ueber unendliche, lineare Punktmannigfaltigkeiten," Part 3, *Mathematische Annalen 20,* 113–121. In Cantor (1932), 149–157. Translated as Cantor (1882b/1883).

——— 1882b/1883. "Sur les ensembles infinis et linéares de points," *Acta Mathematica 2,* 361–371.

——— 1883a. "Ueber unendliche, lineare Punktmannigfaltigkeiten," Part 4, *Mathematische Annalen 21,* 51–58. In Cantor (1932), 157–164. Translated as Cantor (1883a/1883).

——— 1883a/1883. "Sur les ensembles infinis et linéares de points," *Acta Mathematica 2,* 372–380.

——— 1883b. "Ueber unendliche, lineare Punktmannigfaltigkeiten," Part 5, *Mathematische Annalen 21,* 545–586. In Cantor (1932), 165–208. See as well Cantor (1883c).

——— 1883c. *Grundlagen einer allgemeinen Mannigfaltigkeitslehre. Ein mathematisch-philosophischer Versuch in der Lehre des Unendlichen.* Leipzig: B. G. Teubner. In Cantor (1932), 165–208. Translated, in part, into French as Cantor (1883d).An English translation has recently appeared: Cantor (1976). See the warning about both translations in note 8 to Chapter 5.

——— 1883d. "Fondaments d'une théorie générale des ensembles," *Acta Mathematica 2,* 381–408.

——— 1883e. "Sur divers théorèmes de la théorie des ensembles de points situés dans un espace continu à *n* dimensions" (Première Communication) *Acta Mathematica 2,* 409–414. In Cantor (1932), 247–251.

——— 1884a. "Ueber unendliche, lineare Punktmannigfaltigkeiten," Part 6, *Mathematische Annalen 23,* 453–488. In Cantor (1932), 210–244.

——— 1884b. "De la puissance des ensembles parfaits de points," *Acta Mathematica 4,* 381–392.

——— 1884c. (A review of Cohen (1883)) *Deutsche Literaturzeitung* No. 8 (Berlin: February 23), 266–268.

——— 1885a. "Über verschiedene Theoreme aus der Theorie der Punktmengen in einem *n*-fach ausgedehnten stetigen Raume G_n," (Zweite Mittheilung) *Acta Mathematica 7,* 105–124. In Cantor (1932), 261–276.

——— 1885b. "Über die verschiedenen Ansichten in bezug auf die actualunendlichen Zahlen," *Bihang till Kongl. Svenska Vetenskaps-Akademiens Handlingar 11,* No. 19.

——— 1885c. "Ludwig Scheefer," *Bibliotheca Mathematica 1,* 197–199. In Cantor (1932), 368–369.

——— 1885d. Review of Frege (1884). *Deutsche Litteraturzeitung* No. 20 (Berlin: May 16), 728–729. In Cantor (1932), 440–441.

——— 1886a. "Über die verschiedenen Standpunkte in bezug auf das aktuelle Unendliche," *Zeitschrift für Philosophie und philosophische Kritik 88,* 224–233. In Cantor (1932), 370–376. This paper also appears as part of Cantor (1890).

——— 1887. "Mitteilungen zur Lehre vom Transfiniten," *Zeitschrift für Philosophie und philosophische Kritik 91* (1887), 81–125; *92* (1888), 240–265. In Cantor (1932), 378–439. Both articles were reissued as part of Cantor (1890).

——— 1889. "Bemerkung mit Bezug auf den Aufsatz: Zur Weierstrass-Cantorschen Theorie der Irrationalzahlen," *Mathematische Annalen 33,* 476. In Cantor (1932), 114.

——— 1890. *Gesammelte Abhandlungen zur Lehre vom Transfiniten.* Halle: C. E. M. Pfeffer.

——— 1891. "Über eine elementare Frage der Mannigfaltigkeitslehre," *Jahresbericht der Deutschen Mathematiker-Vereinigung 1,* 75–78. In Cantor (1932), 278–280.

——— 1894. "Vérification jusqu'a 1000 du théorème empirique de Goldbach," *Congrès de Caen, Séance du 10 Août, 1894. Association Françiase pour L'Avancement des Sciences*. Paris: Chaix.

——— 1895a. "Beiträge zur Begrundung der transfiniten Mengenlehre," Part I, *Mathematische Annalen 46,* 481–512. In Cantor (1932), 282–311. Translated as Cantor (1895c). The "Beiträge" was continued in Cantor (1897). For translations of both parts together, see Cantor (1899) and Cantor (1915).

——— 1895b. "Sui numeri transfinite," *Rivista di Matematica 5,* 104–109.

——— 1895c. "Contribuzione al fondamento della teoria degli insiemi transfiniti." Trans. F. Gerbaldi. *Rivista di Matematica 5,* 129–162.

——— 1897. "Beiträge zur Begründung der transfiniten Mengenlehre," Part II, *Mathematische Annalen 49,* 207–246. In Cantor (1932), 312–351. For translations of both Cantor (1895a) and Cantor (1897), see Cantor (1899) and Cantor (1915).

——— 1899. *Sur les fondements de la théorie des ensembles transfinis*. Trans. F. Marotte. Paris: Hermann.

——— 1905a. *Ex Oriente Lux, Gespräche eines Meisters mit seinem Schüler über wesentliche Puncte des urkundlichen Christenthums. Berichtet vom Schüler selbst.* Halle: C. E. M. Pfeffer.

——— 1905b. "Ein Brief von Carl Weierstrass über das Dreikörperproblem," *Rendiconti del Circolo Matematico di Palermo 19,* 305–308.

——— 1915. *Contributions to the Founding of the Theory of Transfinite Numbers*. Trans. P. E. B. Jourdain. Chicago: Open Court.

——— 1932. *Gesammelte Abhandlungen mathematischen und philosophischen Inhalts*. Ed. E. Zermelo. Berlin: J. Springer; reprinted Hildesheim: Olms, 1966.

——— 1970. "Principien einer Theorie der Ordnungstypen," (Erste Mittheilung). Ed. I. Grattan-Guinness: *Acta Mathematica 124,* 65–107. The paper was originally dated November 6, 1884.

——— 1976."Foundations of the Theory of Manifolds." Trans. U. Parpart. *The Campaigner* (The Theoretical Journal of the National Caucus of Labor Committees) *9* (January and February), 69–96.

Cantor/Dedekind. 1937. *Briefwechsel Cantor-Dedekind*. Eds. E. Noether and J. Cavaillès. Paris: Hermann.

Cantor, Moritz. 1879. "Hankel," *Allgemeine Deutsche Biographie 10,* 516–519. Reprinted Berlin: Duncker und Humblot, 1968.

Carathéodory, C. 1943. *Vorlesungen über reele Funktionen*. 2nd ed. New York: Chelsea, 1948.

Cassirer, E. 1950. *The Problem of Knowledge. Philosophy, Science, and History since Hegel*. Trans. W. H. Woglom and C. W. Hendel. New Haven: Yale University Press.

Cauchy, A. L. 1823a. "Mémoire sur les développements des fonctions en séries périodiques," Mémoires de l'Académie des Sciences 6 (1823: published 1827), 603–612; in *Oeuvres complètes 2* (1). Paris: Gauthier-Villars, 1908, 12–19.

——— 1823b. *Résumé des leçons données à l'École Royale Polytechnique sur le calcul infinitésimal*. Vol. I. Paris: Debure. See also Cauchy's *Oeuvres complètes 4* (2). Paris: Gauthier-Villars, 1899, 5–261.

——— 1868. *Sept leçons de physique générale*. Paris: Gauthier-Villars.

Cavaillès, J. 1962. *Philosophie mathématique*. Paris: Hermann.

Cohen, H. 1883. *Das Princip der infinitesimal Methode und seine Geschichte*. Berlin: F. Dümmlers Verlag; reprinted Frankfurt/M.: Suhrkamp Verlag, 1968.

Cohen, P. J. 1963. "The Independence of the Continuum Hypothesis; I, II," *Proceedings of the National Academy of Sciences, U.S.A. 50* (1963), 1143–1148; *51* (1964) 105–110.

——— 1966. *Set Theory and the Continuum Hypothesis*. New York: W. A. Benjamin, Inc.

Collins, J. 1961. "Leo XIII and the Philosophical Approach to Modernity." In E. T. Gargan, ed., *Leo XIII and the Modern World*. New York: Sheed and Ward, 181–209.

Congress. 1893. *Mathematical Papers Read at the International Mathematical Congress Held in Connection with the World's Columbian Exposition. Chicago*. New York: Macmillan and Co., 1896.

Congress. 1897. *Verhandlungen des ersten internationalen Mathematiker-Kongresses in Zürich vom 9. bis 11. August*. Ed. F. Rudio. Leipzig: B. G. Teubner, 1898.

Congress. 1900. *Compte rendu du deuxième congrès international des mathématiciens tenu à Paris du 6 au 12 aout*. Paris: Gauthier-Villars, 1902.

Congress. 1904. *Verhandlungen des dritten internationalen Mathematiker-Kongresses in Heidelberg vom 8. bis 13. August*. Ed. A. Krazer. Leipzig: B. G. Teubner, 1905.

Congress. 1908. *Atti del IV Congresso Internazionale dei Matematici. Rome, 6–11 April*. Rome: Tipografia della R. Accademia dei Lincei, C. V. Salviucci, 1909.

Cornoldi, G. M. 1889. *La filosofia scolastica di san Tommaso e di Dante*. Rome: A. Befani.

Courant, R. 1926. "Bernhard Riemann und die Mathematik der letzten hundert Jahre," *Die Naturwissenschaften 14*, 813–818; 1265–1277.

Crossley, J. 1973. "A Note on Cantor's Theorem and Russell's Paradox." *Australian Journal of Philosophy 51*, 70–71.

Dauben, J. 1971a. "The Trigonometric Background to Georg Cantor's Theory of Sets," *Archive for History of Exact Sciences 7*, 181–216.

——— 1971b. "The Irrational and Transfinite Numbers: Georg Cantor's Philosophy of Mathematics," *Proceedings of the XIIIth International Congress for the History of Science, August, 1971. 5* (Moscow: Nauka, 1974), 86–93.

——— 1974a. "Denumerability and Dimension: The Origins of Georg Cantor's Theory of Sets," *Rete 2*, 105–134.

——— 1974b. " 'Hypotheses non fingo': Theological Dimensions of Cantorian Set Theory," *Proceedings of the XIVth International Congress of the History of Science, August, 1974*. Tokyo, in press.

——— 1975. "The Invariance of Dimension: Problems in the Early Development of Set Theory and Topology," *Historia Mathematica 2*, 273–288.

——— 1977a. "Georg Cantor and Pope Leo XIII: Mathematics, Theology, and the Infinite," *Journal of the History of Ideas 38*, 85–108.

——— 1977b. "C.S. Peirce's Philosophy of Infinite Sets," *Mathematics Magazine 50*. 3 (May), 123–135.

Dedekind, R. 1872. *Stetigkeit und irrationale Zahlen*. 2nd ed. Braunschweig: Vieweg, 1892; in Dedekind (1930–1932) *3,* 315–334.

——— 1876. "Bernhard Riemann." In Riemann (1876b), 539–558.

——— 1888. *Was sind und was sollen die Zahlen?* Braunschweig: Vieweg. In Dedekind (1930–1932) *3,* 335–391. Translated as Dedekind (1901).

——— 1901. *Essays on the Theory of Numbers, Continuity of Irrational Numbers, the Nature and Meaning of Numbers*. Trans. W. W. Beman. Chicago: Open Court. Reprinted, New York: Dover, 1963.

——— 1930–1932. *Gesammelte mathematische Werke,* eds. R. Fricke, E. Noether, O. Ore. Vieweg: Braunschweig, in 3 volumes.

Deiters, H. 1960. "Wilhelm von Humboldt als Gründer der Universität Berlin," *Forschen und Wirken, Festschrift zur 150-Jahr-Feier der Humboldt-Universität zu Berlin. 1810–1960*. Vol. I, Berlin, 15–39.

——— 1967. "Wilhelm von Humboldts Ansichten vom Wesen der Universität," *Wilhelm von Humboldt, Beiträge vorgelegt von der Humboldt-Universität zu Berlin anlösslich der Feier des zweihundertsten Geburtstages ihres Gründers*. Halle: M. Niemeyer Verlag, 129–161.

Désiré, N. Y. S. 1908. "La Nature de l'espace d'après les théories modernes depuis Descartes," *Memoires, Académie Royale de Belgique: Classe des Lettres et des Sciences morales et politiques et Classe des Beaux-Arts, III*.

Dini, U. 1878. Fondamenti per la Teorica delle Funzioni di Variabili Reali. Pisa: T. Nistri. Trans. as Dini (1892).

——— 1880. *Serie di Fourier e altre rappresentazioni analitiche delle funzioni di una variabili reale*. Pisa: T. Nistri.

——— 1892. *Grundlagen für eine Theorie der Functionen einer veränderlichen reellen Grösse*. Trans. J. Lüroth and A. Schepp. Leipzig: B. G. Teubner.

Dirichlet, P. G. L. 1825. "Mémoire sur l'impossibilité de quelques équations indéterminées du cinquième degré," *Journal für die reine und angewandte Mathematik* (Crelle's *Journal*) *3* (1828) 354–375. In vol. I of Dirichlet (1899), 23–28; with additions, 29–46.

——— 1829. "Sur la convergence des séries trigonométriques qui servent à représenter une fonction arbitraire entre des limites données," *Journal für die reine und angewandte Mathematik 4,* 157–169; also Dirichlet (1889), 117–132.

——— 1837a. "Sur les séries dont le terme général dépend de deux angles, et qui servent à exprimer des fonctions arbitraires entre des limites données," *Journal für die reine und angewandte Mathematik 17,* 35–56; also Dirichlet (1889), 283–306.

——— 1837b. "Ueber die Darstellung ganz willkuhrlicher Funktionen durch Sinus– und Cosinusreihen." In Dove (1837), 152–174; also Dirichlet (1889), 133–160.

——— 1837c. "Beweis des Satzes, dass jede unbegrenzte arithmetische Progression, deren erstes Glied und Differenz ganze Zahlen ohne gemeinschaftlichen Factor sind, unendlich viele Primzahlen enthält," *Abhandlungen der Königlich Preussischen Akademie der Wissenschaften* 45–81; in Dirichlet (1889), 315–342.

——— 1889 and 1897. *G. Lejeune Dirichlet's Werke*. Ed. L. Kronecker and L. Fuchs. Vol. I. Berlin: Georg Reimer, 1889; Vol. II. Berlin: Georg Reimer, 1897.

Dorlodot, Henry de. 1921. *Le Darwinisme au point de vue de l'orthodoxie catholique*. Bruxelles: Vromant & Co.; Trans. E. Messenger, *Darwinism and Catholic Thought*. London: Burns, Oates and Washbourne Ltd., 1922.

Dove, H. W. 1837 and L. Moser, eds. *Repertorium der Physik, Enthaltend eine volständige Zusammenstellung der neuern Fortschritte dieser Wissenschaft*. Vol. I. Berlin: Viet and Co.

Du Bois-Reymond, P. 1871. "Sur la grandeur relative des infinis des fonctions," *Annali di Matematici 4*, 338–353.

——— 1872. "Théorème général concernant la grandeur relative des infinis des fonctions et de leurs dérivées," *Journal für die reine und angewandte Mathematik* (Crelle's *Journal*) *74*, 294–304.

——— 1875a. "Versuch einer Classification der willkürlichen Functionen reeller Argumente nach ihren Aenderungen in den kleinsten Intervallen," *Journal für die reine und angewandte Mathematik 79*, 21–37.

——— 1875b. "Ueber asymptotische Werthe, infinitäre Approximationen und infinitäre Auflösung von Gleichungen," *Mathematische Annalen 8*, 363–414; 574–576.

——— 1876. "Notiz über infinitäre Gleichheiten," *Mathematische Annalen 10*, 576–578.

——— 1877. "Ueber die Paradoxen des Infinitärcalcüls," *Mathematische Annalen 11*, 149–167.

——— 1879. "Erläuterungen zu den Anfangsgründen der Variationsrechnung," *Mathematische Annalen 15*, 283–314.

——— 1880a. "Der Beweis des Fundamentalsatzes der Integralrechnung: $\int_a^b F'(x)\,dx = F(b) - F(a)$," *Mathematische Annalen 16*, 115–128.

——— 1880b. *Zur Geschichte der trigonometrische Reihen, eine Entgegnung*. Tübingen: H. Laupp.

——— 1882. *Die allgemeine Functionentheorie*. Tübingen: H. Laupp.

Dugac, P. 1970. "Charles Méray (1835–1911) et la notion de limite." *Revue d'histoire des sciences et de leurs applications 23*, 333–350.

——— 1973. "Éléments d'analyse de Karl Weierstrass," *Archive for History of Exact Sciences 10*, 41–176.

——— 1976a. *Richard Dedekind et les fondements des mathématiques*. Collection des travaux de l'Académie internationale d'Histoire des Sciences No. 24, Paris: J. Vrin.

——— 1976b. "Problèmes d'histoire de l'analyse mathématique au XIXème siècle. Cas de Karl Weierstrass et de Richard Dedekind," *Historia Mathematica 3*, 5–19.

Eccarius, W. 1976. "August Leopold Crelle als Herausgeber wissenschaftlicher Fachzeitschriften," *Annals of Science 33*, 229–261; also printed as "August Leopold Crelle als Herausgeber des Crelleschen Journals," *Journal für die reine und angewandte Mathematik* (Crelle's *Journal*) *286/287*, 5–25.

Ecker, A. 1883. *Lorenz Oken: A Biographical Sketch*. Trans. A. Tulk. London: Kegan Paul, French & Company.

Ehrle, F. 1954. *Zur Enzyklika "Aeterni Patris": Text und Kommentar*. Ed. F. Pelster. Rome: Edizioni di Storia e Literatura.

Eisele, C, ed. 1976. *The New Elements of Mathematics by Charles S. Peirce*. The Hague: Mouton.

Esser, T. 1895. *Die Lehre des hl. Thomas von Aquino über die Möglichkeit einer anfangslosen Schöpfung*. Münster: Aschendorffschebuchhandlung.

Essler, W. K. 1964. *Aufzählbarkeit und Cantorsches Diagonalverfahren: Unter-*

suchungen zu Grundfragen der Logik. München: *Inaugural Dissertation*.

Euler, L. 1748. *Introductio in analysin infinitorum*. Lausanne: Bousquet. In *Leonhardi Euleri Opera Omnia*. Eds. A. Krazer and F. Rudio, *8* and *9*.1. Leipzig: B. G. Teubner, 1922.

Fang, J. 1976. *The Illusory Infinite. A Theology of Mathematics*. Memphis, Tennessee: Paideia.

Fechner, G. T. 1855. *Über die physikalische und philosophische Atomenlehre*. Leipzig: H. Mendelssohn. Rev. ed., Leipzig: H. Mendelssohn, 1864.

Fraenkel, A. 1930. "Georg Cantor," *Jahresbericht der Deutschen Mathematiker-Vereinigung 39,* 189–266. This biography also appears separately as *Georg Cantor*. Leipzig: B. G. Teubner, and is reprinted in an abridged version in Cantor (1932), 452–483.

——— 1935. "Zum Diagonalverfahren Cantors," *Fundamenta Mathematicae 25,* 45–50.

——— 1953. *Abstract Set Theory*. Amsterdam: North-Holland Publishing Company.

——— 1960. "Jewish Mathematics and Astronomy," *Scripta Mathematica 25,* 33–47.

——— 1967. *Lebenskreise: Aus den Erinnerungen eines jüdischen Mathematikers*. Stuttgart: Deutsche Verlags-Anstalt.

Frege, G. 1879. *Begriffsschrift, eine der arithmetischen nachgebildete Formelsprache des reinen Denkens*. Halle: L. Nebert. Reprinted, Hildesheim: Olms, 1964.

——— 1884. *Die Grundlagen der Arithmetik*. Breslau: Wilhelm Koebner.

——— 1884/1950. *The Foundations of Arithmetic*. Trans. J. Austin. Oxford: B. Blackwell.

——— 1885a. "Über formale Theorien der Arithmetik," *Sitzungsberichte der Jenaischen Gesellschaft für Medizin und Naturwissenschaften für d. Jahr 1885. 17. Juli*. Jena: G. Fischer, 94–104. In Frege (1967), 103.

——— 1885b. "Rezension von: H. Cohen, Das Prinzip der Infinitesimal-Methode und seine Geschichte," *Zeitschrift für Philosophie und philosophisches Kritik 87,* 324–329. In Frege (1967), 99–102.

——— 1885c. "Erwiderung auf Cantor's *Rezension,*" *Deutsche Literaturzeitung 6,* No. 28, 1030. In Frege (1967).

——— 1890. "Entwurf zu einer Besprechung von Cantor's *Gesammelten Abhandlungen zur Lehre vom Transfiniten*." Dated between 1890–1892. In Hermes (1969), 76–80.

——— 1892. "Rezension von: Georg Cantor. Zum Lehre vom Transfiniten," *Zeitschrift für Philosophie und philosophische Kritik 100,* 269–272. In Frege (1967), 163–166.

——— 1893. *Grundgesetze der Arithmetik, begriffsschriftlich abgeleitet,* I. Jena: Verlag Hermann Pohle. Reprinted, Hldesheim: Olms, 1962.

——— 1893/1964. *The Basic Laws of Arithmetic*. Trans. M. Furth. Berkeley, California: University of California Press.

——— 1903. *Grundgesetze der Arithmetik, begriffsschriftlich abgeleitet,* II. Jena: Verlag Hermann Pohle. Reprinted, Hildesheim: Olms, 1962.

——— 1967. *Kleine Schriften*. Ed. I. Angelelli. Hildesheim: Olms.

——— 1969. *Nachgelassene Schriften*. Ed. H. Hermes, F. Kambartel, F. Kaulbach. Hamburg: Felix Meiner Verlag.

Freudenthal, H. 1975. "The cradle of modern topology, according to Brouwer's inedita." *Historia Mathematica 2,* 495–502. Note that the pages are incorrectly numbered, and should be read in the sequence 495, 500, 496, 501, 502.

Galland, J. 1880. *Papst Leo XIII. Festschrift zum goldenen Priester-Jubiläum des h. Vaters.* Paderborn: Schöningh.

Gauss, K. F. 1860. *Briefwechsel zwischen C. F. Gauss und H. C. Schumacher.* Ed. C. A. F. Peters. Vol. II, Altona: G. Esch.

Gericke, H. 1966. "Aus der Chronik der Deutschen Mathematiker-Vereinigung," *Jahresbericht der Deutschen Mathematiker-Vereinigung 68,* 46–74.

——— 1970. *Geschichte des Zahlbegriffs.* Mannheim: Hochschultaschenbücher Verlag.

Gödel, K. 1931. "Ueber formal unentscheidbare Sätze der Principia Mathematica und verwandter Systeme," *Monatshefte für Mathematik und Physik 38,* 173–198.

——— 1936. *The Consistency of the Axiom of Choice and of the Generalized Continuum Hypothesis with the Axioms of Set Theory.* 4th printing, Princeton: Princeton University Press, 1958.

——— 1940. *The Consistency of the Continuum Hypothesis.* Princeton: Princeton University Press.

——— 1947. "What is Cantor's Continuum Problem?" *American Mathematical Monthly 54,* 515–525. Rev. and expanded in Benacerraf (1964).

Grattan-Guinness, I. 1969. "Joseph Fourier and the Revolution in Mathematical Physics," *Journal of the Institute of Mathematics and its Applications 5,* 230–253.

——— 1970a. "An Unpublished Paper by Georg Cantor: 'Principien einer Theorie der Ordnungstypen. Erste Mittheilung.' " *Acta Mathematica 124,* 65–107.

——— 1970b. *The Development of the Foundations of Mathematical Analysis from Euler to Riemann.* Cambridge, Mass.: MIT Press.

——— 1971a. "Towards a Biography of Georg Cantor," *Annals of Science 27,* 345–391 and plates xxv–xxviii.

——— 1971b. "The Correspondence between Georg Cantor and Philip Jourdain," *Jahresbericht der Deutschen Mathematiker-Vereinigung 73,* 111–130.

——— 1971c. "Some Remarks on Cantor's Published and Unpublished Work on Set Theory," *NTM-Schriftenreihe für Geschichte der Naturwissenschaften, Technik, und Medizin 8,* 1–8.

——— 1971d. "Missing Materials Concerning the Life and Work of Georg Cantor," *Isis 62,* 516–517.

——— 1972a. "Bertrand Russell on His Paradox and the Multiplicative Axiom. An Unpublished Letter to Philip Jourdain," *Journal of Philosophical Logic 1,* 103–110.

——— 1972b. "University Mathematics at the Turn of the Century. Unpublished Recollections of W. H. Young," *Annals of Science 28,* 369–384.

——— 1972c. "A Mathematical Union: William Henry and Grace Chisholm Young," *Annals of Science 29,* 105–186.

——— 1972d. *Joseph Fourier: 1768–1830.* Written with J. Ravetz. Cambridge, Mass.: MIT Press.

——— 1974. "The Rediscovery of the Cantor-Dedekind Correspondence," *Jahresbericht der Deutschen Mathematiker-Vereinigung 76,* 104–139.

Greenwood, T. 1944. "La nature du transfini," *Revue de l'Université d'Ottawa 14,* 109–134.

——— 1945. "La nature du transfini," *Revue de l'Université d'Ottawa 15,* 147–185.

Gutberlet, C. 1878. *Das Unendliche metaphysisch und mathematisch betrachtet*. Mainz: F. Kirchheim.

——— 1886. "Das Problem des Unendlichen," *Zeitschrift für Philosophie und philosophische Kritik 88*, 179–223.

——— 1919. "Rezension von: A. Fraenkel. Einleitung in die Mengenlehre," *Philosophisches Jahrbuch der Görres-Gesellschaft 32*, 364–370.

Gutzmer, A. 1904. *Geschichte der Deutschen Mathematiker-Vereinigung von ihrer Begründung bis zur Gegenwart dargestellt*. Leipzig: B. G. Teubner.

——— 1909. "Geschichte der Deutschen Mathematiker-Vereinigung," *Jahresbericht der Deutschen Mathematiker-Vereinigung 10*, 1–49. Gutzmer's report, made to the Third International Congress in Heidelberg, 1903 [See Congress (1903), 34–35], was printed in full by Teubner: Gutzmer (1904).

——— 1918. "Georg Cantor," *Illustrierte Zeitung 150*, 86.

Hadamard, J. 1897. "Sur certaines applications possibles de la théorie des ensembles." In Congress (1897), 201–202.

Hahn, H. 1934. "Gibt es Unendliches?" *Alte Probleme—Neue Lösungen in den exakten Wissenschaften*. Leipzig: F. Denticke.

Hankel, H. 1867. *Vorlesungen über die complexen Zahlen und ihre Functionen. 1. Theil: Theorie der complexen Zahlensysteme*. Leipzig: L. Voss.

——— 1870 and 1905. *Untersuchungen über die unendlich oft oszillierenden und unstetigen Funktionen*. Tübingen: *Universitätsprogramm*, F. Fues. Also ed. P. E. B. Jourdain for *Ostwald's Klassiker der exakten Wissenschaften* No. 153. Leipzig: W. Engelmann, 1905, 44–102.

——— 1871. "Grenze," *Allgemeine Encyklopädie der Wissenschaften und Künste*. Eds. J. S. Ersch and J. G. Gruber. Leipzig: F. A. Brocthaus, 185–211.

Hardy, G. H. 1954. *Orders of Infinity. The "Infinitärcalcül" of Paul du Bois-Reymond*. 1st ed. 1910; rev. 1924; reprinted, Cambridge, England: at the University Press, 1954.

Harnack, A. 1880. "Ueber die trigonometrische Reihe und die Darstellung willkürlicher Functionen," *Mathematische Annalen 17*, 123–132.

——— 1881. *Die Elemente der Differenzial- und Integralrechnung*. Leipzig: B.G. Teubner; translated as Harnack (1891).

——— 1882a. "Vereinfachung der Beweise in der Theorie der Fourier'schen Reihe," *Mathematische Annalen 19*, 235–279.

——— 1882b. "Berichtigung zu dem Aufsatze: 'Ueber die Fourier'sche Reihe,'" *Mathematische Annalen 19*, 524–528.

——— 1885. "Ueber den Inhalt von Punktmengen," *Mathematische Annalen 25*, 241–250.

——— 1886. "Bemerkung zur Theorie des Doppelintegrales," *Mathematische Annalen 26*, 566–568.

——— 1891. *An Introduction to the Study of the Elements of the Differential and Integral Calculus*. Trans. G. Cathcart. London: Williams and Norgate.

Hauch-Tausböll, T. 1937. A German translation of the Cantor's family history, prepared by the Danish Genealogical Institute, Copenhagen, and preserved in *Nachlass Cantor I*. See Grattan-Guinness (1971a), 351.

Hausdorff, F. 1907. "Über dichte Ordnungstypen," *Jahresbericht der Deutschen Mathematiker-Vereinigung 16*, 541–546.

——— 1908. "Grundzüge einer Theorie der geordneten Mengen," *Mathematische Annalen 65,* 435–505.

——— 1914. *Grundzüge der Mengenlehre.* Leipzig: von Veit.

Hawkins, T. 1970. *Lebesgue's Theory of Integration: Its Origins and Development.* Madison: University of Wisconsin Press. Reprinted New York: Chelsea Publishing Company, 1975.

Hayes, C. J. H. 1940. *A Generation of Materialism.* New York: Harper and Row.

Heijenoort, J. van, ed. 1967. *From Frege to Gödel: A Source Book in Mathematical Logic, 1879–1931.* Cambridge, Mass.: Harvard University Press.

Heine, E. 1870. "Ueber trigonometrische Reihen," *Journal für die reine und angewandte Mathematik* (Crelle's *Journal*) *71,* 353–365.

——— 1872. "Die Elemente der Functionenlehre," *Journal für die reine und angewandte Mathematik 74,* 172–188.

Helmholtz, H. 1887. "Zählen und Messen erkenntnisstheoretisch betrachtet." In *Philosophische Aufsätze. Eduard Zeller zu seinem fünfzigjährigen Doctor-Jubiläum gewidmet.* Leipzig: Fues's Verlag, 14–52; also reprinted in *Hermann v. Helmholtz. Schriften zur Erkenntnistheorie.* Eds. P. Hertz and M. Schlick. Berlin: Springer Verlag, 1921, 70–108.

Herbart, J. F. 1850. *Johann Friedrich Herbart's sämmtliche Werke.* Ed. G. Hartenstein. Leipzig: L. Voss, 1850–1852.

Hermes, H., *et al.* 1969. *Gottlob Frege: Nachgelassene Schriften.* Hamburg: Felix Meiner.

Hermite, C. 1873. "Sur la fonction exponentielle," *Comptes rendus de l'Académie des Sciences 77,* 18–24; 74–79; 226–233; 285–293. Published separately, Paris: Gauthier-Villars, 1874. Also in Hermite (1912), volume *3,* 150–181.

——— 1912. *Oeuvres de Charles Hermite.* Ed. É. Picard. Paris: Gauthier-Villars.

Hessenberg, G. 1906. *Grundbegriffe der Mengenlehre.* Göttingen: Banderhoeck and Ruprecht.

——— 1908. "Willkürliche Schöpfungen des Verstandes?" *Jahresbericht der Deutschen Mathematiker-Vereinigung 17,* 145–162.

Hilb, E. 1923, and M. Riesz: "Neuere Untersuchungen über trigonometrische Reihen," *Encyklopädie der mathematischen Wissenschaften 2,* Part 3, second half, section 10. Leipzig: B. G. Teubner, 1923–1927, 1189–1228.

Hilbert, D. 1899. *Grundlagen der Geometrie.* Leipzig: Teubner.

——— 1900a. "Sur les Problèmes futurs des mathématiques." Trans. M.L. Laugel. See Congress (1900).

——— 1900b. "Über den Zahlbegriff," *Jahresbericht der Deutschen Mathematiker-Vereinigung 8,* 180–184.

——— 1909. "Hermann Minkowski," *Nachrichten von der Könglichen Gesellschaft der Wissenschaften zu Göttingen. Geschäftliche Mitteilungen,* 72–101.

Hill, L. 1933. "Fraenkel's Biography of Georg Cantor," *Scripta Mathematica 2,* 41–47.

Hofmann, J. E. 1959. "Alexander von Humboldt in seiner Stellung zur reinen Mathematik und ihrer Geschichte." In *Alexander von Humboldt. Gedenkenschrift zur 100. Wiederkehr seines Todestages.* Berlin: Akademie Verlag, 237–287.

Hontheim, J. 1893. *Institutiones theodicaeae, sive Theologiae naturalis secundum principia S. Thomae Aquinatis ad usum scholasticum.* Friburgi Brisgoviae: Herder.

——— 1895. *Der logische Algorithmus in seinem Wesen, in seiner Anwendung und in seiner philosophischen Bedeutung.* Berlin: F. L. Dames.

Hurewicz, W. 1948, and H. Wallman. *Dimension Theory.* Princeton: Princeton University Press.

Hurwitz, A. 1897. "Über die Entwickelung der allgemeinen Theorie der analytischen Funktionen in neuerer Zeit." See Congress (1897).

Hugon, E. 1924. "Les Services rendus à la cause Thomiste par son Eminence le Cardinal Mercier," *Revue Thomiste 7* (July–August), 333–339.

Isenkrahe, C. 1885. "Das Unendliche in der Ausdehnung. Sein Begriff und seine Stützen," *Zeitschrift für Philosophie und philosophische Kritik 86,* 73–111, 145–198.

——— 1920. *Untersuchungen über das Endliche und das Unendliche, mit Ausblicken auf die philosophische Apologetik.* Bonn: Marcus and Weber.

Jacobi, F. H. 1785. *Ueber die Lehre des Spinoza in Briefen an den Herrn Moses Mendelssohn.* Breslau: G. Löwe.

——— 1786. *Wider Mendelssohns beschuldigungen betreffend die Briefe über die Lehre des Spinoza.* Leipzig: G. J. Goeschen.

Jacobsthal, E. 1907. "Vertauschbarkeit transfiniter Ordnungszahlen," *Mathematische Annalen 64,* 475–488.

——— 1908. "Über den Aufbau der transfiniten Arithmetik," *Mathematische Annalen 66,* 145–194.

Jahresbericht. 1891. *Jahresbericht der Deutschen Mathematiker-Vereinigung 1* (1890–1891).

Jahresbericht. 1903. *Jahresbericht der Deutschen Mathematiker-Vereinigung 13.*

Jahresbericht. 1911. *Jahresbericht der Deutschen Mathematiker-Vereinigung 20.*

Jansen, B. 1926. "The Neo-Scholastic Movement in Germany." In Zybura (1926), 250–275.

Jeiler, I. 1897. *S. Bonaventurae principia de concursu Dei generali ad actiones causarum secundarum collecta et S. Thomae doctrina confirmata.* Ad Claras Aquas: Collegii S. Bonaventurae.

Jentsch, W. 1976. "Über ein Hallenser Manuskript der Dissertation Georg Cantors," *Historia Mathematica 3,* 449–462.

John, H. T. 1966. *The Thomist Spectrum.* New York: Fordham University Press.

Johnson, P. E. 1968. *A History of Cantorian Set Theory.* Ph.D. dissertation, George Peabody College for Teachers. Available through University Microfilms Inc., Ann Arbor, Michigan, Film #69-13827.

——— 1972. *A History of Set Theory.* Boston: Prindle, Weber and Schmidt; volume 16 in the Complementary Series in Mathematics.

Jourdain, P. E. B. 1904. "On the Transfinite Cardinal Numbers of Well-Ordered Aggregates," *Philosophical Magazine 7,* 61–75.

——— 1905. "On a Proof that Every Aggregate Can Be Well-Ordered," *Mathematische Annalen 60,* 465–470.

——— 1906. "The Development of the Theory of Transfinite Numbers. The Growth of the Theory of Functions up to the year 1870," *Archiv der Mathematik und Physik* (Grunnert's *Archiv*) *10,* 254–281.

——— 1909. "The Development of the Theory of Transfinite Numbers (Part 2). Weierstrass (1840–1880)," *Archiv der Mathematik und Physik 14,* 289–311.

——— 1910. "The Development of the Theory of Transfinite Numbers (Part 3). Georg Cantor's Work on Trigonometrical Series and his Theory of Irrational Numbers (1870–1871). The Other Theories of Irrational Numbers," *Archiv der Mathematik und Physik 22,* 1–21.

——— 1913. "The Development of the Theory of Transfinite Numbers (Part 4)," *Archiv der Mathematik und Physik 22,* 1–21.

Jürgens, E. 1878. "Ueber eindeutige und stetige Abbildungen von Mannigfaltigkeiten," *Tageblatt der Versammlung Deutscher Naturforscher und Aerzte.* Cassel.

——— 1879. *Allgemeine Sätze über Systeme von zwei eindeutigen und stetigen reellen Functionen von zwei reellen Veränderlichen.* Leipzig: B. G. Teubner.

——— 1898. "Der Begriff der n-fachen stetigen Mannigfaltigkeit," *Jahresbericht der Deutschen Mathematiker-Vereinigung 7,* 50–55.

Juškevič, A. 1966. "Georg Cantor und Sof'ja Kovalevskaja," *Ost und West in der Geschichte des Denkens und der kulturellen Beziehungen. Festschrift für Eduard Winter zum 70. Geburtstag.* Eds. H. Mohr and C. Grau. Berlin: Akademie-Verlag.

Kerry, B. 1885. "Ueber G. Cantor's Mannigfaltigkeitsuntersuchungen," *Vierteljahrsschrift für wissenschaftliche Philosophie 9,* 191–232.

Kertész, A. 1970. "The Significance of Cantor's Ideas for the Development of Algebra," *Scientia 105,* 203–209.

——— 1976. *Leben und Wirken Georg Cantors.* Ed. M. Stern. Halle: in press, expected 1976.

Kimberling, C. 1972. "Emmy Noether," *American Mathematical Monthly 79,* 136–149.

Klein, F. 1893. "The Present State of Mathematics." See Congress (1893), 133–135.

——— 1894. "Riemann und seine Bedeutung für die Entwickelung der modernen Mathematik," *Gesellschaft Deutscher Naturforscher und Aerzte. Verhandlungen 1894. Allgemeiner Theil.* Leipzig: F. C. W. Vogel, 3–18.

——— 1926. *Vorlesungen über die Entwicklung der Mathematik im 19. Jahrhundert.* Berlin: Springer.

Kneser, A. 1925. "Leopold Kronecker," *Jahresbericht der Deutschen Mathematiker-Vereinigung 33,* 210–228.

König, G. 1904. "Zum Kontinuum-Problem." See Congress (1904), 144–147.

——— 1905a. "Zum Kontinuum-Problem," *Mathematische Annalen 60,* 177–180.

——— 1905b. "Über die Grundlagen der Mengenlehre und das Kontinuumproblem," *Mathematische Annalen 61,* 156–160.

Koenigsberger, L. 1904. *Carl Gustav Jacob Jacobi: Festschrift zur Feier der 100. Wiederkehr seines Geburtstages.* Leipzig. B. G. Teubner.

——— 1919. *Mein Leben.* Heidelberg: Winter.

Kortum, H. 1906. "Rudolf Lipschitz," *Jahresbericht der Deutschen Mathematiker-Vereinigung 15,* 56–59.

Kossak, E. 1872. *Die Elemente der Arithmetik.* Berlin: Nicolai'sche Verlagsbuchhandlung.

Kowalewski, G. 1950. *Bestand und Wandel.* München: R. Oldenbourg.

Koyré, A. 1957. *From the Closed World to the Infinite Universe.* Baltimore: Johns Hopkins Press.

Kronecker, L. 1882. *Grundzüge einer arithmetischen Theorie der algebraischen Grössen. Festschrift zu Herrn Ernst Eduard Kummer's Fünfzigjährigem Doctor-*

Jubiläum, 10. September, 1881. Berlin: G. Reimer. Also published in *Journal für die reine und angewandte Mathematik* (Crelle's *Journal*) *92* (1882), 1–122; and in Kronecker (1897), 237–387.

——— 1886. "Ueber einige Anwendungen der Modulsysteme auf elementare algebraische Fragen," *Journal für die reine und angewandte Mathematik 99*, 329–371. Also in Kronecker (1899), 145–208.

——— 1887. "Ueber den Zahlbegriff," *Philosophische Aufsätze. Eduard Zeller zu seinem fünfzigjährigen Doctor-Jubiläum gewidmet*. Leipzig: Fues's Verlag, 263–274. Also in *Journal für die reine und angewandte Mathematik 101* (1887), 337–355; and in Kronecker (1899), 249–274.

——— 1894. *Vorlesungen über die Theorie der einfachen und der vielfachen Integrale*. Ed. E. Netto. Leipzig: B. G. Teubner.

——— 1897 and 1899. *Leopold Kroneckers Werke*. Ed. K. Hensel. Leipzig: B. G. Teubner, vol. II: 1897; vol. III: 1899.

Kummer, E. E. 1860. "Gedächtnissrede auf Gustav Peter Lejeune-Dirichlet," *Abhandlungen der Königlichen Akademie der Wissenschaften zu Berlin*, 1–36. Also in Dirichlet (1897), 311–344.

Landau, E. 1917. "Richard Dedekind," *Nachrichten von der Königlichen Gesellschaft der Wissenschaften zu Göttingen. Geschäftliche Mitteilungen*, 50–70.

Langer, R. E. 1947. "Fourier's Series. The Genesis and Evolution of a Theory," *The American Mathematical Monthly 54*, 1–86.

Lebesgue, H. 1906. *Leçons sur les séries trigonométriques*. Paris: Gauthier-Villars.

Leibniz, G. W. F. 1840. *God. Guil. Leibnitii Opera philosophica quae exstant latina, gallica, germanica omnia*. Ed. J. E. Erdmann. Berlin: Stumtibus G. Eichler.

Leo XIII. 1877. *La Chiesa e la Civiltà; lettera pastorale per la quaresima 1877, diocesi di Perugia*. Perugia: V. Santucci. Trans. *The Church and Civilization. Pastoral Letters for Lent 1877–1878*. New York: O'Shea, 1878.

——— 1878. *Inscrutabili*. Trans. "On the Evils Affecting Modern Society. *Inscrutabili*. April 21, 1878." In Wynne (1903), 9–21.

——— 1879. *Aeterni Patris*. Trans. "The Study of Scholastic Philosophy. *Aeterni Patris*. August 4, 1879." In Wynne (1903), 34–57.

Lindemann, F. 1882. "Ueber die Zahl π," *Mathematische Annalen 20*, 213–225.

Lingua, P. 1966. "Il significato topologico della dimensione, nella corrispondenza tra G. Cantor e R. Dedekind," *Periodico di matematiche 44*, 169–188.

Liouville, J. 1851. "Sur des classes très-étendues de quantités dont la valeur n'est ni algébrique, ni même réductible à des irrationnelles algébriques," *Journal de mathématiques pures et appliquées* (Liouville's *Journal*) *16*, 133–142.

Lipschitz, R. 1864. "De explicatione per series trigonometricas instituenda functionum unius variabilis arbitrariarum, et praecipue earum, quae per variabilis spatium finitum valorum maximorum et minimorum numerum habent infinitum, disquisitio," *Journal für die reine und angewandte Mathematik* (Crelle's *Journal*) *63*, 296–308. Trans. Lipschitz (1913).

——— 1877. *Lehrbuch der Analysis*. Bonn: Cohen and Son, 1877–1880.

——— 1913. "Recherches sur le développement en séries trigonométriques des fonctions arbitraires d'une variable et principalement de celles qui, dans un intervalle fini, admettent une infinité de maxima et de minima." Trans. P. Montel. *Acta Mathematica 36*, 281–295.

Lorey, W. 1915. "Der 70. Geburtstag des Mathematikers Georg Cantor," *Zeitschrift für mathematischen und naturwissenschaftlichen Unterricht 46*, 269–274.

——— 1916. *Das Studium der Mathematik an den Deutschen Universitäten seit Anfang des 19. Jahrhunderts*. Leipzig: B. G. Teubner.

——— 1918. "Max Simon zum Gedächtnis," *Zeitschrift für mathematischen und naturwissenschaftlichen Unterricht aller Schulgattungen 49*, 268–271.

Lüroth, J. 1878. "Ueber gegenseitig eindeutige und stetige Abbildung von Mannigfaltigkeiten verschiedener Dimensionen aufeinander," *Sitzungsberichte der physikalisch-medizinischen Societät zu Erlangen 10*, 190–195.

——— 1907. "Über Abbildungen von Mannigfaltigkeiten," *Mathematische Annalen 63*, 222–238.

Manheim, J. 1964. *The Genesis of Point Set Topology*. New York: Pergamon.

Manning, K. 1975. "The Emergence of the Weierstrassian Approach to Complex Analysis," *Archive for History of Exact Sciences 14*, 297–383.

Maritain, J. 1958. *St. Thomas Aquinas*. New York: Meridian Books.

McWilliams, J. A. 1945. *Physics and Philosophy: A Study of Saint Thomas' Commentary on the Eight Books of Aristotle's Physics*. Washington: American Catholic Philosophical Association.

Medvedev, F. A. 1965. *Razvitie teorii mnozhestv v deviatnadtsatom veke*. Moscow: Nauka.

Meschkowski, H. 1956. *Wandlungen des mathematischen Denkens*. Braunschweig: Vieweg.

——— 1961. *Denkweisen grosser Mathematiker*. Braunschweig: Vieweg. Trans. *Ways of Thought of Great Mathematicains*. San Francisco, 1964.

——— 1965. "Aus den Briefbüchern Georg Cantors," *Archive for History of Exact Sciences 2*, 503–519.

——— 1967. *Probleme des Unendlichen. Werk und Leben Georg Cantors*. Braunschweig: Vieweg.

——— 1971. "Zwei unveröffentlichte Briefe Georg Cantors," *Der Mathematikunterricht 4*, 30–34.

Metrios, G. 1968. *Cantor a tort*. Puteaux, France: Sival-Presse.

Meyer, G. G. 1873. "Bemerkungen über den Du Bois-Reymond'schen Mittelwerthsatz," *Mathematische Annalen 6*, 313–318.

Minkowski, H. 1905. "Peter Gustav Lejeune Dirichlet und seine Bedeutung für die heutige Mathematik," *Jahresbericht der Deutschen Mathematiker-Vereinigung 14*, 149–163.

Minnigerode, B. 1871. "Bemerkung über irrationale Zahlen," *Mathematische Annalen 4*, 497–498.

Mittag-Leffler, G. 1884a. "Sur la représentation analytique des fonctions monogènes uniformes d'une variable indépendante," *Acta Mathematica 4*, 1–79.

——— 1884b. "Démonstration nouvelle du théorème de Laurent," *Acta Mathematica 4*, 80–88.

——— 1900. "Une page de la vie de Weierstrass." See Congress (1900).

——— 1920. "Die Zahl: Einleitung zur Theorie der analytischen Funktionen," *The Tôhoku Mathematical Journal 17* (May), 157–209.

Moigno, F. N. M. 1868. *Sur l'impossibilité du nombre actuellement infini*. Paris: Gauthier-Villars.

Molk, J. 1904. "Nombres irrationnels et notion de limite." Trans. from the German article by A. Pringsheim (1898–1904), *Encyclopédie des sciences mathématiques 1* (3) (Paris), 133–208.

Molodshii, V. N. 1969. *Ocherki po filosofskim voprosam matematiki*. Moscow: Prosveschchenie.

Monna, A. 1972. "The Concept of Function in the 19th and 20th Centuries, in Particular with Regard to the Discussions between Baire, Borel and Lebesgue," *Archive for History of Exact Sciences 9*, 57–85.

Muldoon, P. J. 1903. *The Great White Shepherd of Christendom: His Holiness Pope Leo XIII*. Chicago: Hyland.

Müller, F. 1873. "Die erste deutsche Mathematiker-Versammlung," *Sonntags-Beilage Nr. 20. zur Königl. privilegierten Berlinischen Zeitung (Vossische Zeitung)* 18. May.

——— 1905. "Karl Schellbach. Rückblick auf sein wissenschaftliches Leben," *Abhandlungen zur Geschichte der mathematischen Wissenschaften 20*.

Müller, R. 1955. "Aus den Ahnentafeln deutscher Mathematiker," *Familie und Volk. Zeitschrift für Genealogie und Bevölkerungskunde 4*, 172.

Murata, T. 1966. "On the meaning of '*virtualité*' in the history of the set theory," *Japanese Studies in the History of Science 5*, 119–139.

——— 1967. "A Few Remarks on the Atomistic Way of Thinking in Mathematics," *Japanese Studies in the History of Science 6, 47–59*.

——— 1973. "Sur l'évolution de l'idée d'"effectif" dans l'histoire de la théorie des ensembles," *Revue d'histoire des sciences 26*. 4 (October), 365–368.

——— 1974. *L'évolution des principes philosophico-mathématiques de la théorie des ensembles chez Georg Cantor et leur diffusion en France jusqu'en 1905*. Dissertation, University of Paris.

Netto, E. 1878. "Beitrag zur Mannigfaltigkeitslehre," *Journal für die reine und angewandte Mathematik* (Crelle's *Journal*) *86*, 263–268.

——— 1893. "Ueber die arithmetisch-algebraischen Tendenzen Leopold Kronecker's." See Congress (1893).

Newton, I. 1713. *Philosophiae naturalis principia mathematica*. 2nd ed., Cambridge, England: Cornelius Crownfield.

Ong. W. 1960. *Darwin's Vision and Christian Perspectives*. New York: Macmillan.

O'Reilly, B. 1887. *Life of Leo XIII*. New York: Webster and Co.

Paplauskas, A. 1966. *Trigonometricheskie rîady ot Eĭlera do Lebega*. Moscow: Nauka.

——— 1968. "L'Influence de la théorie des séries trigonométriques sur le développement du calcul intégrale," *Archives internationales d'histoire des sciences 84–85*, 249–260.

Peano, G. 1889. *Arithmetices principia, nova methodo exposita*. Turin: Fratres Bocca.

——— 1892. "Dimostrazione dell'impossibilità di segmenti infinitesimi constanti," *Rivista di Matematica 2*, 58–62.

——— 1894–1908. *Formulaire de mathématiques*. Vols. I-V. Turin: Bocca.

Peirce, C. S. 1881. "On the Logic of Number," *American Journal of Mathematics 4*, 85–95; in *Collected Papers of Charles Sanders Peirce*, Eds. C. Hartshorne and P. Weiss. Cambridge, Mass.: Harvard University Press, 1960, 158–170.

——— 1976. *The New Elements of Mathematics by Charles S. Peirce*. Ed. C. Eisele. The Hague: Moulton.

Pelster, F. 1954. *Zur Enzyklika "Aeterni Patris": Text und Kommentar*. Rome: Edizioni di Storia e Litteratura.

Perrier, J. L. 1909. *The Revival of Scholastic Philosophy in the Nineteenth Century*. New York: The Columbia University Press.

Pesch, T. 1880. *Institutiones philosophiae naturalis secundum principia S. Thomae Aquinatis*. Friburgi Brisgoviae: Sumptibus Herder.

——— 1897. *Institutiones philosophiae naturalis*. Freiburg: Herder.

Pesin, I. N. 1970. *Classical and Modern Integration Theories*. Trans. S. Kotz. New York: Academic Press.

Peters, M. 1961. *Lied eines Lebens* (A biography of Else Cantor). Privately printed. Halle.

Pfannenstiel, M. 1953. *Lorenz Oken: sein Leben und Wirken*. Freiburger Universitätsreden, neue Folge, *14*. Freiburg: H. F. Schulz.

——— 1958. *Kleines Quellenbuch zur Geschichte der Gesellschaft Deutscher Naturforscher und Ärzte. Gedächtnisschrift für die hundertste Tagung der Gesellschaft*. Berlin: Springer.

Phillips, R. P. 1941. *Modern Thomistic Philosophy*. London: Birns, Oates and Washbourne.

Phragmén, E. 1884. "Beweis eines Satzes aus der Mannigfaltigkeitslehre," *Acta Mathematica 5*, 47–48.

Pierpont, J. 1904. "The History of Mathematics in the Nineteenth Century," *Bulletin of the American Mathematical Society 11*, 136–159.

——— 1928. "Mathematical Rigor, Past and Present," *Bulletin of the American Mathematical Society 34*, 23–53.

Pincherle, S. 1880. "Saggio di una introduzione alla teoria delle funzioni analitiche secondo i principii del Prof. C. Weierstrass," *Giornale di Matematiche 18*, 178–254, 317–357.

Pittard-Bullock, H. 1905. *The Power of the Continuum. Inaugural-Dissertation*, Rostock. Berlin: E. Ebering.

Pius IX. 1864. *Die Encyclica Papst Pius IX vom 8. Dez. 1864 (Syllabus Errorum)*. Stimmen aus Maria-Laach, Freiburg in Breisgau: Herder, 1866–1899.

Poincaré, H. 1898. "L'Oeuvre mathématique de Weierstrass," *Acta Mathematica 22* (1898–1899), 1–18.

——— 1905. "Les Mathématiques et la logique," *Revue de metaphysique et de morale 13*, 815–835; *14* (1906) 17–34, 294–317. See also Russell (1905) and Boutroux (1905).

——— 1908. "L'Avenir des mathématiques." See Congress (1908), 167–182.

Prasad, G. 1933. *Some Great Mathematicians of the Nineteenth Century*. In 3 volumes. India: Mahamandal Press, by the Benares Mathematical Society.

Pringsheim, A. 1898–1904. "Irrationalzahlen und Konvergenz unendlicher Prozesse," *Encyclopädie der mathematischen Wissenschaften 1* (1). Leipzig: B. G. Teubner, 47–62.

——— 1899. "Grundlagen der allgemeinen Funktionenlehre," *Encyklopädie der mathematischen Wissenschaften 2*. Leipzig: B. G. Teubner, 1–53.

Raeymaeker, L. de, 1952. *Le Cardinal Mercier et l'Institut Superieur de Louvain*. Louvain: Publications universitaires.

Ravetz, J. 1971. See Grattan-Guinness (1971b).

Renouvier, C. 1885. *Esquisse d'une classification systématique des doctrines philosophiques*. Paris: Au Bureau de la critique philosophique, 1885, 1886.

Riemann, B. 1854. "Ueber die Darstellbarkeit einer Function durch eine trigonometrische Reihe." *Habilitationsschrift,* Göttingen. First printed in *Abhandlungen der mathematischen Classe der Königlichen Gesellschaft der Wissenschaften zu Göttingen 13* (1867) 87–131. Note that all page references keyed to Riemann (1854) are taken from the second edition of 1902 as reprinted by Dover, 1953. See also Riemann (1876b), 227–265.

——— 1876a. "Neue mathematische Principien der Naturphilosophie," in Riemann (1876b), 528–532.

——— 1876b. *Bernhard Riemann's Gesammelte mathematische Werke und wissenschaftlicher Nachlass*. Ed. H. Weber and R. Dedekind. Leipzig: B. G. Teubner. Reprinting of the second edition of 1892, including the supplement prepared by M. Noether and W. Wirtinger in 1902, New York: Dover, 1953. Unless otherwise noted, all citations from Riemann's works are taken from this reprinted version.

Riesz, M. 1908. "Stetigkeitsbegriff und abstrakte Mengenlehre." See Congress (1908), vol. II, 18–24.

——— 1923. See Hilb (1923).

Ringer, F. K. 1969. *The Decline of the German Mandarins: the German Academic Community, 1890–1933*. Cambridge, Mass.: Harvard University Press.

Rühle, O. 1966. *Idee und Gestalt der deutschen Universität*. Berlin: VEB Deutscher Verlag der Wissenschaften.

——— 1967. "Humboldts Universitätsidee—Tradition und Aufgabe." In *Wilhelm von Humboldt, Beiträge vorgelegt von der Humboldt-Universität zu Berlin anlösslich der Feier des zweihundertsten Geburtstages ihres Gründers*. Halle: Max Niemeyer Verlag.

Russell, B. 1903. *Principles of Mathematics*. Cambridge, England: At the University Press.

——— 1905. "Sur la relation des mathématiques à la logistique," *Revue de métaphysique et de morale 13,* 906–916. See Poincaré (1905) and Boutroux (1905).

——— 1907. "On Some Difficulties in the Theory of Transfinite Numbers and Order Types," *Proceedings of the London Mathematical Society 4,* 29–53.

——— 1956. *Portraits from Memory*. London: George Allen and Unwin Ltd.

——— 1967. *The Autobiography of Bertrand Russell*. New York: Bantam, 1968. Originally published in hardcover, in London: 1967–1969, in three volumes. All citations and references are taken from the Bantam edition of 1968.

Russell/Whitehead (1925–1927). B. Russell, and A. N. Whitehead. *Principia Mathematica*. Cambridge, England: vol. I, 1910; vol. II, 1912; vol. III, 1913. All citations are from the 2nd ed. 1925–1927.

Rust, W. 1934. "An Operational Statement of Cantor's *Diagonalverfahren,*" *Scripta Mathematica 2,* 334–336.

Sachse, A. 1879. *Versuch einer Geschichte der Darstellung willkürlicher Functionen einer Variabeln durch trigonometrische Reihen. Inaugural-Dissertation*. Göttingen. Trans. Sachse (1880).

——— 1880. "Essai historique sur la présentation d'une fonction arbitraire d'une seule variable par une série trigonométrique," *Bulletin des sciences mathématiques et astronomiques 15,* 43–64, 83–112.

Scheefer, L. 1884a. "Allgemeine Untersuchungen über Rectification der Curven," *Acta Mathematica 5,* 49–82.

——— 1884b. "Zur Theorie der stetigen Funktionen einer reellen Veranderlichen," *Acta Mathematica 5,* 279–296.

Schering, E. 1866. "Zum Gedächtniss an B. Riemann." In Schering (1909), 166–168 and 367–383.

——— 1902 and 1909. *Ernst Schering: Gesammelte Mathematische Werke.* Vol. I, Berlin: Mayer and Müller. Vol. II, Berlin: Mayer and Müller, 1909.

Schoenflies, A. 1896/1897. "Transfinite Zahlen, das Axiom des Archimedes und die projective Geometrie," *Jahresbericht der Deutschen Mathematiker-Vereinigung 5,* 75–81.

——— 1900. *Entwickelung der Mengenlehre.* Leipzig: B.G. Teubner. 2nd ed. 1913.

——— 1903. "Beiträge zur Theorie der Punktmengen (Part I)," *Mathematische Annalen 58,* 195–234.

——— 1904a. "Struktur der perfekten Mengen." See Congress (1904) and Schoenflies (1904b).

——— 1904b. "Beiträge zur Theorie der Punktmengen (Part 2)," *Mathematische Annalen 59,* 129–160.

——— 1905. "Über wohlgeordnete Mengen," *Mathematische Annalen 60,* 181–186.

——— 1906. "Über die logischen Paradoxien der Mengenlehre," *Jahresbericht der Deutschen Mathematiker-Vereinigung 15,* 19–25.

——— 1911. "Über die Stellung der Definition in der Axiomatik," *Jahresbericht der Deutschen Mathematiker-Vereinigung 20,* 250–255.

——— 1913. *Entwickelung der Mengenlehre und Ihrer Anwendungen. Umarbeitung des im VIII Bande der Jahresbericht der Deutschen Mathematiker-Vereinigung erstatteten Berichts. Erste Hälfte: Allgemeine Theorie der Unendlichen Mengen und Theorie der Punktmengen.* Leipzig: B. G. Teubner.

——— 1922. "Zur Erinnerung an Georg Cantor," *Jahresbericht der Deutschen Mathematiker-Vereinigung 31,* 97–106.

——— 1927. "Die Krisis in Cantor's mathematischem Schaffen," *Acta Mathematica 50,* 1–23.

——— 1928. "Georg Cantor," *Mitteldeutsche Lebensbilder 3.* Magdeburg: Selbstverlag der Historischen Kommission.

Scholz, H. 1935, and H. Schweitzer. *Die sogenannte Definitionen durch Abstraktion.* Leipzig: Forschungen zur Logistik No. 3.

Schröder, E. 1896. "Über G. Cantorsche Sätze," *Jahresbericht der Deutschen Mathematiker-Vereinigung 5,* 81–82.

——— 1898. "Ueber zwei Definitionen der Endlichkeit und G. Cantor'sche Sätze," *Nova Acta Abh. der Kaiserl. Leop.-Carol. Deutschen Akademie der Naturforscher 71.* Halle, 301–362.

Schwarz, H. A. 1890. *Gesammelte mathematische Abhandlungen.* Berlin: Springer. 2 vols.

Schwarz, H. C. 1888. *Ein Beitrag zur Theorie der Ordnungstypen.* Halle: Schmidt.

Scriba, C. J. 1968. *The Concept of Number*. Manheim: Hochschulschriften Bibliographisches Institut.

Sierpiński, W. 1934. *Hypothèse du continu*. Warsaw-Lwów: Monografie matematyczne, no. 4.

——— 1958. *Cardinal and Ordinal Numbers*. Warsaw: Monografie matematyczne, no. 34.

Simon, M. 1883. "Review of G. Cantor. *Grundlagen einer allgemeinen Mannigfaltigkeitslehre*" in *Deutsche Litteraturzeitung* (5. Mai), no. 18, 641–643.

Smith, H. J. S. 1875. "On the Integration of Discontinuous Functions," *Proceedings of the London Mathematical Society 6* (June 10), 140–153.

Spinoza, B. 1928. *The Correspondence of Spinoza*. Trans. A. Wolf. London: G. Allen and Unwin.

Stolz, O. 1881. "B. Bolzano's Bedeutung in der Geschichte der Infinitesimalrechnung," *Mathematische Annalen 18*, 255–279.

——— 1884. "Ueber einen zu einer unendlichen Punktmenge gehörigen Grenzwerth," *Mathematische Annalen 23*, 152–156.

——— 1885. *Vorlesungen über allgemeine Arithmetik*. Leipzig: B. G. Teubner, 1885–86.

——— 1891. *Grössen und Zahlen. Rede bei Gelegenheit der feierlichen Kundmachung der gelösten Preisaufgaben am 2. März, 1891*. Leipzig: B. G. Teubner.

Sudhoff, K. 1922. *Hundert Jahre Deutscher Naturforscher-Versammlungen. Gedächtnisschrift zur Jahrhundert-Tagung der Gesellschaft deutscher Naturforscher und Ärzte*. Leipzig: F. C. W. Vogel.

Tannery, J. 1897. "De l'Infini mathématique," *Revue générale des sciences pures et appliquées 8*, 129–140.

Tannery, P. 1885. "Le Concept scientifique du continu: Zénon d'Elée et G. Cantor," *Revue philosophique 20*, 385–410.

——— 1934. *Mémoires scientifiques 13*. Correspondance. Paris: Gauthier-Villars.

Ternus, J. 1929. "Ein Brief Georg Cantors an P. Joseph Hontheim, S. J.," *Scholastik 4*, 561–571.

Thomae, J. 1870. *Abriss einer Theorie der complexen Functionen und der Thetafunctionen einer Veränderlichen*. Halle: L. Nebert.

——— 1878. "Sätze aus der Functionentheorie," *Nachrichten von der K. Gesellschaft der Wissenschaften und der Georg-Augusts-Universität,* Göttingen, 466–468.

——— 1906. "Gedankenlose Denker. Eine Ferienplauderei," *Jahresbericht der Deutschen Mathematiker-Vereinigung 15*, 434–438.

Thomas Aquinas. 1963. *Commentary on Aristotle's Physics*. Trans. R. J. Blackwell *et al*. London: Routledge.

——— 1961. *Commentary on the Metaphysics of Aristotle*. Trans. J. P. Rowan. Chicago: Regnery.

——— 1964a. *On the Eternity of the World (De aeternitate mundi)*. Trans. C. Vollert *et al*. Milwaukee: Marquette University Press.

——— 1964b. *Summa theologiae*. Cambridge, England: Blackfriars.

van Heijenoort, J. 1967. *From Frege to Gödel. A Source Book in Mathematical Logic, 1879–1931*. Cambridge, Mass.: Harvard University Press.

Van Riet, G. 1946. *L'Epistémologie Thomiste*. Louvain: Éditions de l'Institut Supérieur de Philosophie.

Van Vleck, E. B. 1914. "The Influence of Fourier's Series upon the Development of Mathematics," *Science 39,* 113–124.

Verhandlungen. 1890. *Verhandlungen der Gesellschaft Deutscher Naturforscher und Ärzte. 63. Versammlung zu Bremen. 15–20. September.* Ed. O. Lassar. Zweiter-Theil: Abtheilungs-Sitzungen. Leipzig: F. C. W. Vogel, 1891.

Verhandlungen. 1891. *Verhandlungen der Gesellschaft Deutscher Naturforscher und Ärzte. 64. Versammlung zu Halle A. S. 21.–25. September.* Ed. A. Wangerin and F. Krause. Erster Theil: Die allgemeinen Sitzungen, Leipzig: F. C. W. Vogel, 1891; Zweiter Theil: Abtheilungs-Sitzungen, Leipzig: F. C. W. Vogel, 1892.

Veronese, G. 1894. *Grundzüge der Geometrie von mehreren Dimensionen und mehreren Arten gradliniger Einheiten in elementarer Form entwickelt.* Trans. A. Schepp. Leipzig: B. G. Teubner.

Virchow, R. 1893. *Die Gründung der Berliner Universität und der Uebergang aus dem philosophischen in das naturwissenschaftliche Zeitalter. Rede am 3. August 1893 in der Aula der Königlichen Friedrich-Wilhelms-Universität zu Berlin.* Berlin: Julius Becker.

Vivanti, G. 1889. "Fondamenti della teoria dei tipi ordinati," *Annali di Matematica pura ed applicata 17,* 1–35.

——— 1891a. "Sull'infinitesimo attuale," *Rivista di Matematica 1,* 135–153.

——— 1891b. "Ancora sull'infinitesima attuale," *Rivista di Matematica 1,* 248–255.

——— 1892. "Sopra una questione elementare della teoria degli aggregati di G. Cantor," *Rivista di Matematica 2,* 165–167. Translation of Cantor (1891).

——— 1893. "Lista bibliografica della teoria degli aggregati," *Rivista di Matematica 3,* 189–192.

——— 1899. *Corso di calcolo infinitesimale.* Messina: Trimarchi.

——— 1901. *Teoria delle funzioni analitiche.* Milan: Holpli.

——— 1906. *Theorie der eindeutigen analytischen Funktionen.* Trans. A. Gutzmer. Leipzig: B. G. Teubner.

Von Neumann, J. 1925. "Eine Axiomatisierung der Mengenlehre," *Journal für die reine und angewandte Mathematik* (Crelle's *Journal*) *154,* 219–240.

——— 1928. "Die Axiomatisierung der Mengenlehre," *Mathematische Zeitschrift 27,* 669–752.

Von Zahn, W. 1874. "Einige Worte zum Andenken an Hermann Hankel," *Mathematische Annalen 7,* 583–590.

Wallace, L. 1966. *Leo XIII and the Rise of Socialism.* North Carolina: Duke University Press.

Wangerin, A. 1918. "Georg Cantor," *Leopoldina 54,* 10–13, 32.

Weber, H. 1893a. "Leopold Kronecker," *Mathematische Annalen 43,* 1–25.

——— 1893b. "Leopold Kronecker," *Jahresbericht der Deutschen Mathematiker-Vereinigung 2,* 5–23.

——— 1906. "Elementare Mengenlehre," *Jahresbericht der Deutschen Mathematiker-Vereinigung 15,* 173–184.

Wehofer, T. 1879. "Die geistige Bewegung im Anschluss an die Thomas-Encyclica Leo XIII vom 4. August 1879," *Vorträge und Abhandlungen herausgegeben von der Leo-Gesellschaft,* no. 7. Vienna.

Weierstrass, K. 1879. *Theorie der Variationsrechnung.* Lectures delivered in the summer semester, 1879, preserved in a typescript copy, Harvard University, from

an original set of handwritten notes in the library of the Johns Hopkins University.
Werner, K. 1858. *Der heilige Thomas von Aquino*. Regensburg: G. J. Manz, 1858–1859.

Weyl, H. 1931. *Die Stufen des Unendlichen*. Jena: Gustav Fischer.

——— 1944a. *Philosophy of Mathematics and Natural Science*. Princeton: Princeton University Press. Rev. ed., 1949.

——— 1944b. "David Hilbert and his Mathematical Work," *Bulletin of the American Mathematical Society 50,* 612–654.

Wilder, R. L. 1965. *Evolution of Mathematical Concepts*. New York: Wiley and Sons.

Wynne, J. J., ed. 1903. *The Great Encyclical Letters of Pope Leo XIII*. New York.

Wyser, P. 1951. "Der Thomismus," *Bibliographische Einführungen in das Studium der Philosophie 15/16*.

Young, W. H. and G. C. 1906. *The Theory of Sets of Points*. Cambridge, England. Reprinted, New York: Chelsea Publishing Company, 1972.

Young, W. H. 1926. "The Progress of Mathematical Analysis in the Twentieth Century," *Proceedings of the London Mathematical Society 24,* 421–434.

Yushkevich, *see* Juškevič.

Zermelo, E. 1901. "Ueber die Addition transfiniter Cardinalzahlen," *Nachrichten von der Königl. Gesellschaft der Wissenschaften zu Göttingen. Mathematisch-physikalische Klasse,* 34–38.

——— 1904. "Beweis, dass jede Menge wohlgeordnet werden kann," *Mathematische Annalen 59,* 514–516.

——— 1908a. "Neuer Beweis für die Möglichkeit einer Wohlordnung," *Mathematische Annalen 65,* 107–128.

——— 1908b. "Untersuchungen über die Grundlagen der Mengenlehre. I." *Mathematische Annalen 65,* 261–281.

——— 1927. "Über das Mass und die Diskrepanz von Punktmengen," *Journal für die reine und angewandte Mathematik* (Crelle's *Journal*) *158,* 154–167.

Zincke, H. 1916. "Erinnerungen an Richard Dedekind," *Braunschweigisches Magazin 7* (July), 73–81.

Zybura, S. J., ed. 1926. *Present-Day Thinkers and the New Scholasticism*. London: Herder.

Zygmund, A. 1959. *Trigonometric Series*. Cambridge, England: Cambridge University Press. 2nd ed. in 2 vols.

Index